VOLUME ONE HUNDRED AND SEVENTY FIVE

PROGRESS IN MOLECULAR BIOLOGY AND TRANSLATIONAL SCIENCE

Prions and Neurodegenerative Diseases

VOLUME ONE HUNDRED AND SEVENTY FIVE

PROGRESS IN MOLECULAR BIOLOGY AND TRANSLATIONAL SCIENCE

Prions and Neurodegenerative Diseases

Edited by

GIUSEPPE LEGNAME

Laboratory of Prion Biology, Department of Neuroscience, Scuola Internazionale Superiore Di Studi Avanzati (SISSA), Trieste, Italy

ELSEVIER

ACADEMIC PRESS

An imprint of Elsevier

Academic Press is an imprint of Elsevier
50 Hampshire Street, 5th Floor, Cambridge, MA 02139, United States
525 B Street, Suite 1650, San Diego, CA 92101, United States
The Boulevard, Langford Lane, Kidlington, Oxford OX5 1GB, United Kingdom
125 London Wall, London EC2Y 5AS, United Kingdom

First edition 2020

ISBN: 978-0-12-820002-5
ISSN: 1877-1173

For information on all Academic Press publications
visit our website at https://www.elsevier.com/books-and-journals

Publisher: Zoe Kruze
Acquisitions Editor: Ashlie M. Jackman
Editorial Project Manager: Leticia Lima
Production Project Manager: James Selvam
Cover Designer: Matt Limbert

Typeset by SPi Global, India

Contents

Contributors

Hamza Arshad
Tanz Centre for Research in Neurodegenerative Diseases; Department of Biochemistry, University of Toronto, Toronto, ON, Canada

Jason C. Bartz
Department of Medical Microbiology and Immunology, School of Medicine, Creighton University, Omaha, NE, United States

Ilia V. Baskakov
Department of Anatomy and Neurobiology, and Center for Biomedical Engineering and Technology, University of Maryland School of Medicine, Baltimore, MD, United States

Emiliano Biasini
Department of Cellular, Computational and Integrative Biology (CIBIO); Dulbecco Telethon Institute, University of Trento, Trento, TN, Italy

Edoardo Bistaffa
Fondazione IRCCS Istituto Neurologico Carlo Besta, Unit of Neurology 5 and Neuropathology, Milan, Italy

Alyssa J. Block
Department of Medical Microbiology and Immunology, School of Medicine, Creighton University, Omaha, NE, United States

Matthew E.C. Bourkas
Tanz Centre for Research in Neurodegenerative Diseases; Department of Biochemistry, University of Toronto, Toronto, ON, Canada

N.R. Cashman
Djavad Mowafaghian Centre for Brain Health, University of British Columbia, Vancouver, BC, Canada

Marcella Catania
Neurology 5 / Neuropathology Unit, Fondazione IRCCS Istituto Neurologico Carlo Besta, Milan, Italy

Federico Angelo Cazzaniga
Fondazione IRCCS Istituto Neurologico Carlo Besta, Unit of Neurology 5 and Neuropathology, Milan, Italy

Alessandra Consonni
Fondazione IRCCS Istituto Neurologico Carlo Besta, Neurology IV—Neuroimmunology and Neuromuscular Diseases Unit, Milan, Italy

Chiara Maria Giulia De Luca
Fondazione IRCCS Istituto Neurologico Carlo Besta, Unit of Neurology 5 and Neuropathology, Milan, Italy

Giuseppe Di Fede
Neurology 5 / Neuropathology Unit, Fondazione IRCCS Istituto Neurologico Carlo Besta, Milan, Italy

Laura J. Ellett
Department of Microbiology and Immunology, The University of Melbourne, Parkville, VIC, Australia

Juan Carlos Espinosa
Centro de Investigación en Sanidad Animal (CISA-INIA), Madrid, Spain

Giorgio Giaccone
Fondazione IRCCS Istituto Neurologico Carlo Besta, Unit of Neurology 5 and Neuropathology, Milan, Italy

Michel Goedert
MRC Laboratory Molecular Biology, Cambridge, United Kingdom

Sara A.M. Holec
Institute for Applied Life Sciences and Department of Biology, University of Massachusetts Amherst, Amherst, MA; Department of Medical Microbiology and Immunology, School of Medicine, Creighton University, Omaha, NE, United States

Richard Knight
Emeritus Professor of Clinical Neurology, Centre for Clinical Brain Sciences, University of Edinburgh, Edinburgh, United Kingdom

Yong Qian Koo
Department of Microbiology and Immunology, The University of Melbourne, Parkville, VIC, Australia

Kazuo Kuwata
United Graduate School of Drug Discovery and Medical Information Sciences, Gifu University, Gifu, Japan

Victoria A. Lawson
Department of Microbiology and Immunology, The University of Melbourne, Parkville, VIC, Australia

Giuseppe Legname
Laboratory of Prion Biology, Department of Neuroscience, Scuola Internazionale Superiore Di Studi Avanzati (SISSA), Trieste, Italy

Alba Marín-Moreno
Centro de Investigación en Sanidad Animal (CISA-INIA), Madrid, Spain

L. McAlary
Illawarra Health and Medical Research Institute; Molecular Horizons and School of Chemistry and Molecular Bioscience, University of Wollongong, Wollongong, NSW, Australia

Fabio Moda
Fondazione IRCCS Istituto Neurologico Carlo Besta, Unit of Neurology 5 and Neuropathology, Milan, Italy

Jesús R. Requena
CIMUS Biomedical Research Institute & Department of Medical Sciences, University of Santiago de Compostela-IDIS, Santiago, Spain

Zoe T. Revill
Department of Microbiology and Immunology, The University of Melbourne, Parkville, VIC, Australia

Carlo Scialò
Laboratory of Prion Biology, Department of Neuroscience, Scuola Internazionale Superiore Di Studi Avanzati (SISSA), Trieste, Italy

Giovanni Spagnolli
Department of Cellular, Computational and Integrative Biology (CIBIO); Dulbecco Telethon Institute, University of Trento, Trento, TN, Italy

Surachai Supattapone
Department of Biochemistry and Cell Biology and Department of Medicine, Geisel School of Medicine at Dartmouth College, Hanover, NH, United States

Juan María Torres
Centro de Investigación en Sanidad Animal (CISA-INIA), Madrid, Spain

Joel C. Watts
Tanz Centre for Research in Neurodegenerative Diseases; Department of Biochemistry, University of Toronto, Toronto, ON, Canada

J.J. Yerbury
Illawarra Health and Medical Research Institute; Molecular Horizons and School of Chemistry and Molecular Bioscience, University of Wollongong, Wollongong, NSW, Australia

Preface

When I accepted to edit this book I thought that it would have been a timely contribution to the field of prions and prion diseases. I also believed that, since the last book I had the pleasure to co-edit with Silvia Vanni, this new publication would have an optimal schedule to explore new concepts such as prion-like disorders. Indeed, many chapters of this book deal with this emerging concept of prion-like diseases, such as more common forms of neurodegenerative disorders.

This book is titled *Prions and neurodegenerative diseases* and, as you will read throughout every chapter, I invited world-leaders in their field to contribute to this new editorial adventure.

It is also a peculiar time of everyone's life. The pandemic of COVID-19 that has hit the world has reshaped the way we face scientific discovery. This book also reminds us that despite contingent global emergencies, neurodegenerative diseases remain an ongoing worldwide, unresolved crisis.

The book opens with a very comprehensive review on "Clinical diagnosis of human prion disease" by Richard Knight. The author illustrates the various aspects of clinical evaluation and characterization of prion diseases. Although very rare disorders, as you will read, it also comes in many forms. Professor Knight also introduces the advent of new technological advances in *ante mortem* diagnosis, which will be described in more detail later in the book.

The second chapter deals with "Understanding prion structure and conversion" authored by Giovanni Spagnolli, Jesús R. Requena, and Emiliano Biasini. The authors of this chapter embark in detailing a long-standing issue of defining prion structure and molecular mechanisms of conversion and replication. Their contribution in this particular field of prion research has been of great value for the scientific community as a whole.

Next the "Role of sialylation in prion disease pathogenesis and prion structure" authored by Ilia V. Baskakov is explored. In this chapter, prions are currently widely accepted as protein-only infectious pathogens and the role of glycosylation is analyzed in great detail. Professor Baskakov argues that sialylation covers an important role in delineating prion diverse structures, besides their protein structures. Whether glycans hold a unique part in the pathogenesis of prions remains to be fully established.

"Cofactor molecules: Essential partners for infectious prions" is the following chapter authored by Surachai Supattapone. Since the first, seminal publication of formal proofs that prion are indeed protein-only pathogens the search for cofactors involved in the conversion and replication of infectious prions has been restless. Professor Supattapone highlights the various lines of evidence emerged in the last few years supporting such idea.

The following two chapters treat the relevant issues of prion strain diversity and the quest for finding an efficacious treatment and cure for these devastating disorders.

Sara A.M. Holec, Alyssa J. Block, and Jason C. Bartz authored the chapter on "The role of prion strain diversity in the development of successful therapeutic treatments." This contribution is extremely detailed in describing the evolution of the concept of prion strain and their role in identifying an effective therapy. Also in the following chapter, "Strain variation in treatment and prevention of human prion diseases," Laura J. Ellett, Zoe T. Revill, Yong Qian Koo, and Victoria A. Lawson focus on prion strain diversity, but this time on a more human prion disease perspective.

One distinctive advantage in the field of prion diseases over other more common forms of neurodegeneration is the accessibility of excellent tools for the study of these disorders at the molecular level. This is the case for animal models that reproduce in every molecular and pathological detail other animal and human diseases. Marín-Moreno Alba, Espinosa Juan Carlos and Torres Juan María accomplish an excellent contribution delivering a very comprehensive review of all animal models developed for prion research: from several knockout mouse models to more sophisticate transgenic mice for the study of the ample spectrum of prion disorders. One such model that has recently attracted the attention of prion researchers is the Bank vole. The unique value of this rodent model is argued and elaborated in the chapter titled "The utility of bank voles for studying prion disease" by Hamza Arshad, Matthew E.C. Bourkas, and Joel C. Watts.

In the chapter "One or more β-amyloid(s)? New insights into the prion-like nature of Alzheimer's disease" by Marcella Catania and Giuseppe Di Fede: the reader enters the world of prion-like disorders of which Alzheimer's disease is the most prominent. Despite several decades of research in this field many aspects of the disease are still elusive. The findings that β-amyloid (or amyloid-β) in Alzheimer's disease may follow the same mechanism of conversion and replication of prions could facilitate the search for suitable treatments.

In recent years, perhaps the most important contribution in delineating the prion-like nature of neurodegenerative disease is the description of different structures of tau amyloids in diverse forms of taupathies, including Alzheimer's disease. The advent of cryo-electron microscopy (cryoEM) has enabled Michel Goedert and colleagues to describe at the atomic level the different forms of tau in diverse neurodegenerative disorders. Indeed, Michel Goedert describes in much depth these discoveries in the chapter "Tau proteinopathies and the prion concept."

Also "The prion-like nature of amyotrophic lateral sclerosis" constitutes an innovative look at the protein aggregates involved in amyotrophic lateral sclerosis. This chapter authored by L. McAlary, J.J. Yerbury, and N.R. Cashman provides remarkable insights into superoxide dismutase (SOD) and other proteins involved in this disease.

My own laboratory contributes to this book with a chapter titled "The role of the cellular prion protein in the uptake and toxic signaling of pathological neurodegenerative aggregates", with coauthor neurologist Carlo Scialò. It came with great surprise that the cellular form of the prion protein seems to act as a general acceptor/receptor of protein aggregates from other disorders. Together we have attempted to describe the evidence that support such an unprecedented role.

As anticipated the coming of age of novel diagnostic tools is changing the way we study neurodegenerative disorders and together with my colleagues Federico Angelo Cazzaniga, Chiara Maria Giulia De Luca, Edoardo Bistaffa, Alessandra Consonni, Giorgio Giaccone, and Fabio Moda have tried to provide an update on these approaches in the chapter "Cell-free amplification of prions: Where do we stand?".

The book ends with a provocative view on how to develop new therapeutic strategies. Kazuo Kuwata with the chapter titled "Principle of protein surgery: Application to logical drug design for treatment of neurodegenerative disease" provides an innovative perspective on how to achieve this.

It has been a pleasure to serve as an Editor to this book. By doing this I have learnt many new things and the flow of incomparable amount of information provides a fantastic background to new ideas and discoveries. Enjoy reading the book.

Giuseppe Legname
Trieste

CHAPTER ONE

Clinical diagnosis of human prion disease

Richard Knight*
Emeritus Professor of Clinical Neurology, Centre for Clinical Brain Sciences, University of Edinburgh, Edinburgh, United Kingdom
*Corresponding author: e-mail address: r.knight@ed.ac.uk

Contents

Abstract

Human prion disease may present in a non-specific way and is often diagnosed at a relatively late stage of the illness. Until recently, clinical diagnosis has been supported by tests that are mostly non-specific and, sometimes, insensitive. Recent laboratory developments have led to a variety of tests that rely on a disease-specific mechanism. One test, the CSF RT-QuIC (Real-Time Quaking-Induced Conversion) test is very sensitive and specific for sporadic CJD and is now used in routine clinical practice. Other tests, based on other tissues, including blood and urine, have been developed and potentially

Progress in Molecular Biology and Translational Science, Volume 175
ISSN 1877-1173
https://doi.org/10.1016/bs.pmbts.2020.07.006

could improve both clinical diagnostic accuracy and lead to earlier diagnosis. While there are yet no proven treatments for prion disease, any treatment to be developed will almost certainly require earlier diagnosis if therapeutic success is to be realized.

1. Introduction

The human prion diseases are listed in Table 1. Their definitive diagnosis is neuropathological, with standard histological, immunohistopathological and protein analysis features.[1] Naturally, this requires study of brain tissue, usually at autopsy, but occasionally via brain biopsy (the latter being undertaken in only very specific clinical circumstances). Diagnosis in life has rested, for many years, on the clinical picture, the exclusion of other possible diagnoses and positive support from several tests which, while very useful, are generally non-specific, with abnormal results that are not unique to prion disease; the main ones being the EEG (electro-encephalogram), cerebral MR (magnetic resonance) imaging and cerebro-spinal fluid 14-3-3 analysis. Two essentially more specific tests have been used in the diagnosis of two forms of human prion disease: *PRNP* mutation testing in genetic disease and tonsil biopsy in variant CJD. The detection of a known pathogenic *PRNP* mutation or of abnormal prion protein in a tonsil biopsy in an ill individual with an appropriate clinical picture, is not absolutely definitive of genetic prion disease or vCJD, respectively, but surely it removes all reasonable doubt.

Accurate diagnosis in life is important for informing affected patients and families, for disease surveillance and for public health protection measures, especially if autopsies are not performed. In addition, although there are no proven effective treatments at this time, they are being sought actively; if they were to be developed and assessed, then early, reliable clinical diagnosis would become very important.

Initially, the EEG was the single most useful supportive test, followed later by CSF 14-3-3 and the brain MRI, however, over the last few years, a number of highly specific tests have been developed, based on protein amplification techniques. Their specificity rests on the detection of abnormal, disease-related prion protein. One of these (the CSF RT-QuIC (Real-Time Quaking-Induced Conversion) test) has become part of routine clinical practice and others are likely to have important roles. There has been a lot of progress in clinical diagnosis, both in terms of distinguishing

Table 1 The human forms of prion disease.

Causal group	Human diseases	Notes
Sporadic disease	Sporadic CJD	Found world-wide. Cause unknown but regarded as a spontaneously developing disease in affected individuals. Been divided into subtypes according to clinical phenotype/*PRNP*-129 genotype/prion protein type.
	VPSPr	Variably Protease Sensitive Prionopathy. Nosological status a little uncertain. Probably best regarded as a type of sCJD.
Genetic disease	Various phenotypic terms in use: Genetic CJD, Fatal Familial Insomnia, Gerstmann Straussler Scheinker Syndrome.	Caused by underlying *PRNP* mutations. Many now identified. Autosomal Dominant inheritance.
Acquired disease	Kuru	Confined to Papua New Guinea. Related to mourning ritualistic cannibalism.
	Iatrogenic CJD	Accidental human-to-human transmission by surgery or certain medical treatments (use of cadaveric-derived dura mater and human pituitary hormones).
	Variant CJD	Originally caused by BSE contamination of diet; some secondary transmission via blood/blood products. Most cases in UK and France.

prion disease from other neurodegenerative illnesses and also in terms of identifying the different forms of CJD (such as, importantly, variant and sporadic CJD).

2. The diagnostic problem

In themselves, prion diseases have relatively characteristic clinical profiles, however, these profiles are certainly not unique and, moreover, the initial symptoms may be relatively common neurological or neuropsychological ones (for example, unsteadiness, confusion, and mood changes); other neurological diseases (some potentially treatable) need consideration by clinicians in the diagnostic process.

Another clinical problem is distinguishing between different forms of prion disease, especially sporadic and variant CJD (where the distinction has potentially immediate management, as well as longer-term public health and policy, implications). Sporadic CJD itself has been divided into subtypes and this has bearing on clinical diagnosis (for example, diagnostic tests may behave in different ways in different subtypes).

3. The non-specific supportive tests

Three supportive tests have been in routine clinical use for some years: the EEG, Brain MRI and certain CSF protein tests. These are reviewed in summary. Two other particular tests that also have been in routine clinical use (*PRNP* analysis and tonsil biopsy) are essentially specific for certain types of prion disease, and these are discussed in a section below.

3.1 The EEG

The EEG was a very important diagnostic investigation in the pre-MRI and CSF 14-3-3 era. Alongside the loss of normal background rhythms, characteristic bi/tri-phasic periodic activity develops in a significant proportion of cases, especially in sCJD (Steinhoff et al.).[2] Its drawbacks are: a relatively low sensitivity, a tendency for the typical pastern to appear relatively late in the disease course and the fact that such periodic activity is not confined to prion disease. Reported sensitivities for sCJD are around 66%.[2–4] The likelihood of finding the characteristic periodic activity depends partly on how far it is into the illness and the particular sCJD subtype.[4] An additional difficulty is that the EEG is not an always readily available investigation. Therefore, with other tests now available, in many clinical situations its role is now a relatively minor one.

3.2 Brain MRI

MR imaging, or MRI, has, of course, a vital role in the initial investigation of progressive brain disease, because of the possibility of non-prion diseases that are diagnosable through imaging. Its role in the positive diagnosis of prion diseases is based mainly on characteristic features seen in sCJD and vCJD.

In sCJD, high signal change in the caudate/putamen region is a particular feature associated with sCJD. Cortical high signal change is also seen. In vCJD, the key diagnostic feature is high signal in the posterior thalamus (the "pulvinar sign").[5] It is important to note that the "pulvinar sign" is not simply high signal in that region (which may be seen in sCJD) but higher signal there relative to what is seen in other areas of the brain.

The MRI changes are not entirely specific to prion disease and cortical signal change alone has the lowest specificity of the prion-associated findings. The MRI findings vary according to the MRI sequence employed.[4–7]

The characteristic findings in sCJD vary also with the sCJD subtype.[4,7] A study of 211 sCJD patients reported details of the MRI findings in different subtypes.[7] Basal ganglia hyperintensities occurred most frequently in MV2, VV2, and MM1 subtypes (79, 77, and 70%, respectively) Cerebral cortical signal increase was most common in VV1, MM2, and MV1 subtypes (86, 77, and 77%, respectively). Thalamic hyperintensities occurred most often in VV2 (45%) and MV2 (43%). The most consistent finding across most subtypes was high signal in basal ganglia, with these abnormalities found in 63% (FLAIR) and 71% (DWI).

While these characteristic abnormalities are easily seen by radiologists experienced in prion disease, one publication noted that they may not always be recognized on hospital routine reporting.[8]

3.3 CSF protein tests

The development of CSF protein tests was a great help in prion disease diagnosis with 14-3-3 being most important, along with lesser roles for S100b and tau. However, these proteins are normal brain proteins and, unsurprisingly, such tests are therefore not entirely specific for prion disease.

The tests require a lumbar puncture to be performed, to obtain CSF, but this is normally undertaken anyway in the investigation of rapidly progressive brain disease, because of the need to consider a fairly wide differential diagnosis in the initial disease stage, including inflammatory disorders. The sensitivity of a CSF 14-3-3 test for sCJD is high, with reported values being around 90%.[3,4,9] The sensitivity of the 14-3-3 test for sCJD also depends on subtype.[4] The specificity is somewhat lower and, of course, will depend on the testing context; in a big collaborative International study, it was reported to be 84%.[3] The sensitivity for vCJD is significantly lower at around 48–58%.[9,10] A study using a protein-capture step reported an increase in sensitivity for vCJD to 77%, although this is still below the sensitivity for sCJD.[9]

3.4 Test combinations

There have been assessments of the utility of combinations of the above tests. For example, in the international collaborative study mentioned above, a combination of positive findings on both EEG and CSF 14-3-3 test results gave a positive predictive value of 98% for sCJD and, when both tests were negative, the predictive value for the absence of CJD was 79%.[3] In another big collaborative study, it was noted that, in most cases of sCJD, it is very uncommon to find negative results in all three (brain MRI, CSF 14-3-3 test and EEG) investigations.[4]

4. Specific tests: Detection of abnormal prion protein

The detection of abnormal prion protein is the key to a truly specific diagnostic test for prion disease. Methods of detecting and identifying PrP^{Sc} are reasonably routine molecular laboratory procedures, as in the use of Western blots and immunohistochemistry. Such identification in brain tissue is the key to definitive neuropathological diagnosis and, in the tonsil, as a test for vCJD (discussed below). However, like all such methods, they have detection limits and, in non-brain tissues, levels of PrP^{Sc} are usually below these limits, particularly in human prion diseases other than vCJD. Two approaches to this problem have been employed: capture/concentration and what have been termed "amplification" techniques (which can, of course, be used in combination).

4.1 Capture/concentration techniques

Methods to capture prion protein, or to concentrate it, have involved different techniques including phosphotungstic acid precipitation, immunoprecipitation, and capture via beads or particles in various combinations.[11–14] These methods will not be reviewed in detail here.

4.2 The development of protein "amplification" techniques

Underlying the prion disease process, is a post-translational conformational change in the structure of the prion protein. Whatever initiates this, it is clear that abnormal prion protein (PrP^{Sc}) is able to interact with normal prion protein (PrP^{C}) in an auto-catalytic production of more abnormal protein, leading to increasing amounts of PrP^{Sc}, which aggregates and is deposited in tissue.[1] This description of the disease process is an over-simplification in a number of ways including variations in the use of terminology (such as exactly to what the term PrP^{Sc} refers, which often is used interchangeably with the term PrP^{RES} that, strictly speaking, refers to the protease-resistant core) and the existence of other prion protein forms that may be pathogenetically important intermediate protein structural steps between PrP^{C} and the aggregating PrP^{Sc}.[15] However, recognition of this disease process has led to the development of what have been termed protein "amplification" techniques, whereby some form of normal prion protein is, in some way, used to "amplify" low levels of abnormal prion protein. Clearly, the detection of PrP^{Sc} by any means is a potentially specific diagnostic test as it relates to a fundamental underlying disease mechanism. Amplification techniques offer this opportunity with two important caveats. The first is the exact meaning of the phrase "amplification of PrP^{Sc}" and the second is that the technique should not in itself "create" abnormal protein from normal, non-disease, samples. The various techniques have in common the combination of a test sample (which may or may not contain disease-related abnormal prion protein) and some form of normal prion protein. There is then a process whereby the normal prion protein is converted into an abnormal form and then this product is detected and measured by some means. However, it is not necessarily clear exactly what specific abnormal protein is being "amplified" nor the exact nature of the resulting product. What is certain is that normal PrP^{C} is being changed into some detectable and measurable abnormal form in a process that depends on the test sample being derived from an individual with a prion disease. These techniques therefore use some "seeding" property of the disease sample to change the normal prion protein substrate. Sometimes these seeding properties are called "prion seeding" but this is potentially confusing if the term "prion" is taken in its standard sense as referring to the infectious agent. Furthermore, the normal prion protein is being subjected to various challenging processes in these test procedures and, for the test to be reliable, one wants to avoid some sort of "spontaneous" abnormal protein product arising simply from the process itself.

The main "amplification" methods in wide use are PMCA and RT-QuIC. PMCA (Protein Misfolding Cyclic Amplification) uses brain tissue extract as the substrate, containing as it does, PrP^{C} (although other things as well, some of which act as co-factors in the process). Its output is detected by Western blot. It produces an amplification product that has a protease-resistant form with identical Western blot electrophoretic mobility and glycosylation pattern to those of PrP^{RES},the protease-resistant core of PrP^{Sc}, obtained from prion disease brain and which also has infectious properties demonstrated in animal transmission experiments.[16,17] A different technique, RT-QuIC (Real-Time Quaking-Induced Conversion) uses some form of recombinant PrP^C as substrate and its output is detected, and measured, in real-time, by the fluorescence properties of protein aggregates. This product has a different Western blot electrophoretic mobility and glycosylation pattern to that of the protease-resistant core derived directly from disease-associated PrP^{Sc} and it appears to lack infectious properties in animal transmission experiments.[18]

The nature of the prion hypothesis naturally suggests that one should be able to create misfolded prion protein, from PrP^C, that has the structural and infective properties of disease-derived PrP^{Sc}. There are a number of reports of experimental success in this using various techniques.[17,19,20] Of most relevance here, success in generating infectious protein has been reported using PMCA, although, as the authors point out, it cannot be certain that this did not result from replication of a form of abnormal prion protein already present in the brain homogenate they used.[21] Even if this is indeed the generation of a form of PrP^{Sc}, to call it "spontaneous" generation (as has been done) is perhaps misleading, as it is of the result of various technical manipulations. It does appear to be more of a concern with PMCA than with RT-QuIC.[22–24]

However, leaving aside the interesting theoretical considerations of these test techniques, as clinically useful tests, the important point is how they behave in the reliable sensitive and specific diagnosis of disease, that is: how their results correlate with the presence or absence of disease in the tested individual.

There are other described techniques based on seeding activity which will not be detailed here, for example, ASA (the amyloid seeding assay) and SOPHIA (the surround optical fiber immunoassay).[23,25,26] There are also modifications of standard techniques such as the e-QuIC modification of Rt-QuIC.[23,24]

Amplification methods may be combined with concentration steps (such as bead capture) to increase the likelihood of detection of abnormal prion protein techniques (such as bead capture).

5. Current routine clinical practice

The clinical diagnosis of human prion disease consists of four steps:

(1) The recognition of the possibility of a prion disease.

(2) The exclusion of other possible illnesses.

(3) Support from essentially non-specific, but often very useful, tests (EEG, cerebral MRI, CSF 14-3-3), as discussed above.
(4) Specific tests relating to particular prion diseases. These specific tests are discussed below.

5.1 Brain biopsy

As a means of obtaining cerebral tissue, this is clearly a very direct diagnostic technique. However, it is an invasive procedure with potential complications and requires proper handling of associated infection risks. It is, therefore, undertaken rarely. With prion disease being untreatable currently, most clinicians feel that the principal indication for cerebral biopsy is the consideration of another, potentially treatable, disease that cannot be diagnosed by another method.

5.2 *PRNP* analysis

Genetic prion disease will be suspected, of course, if there is a clear family history either of confirmed prion disease or suggestive of prion disease. However, a family history may be absent in many cases; in one big, international study, in 12–88% depending on the specific mutation.[27] The reasons for this relate to factors such as hidden/unrecognized history, death of relatives from other disease before the age of onset of prion disease, non-paternity and variable penetrance.[28] It is reasonable to consider *PRNP* testing (along with other genetic analyses) in cases of neurodegenerative disease of uncertain diagnosis. Additionally, a genetic cause for apparently sporadic prion disease can be absolutely excluded only by such testing. Pre-symptomatic *PRNP* analysis can be undertaken in healthy individuals at risk of genetic prion disease, but this is a matter for careful consideration and, here, the discussion is of diagnostic testing in symptomatic individuals.

PRNP analysis includes 129-codon characterization; this is not diagnostic in itself, representing, as it does, a normal population polymorphism, but it can be useful in considering a diagnosis, for example, in the classification of sCJD subtypes.

5.3 Tonsil biopsy

Tonsil biopsy is a potentially useful test for vCJD because of the deposition of PrP^{Sc} in lymphoreticular tissues in this particular form of prion disease.[29] A positive result makes a diagnosis of vCJD very likely although a negative result cannot completely exclude it. A positive biopsy has not been considered to mean a definite diagnosis as lymphoreticular positivity can occur in asymptomatic or pre-clinically infected individuals. It is, therefore, always at least possible that an individual with another neurological disease could have an incidentally positive biopsy. Having said that, this would be an unlikely co-incidence and, in the right clinical context, a positive biopsy must make vCJD a diagnosis beyond reasonable doubt. Despite its specificity for vCJD, tonsil biopsy is not necessary in most suspected cases, as a very probable clinical diagnosis often can be made on clinical features with other, less invasive, investigations such as MRI brain imaging.

5.4 CSF RT-QuIC

This test is now employed as part of routine clinical diagnosis in relation to sCJD; its sensitivity for this diagnosis is high but there are different technical forms of the test and any reported sensitivity needs to be related to the particular test used. What has been termed "1st generation RT-QuIC" is based on using recombinant full-length Syrian Hamster prion protein (rec SHa(23 – 231)).[18] However, some versions of the test have other substrates.[30,31] What has been termed "2nd generation RT-QuIC" is based on using truncated Syrian Hamster protein (rec SHa(90–231)).[32] Using different substrates may require other, but slight, changes to the test methodology. The terms 1st and 2nd generation can cause confusion and it is probably best to refer to the relevant test in terms of its specific substrate (for example, rec SHa(23–231) Rt-QuIC). The results of Rt-QuIC testing may vary with the substrate used and, therefore, any particular method may require its own evaluation before adoption into clinical practice.[30,33] Another factor potentially affecting sensitivity is test result variation with different sCJD subtypes.[32,34] In the UK CJD surveillance system, using RT-QuIC based on full-length Syrian Hamster prion protein, a sensitivity for sSCD of 92% has been reported.[34] The specificity of this test for sCJD depends not only on the test characteristics but also on the sample population being tested (as for any test). In the circumstances of the UK surveillance system, the reported current specificity for sCJD is 100%.[34]

6. Other tests

A variety of diagnostic tests have been developed, typically employing amplification techniques, but also (as in the case of nasal brushing) based on ways of obtaining neural tissue other than by brain biopsy. Naturally, these tests are usually developed initially in a laboratory context, sometimes in animal (not human) contexts, using particular strains of prion and small-scale methods that may require time and close attention from experienced individuals. There is a potentially big gap between reports of such laboratory success and a reliable, robust diagnostic test suitable for routine human clinical use. In particular, human strains of prion in a real disease setting may behave differently from other prion strains or from human prion strains in an *in vitro* setting. The real utility of potential tests in routine clinical settings has to take three things into particular account: technical aspects, sensitivity and specificity.

1. On the technical side, a test has to be robust, consistent and produce timely results. A test may be reasonable in a small-scale laboratory setting where time is not critical, but which may be difficult to scale-up to use in a large-volume, routine service setting where results are required quickly.
2. Its sensitivity has to be assessed against some established standard (such as eventual autopsy confirmation); for sporadic CJD, this should be a practical possibility, but, for vCJD, this is difficult as, fortunately, there are no living cases at this time and only very limited historical samples available for testing.
3. The assessment of specificity is generally more difficult than that of sensitivity. Firstly, the specificity of a test depends not only on the test but also on the context in which it is used. Initial assessments often depend on testing a set of selected controls such as samples from healthy individuals and those with diseases thought to be appropriate (such as other neurodegenerative dementias). However, specificity cannot really be established accurately until a test has been used in a significant number of instances in real-life clinical practice. Even then, clinicians need to be aware that the quoted specificity for a test is valid only if the test is being used in the same context in which that specificity was assessed.

All of these considerations are complicated by the fact that there are often variations in technique between different laboratories and, therefore, reported assessments are not always directly comparable. For example, as noted above, in RT-QuIC, different prion protein substrates may be used and, in PMCA, the number of amplification cycles employed may vary.

Aside from genetic analysis, EEG, CSF 14-3-3, brain MRI and CSF RT-QuIC, the various reported clinical diagnostic tests have been, so far, subject to relatively limited evaluation.

Some of the tests discussed below have been evaluated only in a research setting and some in only limited clinical or surveillance settings; none have been adopted yet into internationally accepted routine protocols.

6.1 Skin biopsy

A study of skin samples (all taken after death) concerned 21 sCJD, 2 vCJD and 15 non-CJD individuals. Using RT-QuIC to detect PrP^{Sc} all 23 CJD samples were positive, with negative results in the 15 controls.[35]

A further study, concerned skin samples taken from two sites (cervical and thigh areas) in 57 patients referred as potential cases of CJD (35 of whom proved to have sCJD; 22 did not) and 15 patients with other neurodegenerative conditions.[33] Interestingly, positive results were not always obtained in both sample sites from the same individual. Most of the CJD samples were obtained after death, but in three cases they were biopsies and, in a further two cases, samples were taken both in life and after death. They used RT-QuIC as the detection method but used two different substrates (rec SHa(23–231) and rec Bv(22 – 230)). Their overall results gave a sensitivity of 89% and a specificity of 100% for sCJD. They noted that the SHa(23–231) substrate performed less well than the Bv(22–230) substrate, illustrating the potential importance of the specific test methodology employed. They suggested that their test might be more sensitive in some subtypes of sCJD than others; although the small numbers in their study do not allow for a definitive statement, variation in test results with sCJD subtypes is well recognized (for example, as reported with brain MRI).

The precise role of skin biopsy in routine prion disease diagnosis requires further study. Interestingly, in the 2017 study, two sCJD skin samples showed successful disease transmission to transgenic mice, indicating infectivity as well as seeding activity in the skin.[35]

6.2 Nasal brushing/swabbing

Reports concerning olfactory epithelium (obtained by nasal brushing or swabbing) in RT-QuIC testing represent an interesting development not only because of the use of an amplification technique but also because nasal epithelium contains olfactory neurons; the technique therefore represents a relatively non-invasive method of obtaining neurological tissue. The initial publication described testing of nasal brushings from 31 CJD cases (15 definite sCJD, 13 probable sCJD and 2 genetic cases) and 43 controls, reporting a sensitivity for CJD 97% and a specificity of 100%.[36]

Another study described the results in a further set of CJD cases and controls, using an improved, gentler, method of obtaining olfactory epithelium via a designed swab. In this study, the authors evaluated the sensitivity and specificity of combining an RT-QuIC analysis of both CSF and olfactory epithelium as both being 100%.[37] The sensitivity for sCJD from the nasal swab test alone was also high but the authors reported two cases of sCJD with negative olfactory mucosal testing and discuss possible reasons for this, including nasal problems that might interfere with obtaining a satisfactory sample.[37]

6.3 Urine tests

A PMCA-based urine study found positive results in 13/14 vCJD urine samples, and negative results in samples from 68 sCJD cases, 4 genetic prion disease cases, 52 healthy individuals, 50 other neurodegenerative disease cases and 50 other neurological cases.[38]

Another study of a urine test, based on a previously described blood test, analyzed specimens from 125 controls (91 healthy; 34 non-prion neurodegenerative disease) and 37 prion disease cases (20 sCJD, 13 vCJD, 2 genetic, 2 iatrogenic CJD), finding a positive result in 40% of the sCJD cases but in only 1/13 vCJD cases.[39]

6.4 Blood tests

Developing blood tests for prion disease has proved technically difficult, reflecting the problem of detecting low levels of abnormal prion protein in the complex matrix of blood. Aside from possible variations in test methodology, the performance of any test may vary according to the nature of the sample (whole blood, plasma, etc) and possibly on the anticoagulant used in the blood sample (EDTA, citrate, etc). In addition, some experiments have used blood spiked with prion material (such as brain), but prion protein in natural disease blood might not behave in the same way. A number of animal-based studies have been reported. For example, a PMCA/SOFIA test has produced positive results in both clinical and pre-clinical blood samples from sheep and white-tailed deer with CWD.[26] In another, RT-QuIC-based technique study, success was reported in experiments with spiked-plasma and hamster scrapie samples.[24] A recently reported animal study arguably has the greatest possible relevance to the human situation as it reported positive results using a PMCA-based blood test on pre-clinical samples from vCJD-infected macaques.[40]

However, the discussion here will concern those tests that have been evaluated in the human diagnostic context, of which there are three, each of which relates to vCJD. Naturally, in the instance of vCJD, one real problem is the scarcity of blood samples on which to evaluate any proposed test.

1. Edgeworth et al.

The first was reported as a test for vCJD, using a solid capture/concentration step followed by a direct immunodetection method. This report concerned 21 vCJD, 27 sCJD, 42 other neurological disease and 100 normal samples.[14] All were whole blood, using EDTA as the anticoagulant. The reported sensitivity and specificity for vCJD was 71.4% and 100%, respectively. In a further study with 5552 control and 10 vCJD samples, the same research group reported a sensitivity of 70%, with no positive results in the control group, but with positive results in 2 samples from sCJD cases.[41]

2. Concha-Marambio et al.

This study used PMCA following centrifugation of samples in the presence of sarkosyl. They reported positive results for 14 samples from clinically ill vCJD individuals with negative results in 137 controls (62 non-CJD neurodegenerative illnesses, 26 other neurological illnesses, 49 healthy controls) and also negative results in samples from 6 individuals with sCJD.[42]

3. Bougard et al.

This test was reported as a test for vCJD.[43] It is based on a combination of plasminogen-bead capture and PMCA on plasma samples, with results not depending on the use of any particular anticoagulant in the plasma sample. The study had a 100% sensitivity and a diagnostic specificity of 100% for vCJD in a sample set of 256 (67 sCJD, 67, non-CJD neurological illness, 104 healthy controls and 18 clinical cases of vCJD).

In addition, this study reported, for the first time, positive results in presymptomatic vCJD blood samples. Samples had been stored from two blood donors who went on to develop vCJD. Samples taken 1.3 and 2.6 years prior to the onset of illness were positive.

7. Concluding comments

Looking back over the four decades since the prion protein underpinning of these diseases came to light, there has been a big improvement in the clinical diagnosis of human prion disease. There was a relatively limited appreciation of the full range of the clinical phenotype and the EEG was the only supportive diagnostic test. Things moved on with MRI and CSF 14-3-3 testing; these clinical test developments along with the detection of abnormal PrP in tissue as a pathological hallmark, allowed the recognition of less typical clinical pictures. Now, with the development of laboratory techniques based on the underlying protein process characterizing these diseases, there is an ever-increasing number of potentially useful diagnostic tests. As with all tests, they need to be carefully evaluated in appropriate clinical settings and, to date, the most clearly established test in routine use is the CSF RT-QuIC test for sCJD. Earlier, more accurate, clinical diagnosis has obvious importance for those affected in terms of understanding or planning of care and will be crucial if successful treatments are developed. Quite aside from their diagnostic utility in the management of neurologically ill individuals, these tests have potential public health and research functions. At the public health level, an appropriate and properly validated test could answer the question as to the number of asymptomatically infected people there are in BSE-exposed populations or act as a screening test for potential infectivity in, for example, blood donations.[41] However, any routine testing of blood donors requires a robust, rapid test that can deal with large numbers of samples and there are considerations other than technical ones: ethical, practical consequences and cost. For example, would there be a significant effect on the willingness of individuals to donate, what does one tell healthy donors who test positive and how should they be managed if they require medical or surgical treatment? All of these questions needing to be considered in the context of the occurrence of some false-positive tests and the increase in costs of providing blood. On the research side, these test methodologies, with their varying behaviors (such as the differences between PMCA and RT-QuIC) raise important theoretical questions and may give potential insights into prion disease mechanisms. Finally, it is important to note that although the newer tests are finding positive results in tissue such as blood, urine, nasal epithelium and skin, with some successful laboratory transmission experiments using such tissue, this does not imply significant infection risk in ordinary life: there is no evidence for person-to-person transmission of human prion diseases through ordinary (even intimate) contact.

References

1. Head MW, Ironside JW, Ghetti B, Jeffrey M, Piccardo P, Will RG. Prion diseases. In: Love S, Perry A, Ironside JW, Budka H, eds. *Greenfield's Neuropathology*. Boca Raton, Florida: CRC Press, Taylor and Francis Group; 2015:1016–1098.
2. Steinhoff BJ, Racker S, Herrendorf G, et al. Accuracy and reliability of periodic sharp wave complexes in Creutzfeldt-Jakob disease. *Arch Neurol*. 1996;53:162–166.

3. Zerr I, Pocchiari M, Collins S, et al. Analysis of EEG and CSF 14-3-3 proteins as aids to the diagnosis of Creutzfeldt–Jakob disease. *Neurology*. 2000;55(6):811–815.
4. Collins SJ, Sanchez-Juan P, Masters CL, et al. Determinants of diagnostic investigation sensitivities across the clinical spectrum of sporadic Creutzfeldt-Jakob disease. *Brain*. 2006;129(Pt. 9):2278–2287.
5. Collie DA, Sellar RJ, Zeidler M, Colchester AFC, Knight R, Will RG. MRI of Creutzfeldt-Jakob disease: imaging features and recommended MRI protocol. *Clin Radiol*. 2001;56:726–739.
6. Kallenberg K, Shulz-Shaeffer WJ, Jastrow S, et al. Creutzfeldt-Jakob disease: comparative analysis of MR imaging sequences. *AJNR*. 2006;7:1459–1462.
7. Meissner B, Kallenberg K, Sanchez-Juan P, et al. MRI lesion profiles in sporadic CJD. *Neurology*. 2009;72(23):1994–2001.
8. Carswell C, Thompson A, Lukic A, et al. MRI findings are often missed in the diagnosis of Creutzfeldt-Jakob disease. *BMC Neurol*. 2012;12:153. https://doi.org/10.1186/1471-2377-12-153.
9. AJE G, Ramliak S, WEG M, RSG K, Achrooder HC. 14-3-3 in the cerebrospinal fluid of patients with variant and sporadic Creutzfeldt-Jakob disease measured using capture assay able to detect low levels of 14-3-3. *Neurosci Lett*. 2002;324(1):57–60.
10. Green AJE, Thompson EJ, Stewart GE, et al. Use of 14-3-3 and other brain-specific proteins in CSF in the diagnosis of variant Creutzfeldt-Jakob disease. *J Neurol Neurosurg Psychiatry*. 2001;70(6):744–748.
11. Morel N, Simon S, Frobert Y, et al. Selective and efficient immunoprecipitation of the disease-associated form of the prion protein can be mediated by nonspecific interactions between monoclonal antibodies and scrapie-associated fibrils. *J Biol Chem*. 2004;279: 30143–30149.
12. Wenborn A, Terry C, Gros N, et al. A novel and rapid method for obtaining high titre intact prion strains from mammalian brain. *Sci Rep*. 2015;5:10062. https://doi.org/10.1038/srep10062.
13. Miller MB, Supattapone S. Superparamagnetic nanoparticle capture of prions for amplification. *J Virol*. 2011;85(6):2813–2817. https://doi.org/10.1128/JVI.02451-10.
14. Edgeworth JA, Farmer M, Sicilia A, et al. Detection of prion infection in variant Creutzfeldt-Jakob disease: a blood-based assay. *Lancet*. 2011;377(9764):487–493. https://doi.org/10.1016/S0140-6736(10)62308-2.
15. Terry C, Wadsworth JDF. Recent advances in understanding mammalian prion structure: a mini review. *Front Mol Neurosci*. 2019;12:169. https://doi.org/10.3389/fnmol.2019.00169.
16. Saborio GP, Permanne B, Soto C. Sensitive detection of pathological prion protein by cyclic amplification of protein misfolding. *Nature*. 2001;411:810–813.
17. Castilla J, Saá P, Hetz C, Soto C. In vitro generation of infectious scrapie prions. *Cell*. 2005;121:95–206.
18. Atarashi R, Wilham JM, Christensen L, et al. Simplified ultrasensitive prion detection by recombinant PrP conversion with shaking. *Nat Methods*. 2008;5:211–212.
19. Legname G, Baskakov IV, Nguyen HO, et al. Synthetic mammalian prions. *Science*. 2004;305(5684):673–676. https://doi.org/10.1126/science.1100195.
20. Deleault NR, Harris BT, Rees JR, Supattapone S. Formation of native prions from minimal components in vitro [published correction appears in Proc Natl Acad Sci U S A. 2008 Aug 26;105(34):12636]. *Proc Natl Acad Sci U S A*. 2007;104(23): 9741–9746. https://doi.org/10.1073/pnas.0702662104.
21. Barria MA, Mukherjee A, Gonzalez-Romero D, Morales R, Soto C. De novo generation of infectious prions in vitro produces a new disease phenotype. [published correction appears in PLoS Pathog. 2013 Mar;9(3). doi:10.1371/annotation/4b55946a-edb5-4feb-aeb5-2ee160394d17]. *PLoS Pathog*. 2009;5(5):e1000421. https://doi.org/10.1371/journal.ppat.1000421.

22. Atarashi R, Sano K, Satoh K, Nishida N. Real-time quaking-induced conversion: a highly sensitive assay for prion detection. *Prion*. 2011;5(3):150–153. https://doi.org/10.4161/pri.5.3.16893.
23. Orrù CD, Wilham JM, Vascellari S, Hughson AG, Caughey B. New generation QuIC assays for prion seeding activity. *Prion*. 2012;6(2):147–152. https://doi.org/10.4161/pri.19430.
24. McGuire LI, Peden AH, Orrú CD, et al. Real time quaking-induced conversion analysis of cerebrospinal fluid in sporadic Creutzfeldt-Jakob disease. *Ann Neurol*. 2012;72(2):278–285. https://doi.org/10.1002/ana.23589.
25. Colby DW, Zhang Q, Wang S, et al. Prion detection by an amyloid seeding assay. [published correction appears in Proc Natl Acad Sci U S A. 2008 Feb 5;105(5):1774]. *Proc Natl Acad Sci U S A*. 2007;104(52):20914–20919. https://doi.org/10.1073/pnas.0710152105.
26. Rubenstein R, Chang B, Gray P, et al. A novel method for preclinical detection of PrPSc in blood. *J Gen Virol*. 2010;91(Pt. 7):1883–1892. https://doi.org/10.1099/vir.0.020164-0.
27. Kovács GG, Puopolo M, Ladogana A, et al. Genetic prion disease: the EUROCJD experience. *Hum Genet*. 2005;118(2):166–174. https://doi.org/10.1007/s00439-005-0020-1.
28. Minikel EV, Vallabh SM, Lek M, et al. Quantifying prion disease penetrance using large population control cohorts. *Sci Transl Med*. 2016;8(322):322ra9. https://doi.org/10.1126/scitranslmed.aad5169.
29. Collinge J, Rossor MN, Thomas D, Frosh A, Tolley N. Tonsil biopsy helps diagnose new variant Creutzfeldt-Jakob disease. *BMJ*. 1998;317:472.
30. Cheng K, Sloan A, Waitt B, et al. Altered rPrP substrate structures and their influence on real-time quaking induced conversion reactions. *Protein Expr Purif*. 2018;143:20–27. https://doi.org/10.1016/j.pep.2017.10.007.
31. Bistaffa E, Vuong TT, Cazzaniga FA, et al. Use of different RT-QuIC substrates for detecting CWD prions in the brain of Norwegian cervids. *Sci Rep*. 2019;9(1):18595. Published 2019 Dec 9, https://doi.org/10.1038/s41598-019-55078-x.
32. Franceschini A, Baiardi S, Hughson AG, et al. High diagnostic value of second generation CSF RT-QuIC across the wide spectrum of CJD prions. *Sci Rep*. 2017;7(1):10655. Published 2017 Sep 6, https://doi.org/10.1038/s41598-017-10922-w.
33. Mammana A, Baiardi S, Rossi M, et al. Detection of prions in skin punch biopsies of Creutzfeldt-Jakob disease patients. *Ann Clin Transl Neurol*. 2020;7(4):559–564. https://doi.org/10.1002/acn3.51000.
34. Green AJE. RT-QuIC: a new test for sporadic CJD. *Pract Neurol*. 2019;19(1):49–55. https://doi.org/10.1136/practneurol-2018-001935.
35. Orrú CD, Yuan J, Appleby BS, et al. Prion seeding activity and infectivity in skin samples from patients with sporadic Creutzfeldt-Jakob disease. *Sci Transl Med*. 2017;9(417):eaam7785. https://doi.org/10.1126/scitranslmed.aam7785.
36. Orrú CD, Bongianni M, Tonoli G, et al. A test for Creutzfeldt-Jakob disease using nasal brushings. [published correction appears in N Engl J Med. 2014 Nov 6;371(19):1852]. *N Engl J Med*. 2014;371(6):519–529. https://doi.org/10.1056/NEJMoa1315200.
37. Bongianni M, Orrù C, Groveman BR, et al. Diagnosis of human prion disease using real-time quaking-induced conversion testing of olfactory mucosa and cerebrospinal fluid samples. *JAMA Neurol*. 2017;74(2):155–162. https://doi.org/10.1001/jamaneurol.2016.4614.
38. Moda F, Gambetti P, Notari S, et al. Prions in the urine of patients with variant Creutzfeldt-Jakob disease. *N Engl J Med*. 2014;371(6):530–539. https://doi.org/10.1056/NEJMoa1404401.
39. Luk C, Jones S, Thomas C, et al. Diagnosing sporadic Creutzfeldt-Jakob disease by the detection of abnormal prion protein in patient urine. *JAMA Neurol*. 2016;73(12):1454–1460. https://doi.org/10.1001/jamaneurol.2016.3733.

40. Concha-Marambio L, Chacon MA, Soto C. Preclinical detection of prions in blood of nonhuman primates infected with variant Creutzfeldt-Jakob disease. *Emerg Infect Dis.* 2020;26(1):34–43. https://doi.org/10.3201/eid2601.181423.
41. Jackson GS, Burk-Rafel J, Edgeworth JA, et al. Population screening for variant Creutzfeldt-Jakob disease using a novel blood test: diagnostic accuracy and feasibility study. *JAMA Neurol.* 2014;71(4):421–428. https://doi.org/10.1001/jamaneurol.2013.6001.
42. Concha-Marambio L, Pritzkow S, Moda F, et al. Detection of prions in blood from patients with variant Creutzfeldt-Jakob disease. *Sci Transl Med.* 2016;8(370):370ra183. https://doi.org/10.1126/scitranslmed.aaf6188.
43. Bougard D, Brandel JP, Bélondrade M, et al. Detection of prions in the plasma of presymptomatic and symptomatic patients with variant Creutzfeldt-Jakob disease. *Sci Transl Med.* 2016;8(370):370ra182. https://doi.org/10.1126/scitranslmed.aag1257.

CHAPTER TWO

Understanding prion structure and conversion

Giovanni Spagnolli[a,b], Jesús R. Requena[c], and Emiliano Biasini[a,b,*]
[a]Department of Cellular, Computational and Integrative Biology (CIBIO), Trento, TN, Italy
[b]Dulbecco Telethon Institute, University of Trento, Trento, TN, Italy
[c]CIMUS Biomedical Research Institute & Department of Medical Sciences, University of Santiago de Compostela-IDIS, Santiago, Spain
*Corresponding author: e-mail address: emiliano.biasini@unitn.it

Contents

Abstract

Since their original identification, prions have represented enigmatic agents that defy the classical concept of genetic inheritance. For almost four decades, the high-resolution structure of PrP^{Sc}, the infectious and misfolded counterpart of the cellular prion protein (PrP^{C}), has remained elusive, mostly due to technical challenges posed by its high insolubility and aggregation propensity. As a result, such a lack of information has critically hampered the search for an effective therapy against prion diseases. Nevertheless, multiple attempts to get insights into the structure of PrP^{Sc} have provided important experimental constraints that, despite being at limited resolution, are paving the way for the application of computer-aided technologies to model the three-dimensional architecture of prions and their templated replication mechanism. Here, we review the most relevant studies carried out so far to elucidate the conformation of infectious PrP^{Sc} and offer an overview of the most advanced molecular models to explain prion structure and conversion.

1. Introduction

Since its original formulation, the idea of prions as proteinaceous infectious particles causing rare transmissible pathologies of the brain has gradually moved to the broader concept of protein-based inheritance.[1–3]

Progress in Molecular Biology and Translational Science, Volume 175
ISSN 1877-1173
https://doi.org/10.1016/bs.pmbts.2020.07.005

This view has mainly been supported by the identification of prion-like features in several other pathological proteins,[4] as well as by the discovery of a number of physiological prions in a variety of species, from bacteria to low eukaryotes to mammalian cells, propagating biological traits as epigenetic elements.[5] PrP^{Sc}, the original, bona fide prion identified in mammals, is the causative agent of transmissible neurodegenerative pathologies known as prion diseases.[2] These include conditions affecting humans, such as Kuru, Creutzfeldt-Jakob disease (CJD), Gerstmann-Straussler-Scheinker (GSS) and Fatal Familial Insomnia (FFI), but also a wide variety of animals, including, among many others, scrapie in sheep, Bovine Spongiform Encephalopathy (BSE) in cows and Chronic Wasting Disease (CWD) in elks and deer.[6,7] The way PrP^{Sc} propagates involves the templated conversion of a physiological protein, the cellular prion protein (PrP^{C}).[8,9] PrP^{C} is a glycoprotein of uncertain function widely expressed on the surface of several cell types, particularly neuronal cells.[10] Coherently, the central dogma in prion biology is that blocking the ability of PrP^{Sc} to replicate and accumulate into the brain is the most direct way to treat prion diseases.[11,12] However, any serious attempt to rationally achieve this goal would require atomic-level information about PrP^{Sc} architecture as well as all the structural transitions underlying the templated conversion of PrP^{C}.[13–16] Consequently, a large number of strategies have been exploited from many different laboratories worldwide over the last four decades to shed light on the structure of infectious PrP^{Sc} and its replication.[13] However, so far these attempts have defined the mechanism by which PrP^{Sc} interacts with PrP^{C} and catalyzes its conformational conversion only at a macroscopic level, while a deep understanding of the molecular aspects of its structure, replication and pathogenicity are still missing.[17] Such a fundamental lack of information, coupled with the surprising ability of prions to exist under different conformations possessing biochemical and biological distinct features (known as prion strains), likely explain why an effective therapy for prion diseases is still missing.[18]

In this chapter, we focus on previous experimental attempts to elucidate the architecture of infectious PrP^{Sc} and how the information arising from these approaches is being exploited by emerging computational technologies to shed light on the ability of prion to propagate their conformationally-encoded information.

2. Structural analyses of infectious and non-infectious PrP aggregates

An important factor should be taken into account when reviewing studies focused on the structural characterization of PrP^{Sc}. The term "prion," as originally coined by Stanley Prusiner, indicated a pathologically-relevant protein capable of replicating like an infectious agent but lacking nucleic acids.[1] Such a definition implies both biochemical and biological features. From a structural standpoint, PrP^{Sc} is generally conceived as aggregated and rich in β-sheets, in contrast to its physiological counterpart PrP^{C}, which is monomeric and mainly alpha-helical. However, compelling evidence indicates that different aggregated assemblies of PrP may correlate with disease, but not necessarily be infectious.[19,20] This is the case of some inherited forms of prion diseases,[20] but also of synthetic PrP amyloids obtained by various procedures.[21] For example, few naturally occurring mutant PrP aggregates or synthetic PrP amyloids may induce pathology in mice but be unable to transmit infectivity.[22,23] Importantly, infectious and non-infectious aggregates of PrP may share multiple biochemical similarities, implying that the ability of self-propagation in a biologically-effective manner may lie within subtle structural determinants.[19,24] To add further complexity, some non-infectious PrP aggregates show the ability to propagate in vitro, following a "seeding" mechanism shared by many different amyloids.[13] Conversely, several amyloidogenic proteins associated with non-transmissible diseases have been shown to propagate in cells and tissues like prions, possibly explaining their spreading in different brain areas during the course of the disease.[4,25] In light of such a great deal of complexity, one important question remains: What are the structural determinants underlying prion infectivity? The most effective way to answer this question is clearly to solve the structure of both infectious and non-infectious PrP aggregates at high-resolution. Thus, when considering studies aimed at elucidating the structure of prions, it is of fundamental importance to distinguish whether they have been performed on infectious or non-infectious PrP material. For the rest of this chapter, we will mainly focus on the former.

3. Low-resolution approaches to study the structure of infectious prions

A number of low-resolution techniques have provided important experimental constraints about the structure of infectious PrP^{Sc}. Fourier-transform infrared spectroscopy (FTIR) and circular dichroism (CD) spectroscopy were among the first techniques employed to collect a first glance of PrP^{Sc} structure. These methods originally revealed the high β-sheet content of N-terminally truncated PrP^{Sc} molecules.[26–28] However, for long time, an incorrect interpretation of the ~1660 cm^{-1} band in FTIR spectra led to the conclusion that a substantial fraction of PrP^{Sc} folding retained an α-helical conformation, like PrP^C.[29] Later on, it was instead showed that the same band is also present in spectra of recombinant PrP (recPrP) amyloid fibrils folded as parallel in-register β-sheets, which completely lack α-helical regions.[30] Despite being limited by their intrinsic low resolution, these techniques provided a first, important experimental constraint: infectious PrP^{Sc} particles are organized exclusively as β-sheets and random coils.

Additional information regarding the structural organization of PrP^{Sc} arose from the use of chemical probes. For example, cross-linking experiments using the bis-(sulfosuccinimidyl)-suberate (BS3) coupled with mass spectrometry on the 263K Syrian hamster PrP^{Sc} identified inter-molecular proximity of two glycine residues in position 90, possibly reflecting a close proximity of the two residues positioned on two intertwining fibrils.[31] In another study, labeling of the 263K prion with the tyrosine-specific reagent tetranitromethane (TNM) revealed a very substantial rearrangement of the C-terminal domain of PrP^C upon conversion into PrP^{Sc}, as shown by the drastically reduced reactivity of two tyrosine residues (Tyr225 and Tyr226).[32] This observation further refuted the hypothesis that the C-terminal α-helices of PrP^C are conserved in PrP^{Sc}.

Another technique employed to visualize mature PrP^{Sc} fibrils is scanning probe microscopy (SPM), also known as atomic force microscopy (AFM). In this approach, a nano tip attached to a cantilever was employed to scan a silica surface where PrP aggregates have previously been immobilized, giving rise to unique 3D scanning profiles for each aggregate.[33] Because of its simplicity, SPM has been employed in several other studies aimed at characterizing PrP^{Sc}. For example, fibrillar structures likely corresponding to large PrP^{Sc} amyloids were observed by SPM in neuroblastoma cells persistently infected with mouse prions.[34] Moreover, the technique was used to estimate the height and helical periodicity of individual amyloid fibrils derived from purified preparations of 22L and RML mouse prions, as well as to compare the morphology of in vitro generated PrP^{Sc} fibrils containing various co-factors, such as palmitoyloleoylphosphatidylglycerol (POP-G) or RNA molecules.[35]

Two additional techniques, limited proteolysis and hydrogen/deuterium (H/D) exchange, provided important insights into the structure of PrP^{Sc} by focusing on different aspects of solvent accessibility of the PrP polypeptide. Limited proteolysis is based on the concept that, under limiting conditions, proteases show a net preference for protein regions exhibiting accessible and flexible secondary structure, such as α-helices and loops, while largely sparing β-strands.[36] Similarly, H/D exchange builds on the notion that unstructured protein regions in solution are characterized by a rapid exchange of hydrogen for deuterium ions at backbone amides, as compared with the moderately or largely slower exchange of α-helices and β-sheets, respectively. Limited proteolysis performed with proteinase K (PK) was among the first set of techniques showing that PrP^{Sc} fibrils are strongly resistant to proteolytic cleavage, with the exception of the N-terminal residues (~23–90), which are often found exceedingly labile.[37] Importantly, PrP^{Sc} fibrils N-terminally deleted by treatment with PK (originally referred to as PrP 27–30, following the estimation in kDa of its molecular weight, as observed by SDS PAGE) retained full infectivity.

More recently, the technique was used to obtain a complete PK-sensitivity map of anchorless PrP^{Sc}, defining the approximate location of β-strands and loops.[14] The analysis also identified the C-terminal half, ranging from residue 152 to the end, as the most resistant part of PrP^{Sc}. These results were further corroborated by subsequent H/D exchange analyses, which found that a large portion of PrP^{Sc} C-terminus, from residues ~90 to 224, exhibits relatively low exchange rates, as it would be expected for regions rich in β-strands.[29] The data further reinforced the evidence that the structure of PrP^{Sc} includes β-strands connected by short loops and/or turns but lacks α-helices.

4. Attempts to solve the structure of infectious prions at high-resolution

Even though low-resolution techniques provided fundamental constraints that greatly contribute to our current understanding of the structure of PrP^{Sc}, they still failed to provide the necessary level of detail to allow rational drug discovery schemes against prions. In order to fill this gap, few previous studies attempted to apply high-resolution techniques to solve the structure of PrP^{Sc}. Indeed, the first observation of PrP^{Sc} fibrils was obtained by negative stain electron microscopy (EM) almost 40 years ago, which showed paired fibrils of protofilaments of around 4–6 nm.[38] However, the amount of structural information that could be gathered was greatly limited by the highly aggregated nature of the protein. The discovery of two-dimensional (2D) crystals of PrP^{Sc} represented another promising avenue to directly visualize the structure of prions via electron crystallography, but once again ended up to provide little information.[39] A subsequent comparison of 2D crystals obtained with a deleted PrP mutant lacking residue 141–176 (called PrP106 or miniprion) but retaining prion infectivity allowed the localization of the N-linked oligosaccharides, and the approximate cross-section dimensions of the PrP^{Sc} polypeptide chain, providing the first insights into the molecular architecture of PrP^{Sc}.[40] One key conclusion was that in its PrP^{Sc} conformation, the PrP chain had necessarily to be coiled several times on itself, or otherwise it would not fit into the space assigned to it in averaged reconstructions of the 2D crystal images.[15] These data were subsequently used to build one of the first computational reconstructions of PrP^{Sc}, characterized by a parallel 4-rung β-helical conformation.[41,42]

Application of cryogenic EM (cryo-EM) on purified PrP^{Sc} allowed direct visualization of the PrP^{Sc} fibers, free from negative stain artifacts, which together with reconstruction techniques allowed a reliable measurement of protofilament volumes.[14] Considering the density of PrP^{Sc}, a more accurate calculation of the volume occupied by each PrP^{Sc} molecule in the protofilament stack was possible, and the physical requirement of a 4 × coiling of the long, extended polypeptide to fit in, confirmed. Additional image averaging using boxing and a single particle analysis approach further confirmed the presence of 2 nm "particles" stacked along the protofilaments, which can only correspond to 4 rung β-solenoids.[14]

Several technical advances, including the use of the phosphotungstate (PTA) precipitation for the purification of PrP^{Sc}, improved methods to align the fibril preparations and the use of synchrotron-based X-ray sources, allowed the analysis of PrP^{Sc} fibers using X-ray fiber diffraction techniques.[43] Application of synchrotron-based small-angle X-ray scattering (SAXS) allowed the precise measurement of an ~11 nm diameter for two intertwined PrP^{Sc} fibers.[44] Moreover, X-ray fiber diffraction provided the first biophysical evidence that each monomer of PrP^{Sc} is arranged as a β-solenoid, made by stacks of four β-strands in a cross-β configuration, giving rise to repeated units of 19.2 Å along each fibril. The results were unexpectedly consistent with a later study using solid-state NMR (ssNMR) on a completely different prion, the fungal HET-s, which was found to be arranged in a similar β-solenoid architecture, although composed by only two rungs (consistent with its shorter length).[45]

Encouraged by the success obtained on HET-s, several laboratories have tried to study the structure of PrP^{Sc} by ssNMR. In one of these attempts, the technique was employed to study PrP amyloid fibers prepared by simply incubating recPrP^{Sc} with denaturants and subjecting the mixture to agitation, conditions previously shown to result in amyloid fibers devoid of infectivity.[46] Consistently, the amyloid samples prepared in such a way exhibited modest infectivity, with only 25% of inoculated mice developing symptoms with long incubation times.[46] Even though the analysis unequivocally assigned only few individual signals, these were sufficient to build a coherent structural model, which included a flexible stretch until residue 114, an α-helical structure in the central region (residues 115–154) and a β-sheet-rich core from residue 155 to the end. With the exception of the central α-helical region, the architecture was reminiscent of in-register β-sheet alignments observed in several other amyloids, including non-infectious PrP amyloid fibers.[47] All this suggests that these are standard PrP amyloid fibers with a structure different to that of bona fide PrP^{Sc}, and that their very modest infectivity might be similar to that seen in some of these preparations, ascribed to a special kind of heterologous templating termed "deformed templating."[48] Once again, these studies highlighted the need for a standardized protocol allowing the production of bona fide, truly infectious, ^{13}C- and ^{15}N-labeled $recPrP^{Sc}$ to carry out high-resolution spectroscopic techniques. This goal was partially achieved later on, when a murine recPrP produced in *E. coli* was refolded into an aggregated, protease-resistant form and amplified using a previously described protein misfolding cyclic amplification (PMCA) technique.[49,50] The resulting $recPrP^{Sc}$ was found capable of inducing a prion disease in mice after intracerebral injection. Unfortunately, the protocol still failed to produce the large amounts (milligram quantities) of sample required for ssNMR studies. The task was finally overcome only recently, with the development of a novel amplification method, called protein misfolding shaking amplification (PMSA), which allowed the production of milligram quantities of bona fide, infectious $recPrP^{Sc}$.[51] By taking advantage of this technological advancement, a recent study reported the first series of data obtained by ssNMR on a uniformly labeled (^{13}C, ^{15}N) infectious $recPrP^{Sc}$.[52] Although not definitive, the results were highly consistent with the previously proposed β solenoid architecture.

5. Evolution of computational models for prion structure and conversion

There are several reasons why the currently available data about the structure of infectious PrP^{Sc} and its conversion mechanism derive mainly from low resolution, indirect experimental evidence. Among these, one of the most relevant is the intrinsic propensity of PrP^{Sc} to aggregate, resulting in the formation of insoluble amyloids in many experimental settings. Moreover, infectious PrP^{Sc} particles isolated from several specimen usually appear as widely heterogeneous. As we discussed above, these features make standard protocols of high-resolution techniques such as X-ray crystallography and NMR inapplicable. However, computational resources offer a possibility to build atomistic models coherent with experimental constraints and, when coupled with molecular dynamics (MD) simulations, also challenge their stability. As a consequence, computational models of the PrP^{Sc} structure have been evolving in parallel with the growing amount of experimental constraints.[18] Three main atomistic architectures have been proposed so far (Fig. 1): the Left-handed-β-Helix (LβH),[41] the Parallel In-Register Beta-Sheet (PIRIBS),[53] and the 4-Rung β-Solenoid (4RβS).[54] In the LβH structure, the misfolded core spans from residue 89 to 170, while the two original C-terminal α-helices of PrP^C are retained. As outlined above, this model mainly derives from the electron crystallography of 2D crystals, coupled with an incorrect interpretation of FTIR data, thus resulting incoherent with the more recent, compelling evidence ruling out the presence of α-helices in PrP^{Sc}.[55] The PIRIBS architecture, characterized by intermolecular stacking of aligned PrP monomers, originated from a great deal of data, some of which at high-resolution, but acquired on poorly or non-infectious PrP amyloids. While this model is likely to reflect the structure of several non-infectious PrP assemblies that may be obtained in vitro, it also appears inconsistent with recent cryoEM data on infectious PrP^{Sc} fibrils.[14] Moreover, the parallel in-register alignment could easily be predicted to generate steric clashes of the glycans attached to PrP^C, thus necessarily implying that PrP^{Sc} growth may proceed only by alternatively recruiting glycosylated and non-glycosylated PrP molecules.[56] The 4RβS model was built precisely to solve most of the discrepancies affecting previous models and to accommodate

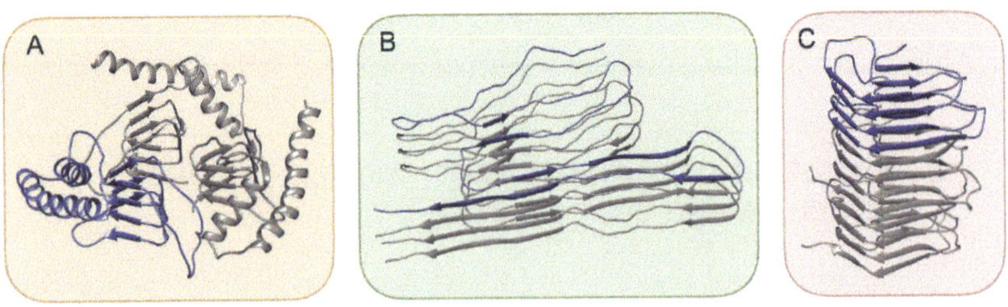

Fig. 1 Computational Models of Infectious PrP^{Sc}. The picture illustrates the three main atomistic architectures proposed so far for the structure of PrP^{Sc}. (A) The Left-handed-β-Helix (LβH), characterized by a misfolded β-sheet-rich core (residues 89–170) and two of the original C-terminal α-helices of PrP^C [41]. (B) The Parallel In-Register Beta-Sheet (PIRIBS), featuring an intermolecular stacking of aligned, extended PrP monomers.[53] (C) The 4-Rung β-Solenoid (4RβS), formed by the stacking of four consecutive rungs, each composed by linker-connected β-strands.[54] In all the models, the first monomer from above is depicted in blue.

the most updated set of available data on the structure of PrP^{Sc}.[54] Its architecture is characterized by the stacking of four consecutive rungs, each composed by linker-connected β-strands to form a 2 nm-high solenoid with a 4 nm diameter. Importantly, when challenged by MD simulations, the stability of the 4RβS architecture was found to be comparable to that of the naturally occurring β-solenoid of the prion forming domain of the fungal prion HET-s. Finally, this new PrP^{Sc} model appears coherent even with latest ssNMR analyses on infectious PrP^{Sc}.[52]

If solving the structure of infectious prions has proven to be an incredibly challenging task, studying at high resolution the highly dynamic molecular transitions underlying the PrP^{C} to PrP^{Sc} templated conversion appears even more ambitious.[18] In fact, in order to approach this problem, one would require to overcome the current spatio-temporal resolution limits of X-ray crystallography and NMR, which instead can only be applied to stable molecular species. At the moment, this possibility is offered exclusively by advanced computer simulations, which have the virtue of allowing the atomistic reconstruction of complex molecular transitions.[57,58] Recently, in two of such attempts, the solenoid architectures of PrP^{Sc} and of the HET-s prion were used as a target structures for the first physically-plausible atomistic reconstructions of the entire sequence of events underlying the templated conversion of PrP^{C} into PrP^{Sc}, and of HET-s monomers into growing fibrils.[54,59] In the case of PrP^{Sc}, rMD simulations yielded a structural transition pathway in which the exposed C-terminal rung of the 4RβS solenoid acts as an primer conversion surface for the unstructured N-terminus of PrP^{C} (residues 89–124). This initial event is followed by a series of conformational transitions in which the formation of each new rung is templated by the previous one, ultimately giving rise to the complete conversion of PrP^{C} into a new 4RβS of the fibril.[54] A very similar conversion pathway was obtained with the HET-s prion, suggesting that common molecular features underlie the ability of evolutionary distant prions to propagate their conformation in a templated fashion.[59] Collectively, while still requiring experimental validation, these computational approaches appear as extremely valuable for formulating new hypotheses that, in turn, may inspire new experiments.[18]

6. Conclusions

A constantly growing number of data derived from experimental improvements and low-resolution approaches are producing a streamline of structural constraints that in turn force a continuous refinement of computational models for PrP^{Sc}. Despite the lack of ideal techniques to study prion structure and replication at high-resolution, such an iterative, stepwise process is taking us increasingly closer to understand the molecular nature of prions.

Acknowledgments

This work was supported by a grant from Fondazione Telethon (Italy, TCP14009). J.R.R. was funded by a Grant (BFU2017-86692-P) from the Spanish Ministry of Economy and Competitiveness, partially funded by FEDER funds. G.S. is a recipient of a fellowship from Fondazione Telethon. E.B. is an Assistant Telethon Scientist at the Dulbecco Telethon Institute (Fondazione Telethon, Italy).

References

1. Prusiner SB. Novel proteinaceous infectious particles cause scrapie. *Science*. 1982;216 (4542):136–144.
2. Prusiner SB. Prions. *Proc Natl Acad Sci U S A*. 1998;95(23):13363–13383.
3. Wickner RB, Taylor KL, Edskes HK, Maddelein ML, Moriyama H, Roberts BT. Prions in Saccharomyces and Podospora spp.: protein-based inheritance. *Microbiol Mol Biol Rev*. 1999;63(4):844–861, [table of contents].
4. Stopschinski BE, Diamond MI. The prion model for progression and diversity of neurodegenerative diseases. *Lancet Neurol*. 2017;16(4):323–332.
5. Harvey ZH, Chen Y, Jarosz DF. Protein-based inheritance: epigenetics beyond the chromosome. *Mol Cell*. 2018;69(2):195–202.
6. Ironside JW, Ritchie DL, Head MW. Prion diseases. *Handb Clin Neurol*. 2017;145: 393–403.
7. Houston F, Andreoletti O. Animal prion diseases: the risks to human health. *Brain Pathol*. 2019;29(2):248–262.
8. Brandner S, Raeber A, Sailer A, et al. Normal host prion protein (PrPC) is required for scrapie spread within the central nervous system. *Proc Natl Acad Sci U S A*. 1996;93 (23):13148–13151.
9. Bueler H, Aguzzi A, Sailer A, et al. Mice devoid of PrP are resistant to scrapie. *Cell*. 1993;73(7):1339–1347.
10. Biasini E, Turnbaugh JA, Unterberger U, Harris DA. Prion protein at the crossroads of physiology and disease. *Trends Neurosci*. 2012;35(2):92–103.
11. Sim VL. Prion disease: chemotherapeutic strategies. *Infect Disord Drug Targets*. 2012;12 (2):144–160.
12. Barreca ML, Iraci N, Biggi S, Cecchetti V, Biasini E. Pharmacological agents targeting the cellular prion protein. *Pathogens*. 2018;7(1).
13. Requena JR, Wille H. The structure of the infectious prion protein and its propagation. *Prog Mol Biol Transl Sci*. 2017;150:341–359.
14. Vazquez-Fernandez E, Young HS, Requena JR, Wille H. The structure of mammalian prions and their aggregates. *Int Rev Cell Mol Biol*. 2017;329:277–301.
15. Wille H, Requena JR. The structure of PrP(Sc) prions. *Pathogens*. 2018;7(1):20.
16. Zweckstetter M, Requena JR, Wille H. Elucidating the structure of an infectious protein. *PLoS Pathog*. 2017;13(4):e1006229.
17. Baskakov IV, Caughey B, Requena JR, Sevillano AM, Surewicz WK, Wille H. The prion 2018 round tables (I): the structure of PrP(Sc). *Prion*. 2019;13(1):46–52.
18. Rigoli M, Spagnolli G, Faccioli P, Requena JR, Biasini E. Ok Google, how could I design therapeutics against prion diseases? *Curr Opin Pharmacol*. 2019;44:39–45.

19. Biasini E, Medrano AZ, Thellung S, Chiesa R, Harris DA. Multiple biochemical similarities between infectious and non-infectious aggregates of a prion protein carrying an octapeptide insertion. *J Neurochem*. 2008;104(5):1293–1308.
20. Chiesa R, Piccardo P, Biasini E, Ghetti B, Harris DA. Aggregated, wild-type prion protein causes neurological dysfunction and synaptic abnormalities. *J Neurosci*. 2008;28 (49):13258–13267.
21. Galkin AP, Velizhanina ME, Sopova YV, Shenfeld AA, Zadorsky SP. Prions and non-infectious amyloids of mammals—similarities and differences. *Biochemistry (Mosc)*. 2018;83(10):1184–1195.
22. Chiesa R, Piccardo P, Quaglio E, et al. Molecular distinction between pathogenic and infectious properties of the prion protein. *J Virol*. 2003;77(13):7611–7622.
23. Requena JR. The protean prion protein. *PLoS Biol*. 2020;18(6):e3000754.
24. Biasini E, Seegulam ME, Patti BN, et al. Non-infectious aggregates of the prion protein react with several PrPSc-directed antibodies. *J Neurochem*. 2008;105(6):2190–2204.
25. Jucker M, Walker LC. Propagation and spread of pathogenic protein assemblies in neurodegenerative diseases. *Nat Neurosci*. 2018;21(10):1341–1349.
26. Caughey BW, Dong A, Bhat KS, Ernst D, Hayes SF, Caughey WS. Secondary structure analysis of the scrapie-associated protein PrP 27-30 in water by infrared spectroscopy. *Biochemistry*. 1991;30(31):7672–7680.
27. Pan KM, Baldwin M, Nguyen J, et al. Conversion of alpha-helices into beta-sheets features in the formation of the scrapie prion proteins. *Proc Natl Acad Sci U S A*. 1993;90 (23):10962–10966.
28. Safar J, Roller PP, Gajdusek DC, Gibbs Jr CJ. Conformational transitions, dissociation, and unfolding of scrapie amyloid (prion) protein. *J Biol Chem*. 1993;268(27): 20276–20284.
29. Smirnovas V, Baron GS, Offerdahl DK, Raymond GJ, Caughey B, Surewicz WK. Structural organization of brain-derived mammalian prions examined by hydrogen-deuterium exchange. *Nat Struct Mol Biol*. 2011;18(4):504–506.
30. Cobb NJ, Sonnichsen FD, McHaourab H, Surewicz WK. Molecular architecture of human prion protein amyloid: a parallel, in-register beta-structure. *Proc Natl Acad Sci U S A*. 2007;104(48):18946–18951.
31. Onisko B, Fernandez EG, Freire ML, et al. Probing PrPSc structure using chemical cross-linking and mass spectrometry: Evidence of the proximity of Gly90 amino termini in the PrP 27-30 aggregate. *Biochemistry*. 2005;44(30):10100–10109.
32. Gong B, Ramos A, Vazquez-Fernandez E, et al. Probing structural differences between PrP(C) and PrP(Sc) by surface nitration and acetylation: evidence of conformational change in the C-terminus. *Biochemistry*. 2011;50(22):4963–4972.
33. Anderson M, Bocharova OV, Makarava N, Breydo L, Salnikov VV, Baskakov IV. Polymorphism and ultrastructural organization of prion protein amyloid fibrils: an insight from high resolution atomic force microscopy. *J Mol Biol*. 2006;358(2):580–596.
34. Wegmann S, Miesbauer M, Winklhofer KF, Tatzelt J, Muller DJ. Observing fibrillar assemblies on scrapie-infected cells. *Pflugers Arch*. 2008;456(1):83–93.
35. Piro JR, Wang F, Walsh DJ, Rees JR, Ma J, Supattapone S. Seeding specificity and ultrastructural characteristics of infectious recombinant prions. *Biochemistry*. 2011;50 (33):7111–7116.
36. Hubbard SJ. The structural aspects of limited proteolysis of native proteins. *Biochim Biophys Acta*. 1998;1382(2):191–206.
37. Sajnani G, Pastrana MA, Dynin I, Onisko B, Requena JR. Scrapie prion protein structural constraints obtained by limited proteolysis and mass spectrometry. *J Mol Biol*. 2008;382(1):88–98.
38. Prusiner SB, McKinley MP, Bowman KA, et al. Scrapie prions aggregate to form amyloid-like birefringent rods. *Cell*. 1983;35(2 Pt 1):349–358.

39. Wille H, Michelitsch MD, Guenebaut V, et al. Structural studies of the scrapie prion protein by electron crystallography. *Proc Natl Acad Sci U S A*. 2002;99(6):3563–3568.
40. Supattapone S, Bosque P, Muramoto T, et al. Prion protein of 106 residues creates an artifical transmission barrier for prion replication in transgenic mice. *Cell*. 1999;96 (6):869–878.
41. Govaerts C, Wille H, Prusiner SB, Cohen FE. Evidence for assembly of prions with left-handed beta-helices into trimers. *Proc Natl Acad Sci U S A*. 2004;101(22):8342–8347.
42. Wille H, Govaerts C, Borovinskiy A, et al. Electron crystallography of the scrapie prion protein complexed with heavy metals. *Arch Biochem Biophys*. 2007;467(2):239–248.
43. Wille H, Bian W, McDonald M, et al. Natural and synthetic prion structure from X-ray fiber diffraction. *Proc Natl Acad Sci U S A*. 2009;106(40):16990–16995.
44. Amenitsch H, Benetti F, Ramos A, Legname G, Requena JR. SAXS structural study of PrP(Sc) reveals ~11 nm diameter of basic double intertwined fibers. *Prion*. 2013;7 (6):496–500.
45. Siemer AB, Ritter C, Ernst M, Riek R, Meier BH. High-resolution solid-state NMR spectroscopy of the prion protein HET-s in its amyloid conformation. *Angew Chem Int Ed Engl*. 2005;44(16):2441–2444.
46. Muller H, Brener O, Andreoletti O, et al. Progress towards structural understanding of infectious sheep PrP-amyloid. *Prion*. 2014;8(5):344–358.
47. Tycko R. Molecular structure of aggregated amyloid-beta: Insights from solid-state nuclear magnetic resonance. *Cold Spring Harb Perspect Med*. 2016;6(8).
48. Makarava N, Baskakov IV. The evolution of transmissible prions: the role of deformed templating. *PLoS Pathog*. 2013;9(12):e1003759.
49. Zhang Z, Zhang Y, Wang F, et al. De novo generation of infectious prions with bacterially expressed recombinant prion protein. *FASEB J*. 2013;27(12):4768–4775.
50. Wang F, Wang X, Yuan CG, Ma J. Generating a prion with bacterially expressed recombinant prion protein. *Science*. 2010;327(5969):1132–1135.
51. Erana H, Charco JM, Di Bari MA, et al. Development of a new largely scalable in vitro prion propagation method for the production of infectious recombinant prions for high resolution structural studies. *PLoS Pathog*. 2019;15(10):e1008117.
52. Martín-Pastor M, Codeseira YB, Spagnolli G, et al. Structural features of an infectious recombinant PrPSc prion using solid state NMR. *bioRxiv*. 2020. https://doi.org/10.1101/2020.04.08.032839.
53. Groveman BR, Dolan MA, Taubner LM, Kraus A, Wickner RB, Caughey B. Parallel in-register intermolecular beta-sheet architectures for prion-seeded prion protein (PrP) amyloids. *J Biol Chem*. 2014;289(35):24129–24142.
54. Spagnolli G, Rigoli M, Orioli S, et al. Full atomistic model of prion structure and conversion. *PLoS Pathog*. 2019;15(7):e1007864.
55. Smirnovas V, Kim JI, Lu X, Atarashi R, Caughey B, Surewicz WK. Distinct structures of scrapie prion protein (PrPSc)-seeded versus spontaneous recombinant prion protein fibrils revealed by hydrogen/deuterium exchange. *J Biol Chem*. 2009;284(36): 24233–24241.
56. Baskakov IV, Katorcha E, Makarava N. Prion strain-specific structure and pathology:a view from the perspective of glycobiology. *Viruses*. 2018;10(12):723.
57. Bartolucci G, Orioli S, Faccioli P. Transition path theory from biased simulations. *J Chem Phys*. 2018;149(7):072336.
58. Orioli S, S AB, Faccioli P. Self-consistent calculation of protein folding pathways. *J Chem Phys*. 2017;147(6):064108.
59. Terruzzi L, Spagnolli G, Boldrini A, Requena JR, Biasini E, Faccioli P. All-atom simulation of the HET-s prion replication. *bioRxiv*. 2020. https://doi.org/10.1101/2020.05.01.072017.

CHAPTER THREE

Role of sialylation in prion disease pathogenesis and prion structure

Ilia V. Baskakov*
Department of Anatomy and Neurobiology, and Center for Biomedical Engineering and Technology, University of Maryland School of Medicine, Baltimore, MD, United States
*Corresponding author: e-mail address: baskakov@som.umaryland.edu

Contents

Abstract

Mammalian prion or PrP^{Sc} is a proteinaceous infectious agent that consists of a misfolded, self-replicating state of a sialoglycoprotein called the prion protein or PrP^C. Sialylation of the prion protein, a terminal modification of N-linked glycans, was discovered more than 30 years ago, yet the role of sialylation in prion pathogenesis is not well understood. This chapter summarizes current knowledge on the role of sialylation of the prion protein in prion diseases. First, we discuss recent data suggesting that sialylation of PrP^{Sc} N-linked glycans determines the fate of prion infection in an organism and control prion lymphotropism. Second, emerging evidence pointing out at the role N-glycans in neuroinflammation are discussed. Thirds, this chapter reviews a mechanism postulating that sialylated N-linked glycans are important players in defining strain-specific structures. A new hypothesis according to which individual strain-specific PrP^{Sc} structures govern selection of PrP^C sialoglycoforms is discussed. Finally, this chapter explain how N-glycan sialylation control the prion replication and strain interference. In summary, comprehensive review of our knowledge on N-linked glycans and their sialylation provided in this chapter helps to answer important questions of prion biology that have been puzzling for years.

Progress in Molecular Biology and Translational Science, Volume 175
ISSN 1877-1173
https://doi.org/10.1016/bs.pmbts.2020.07.004

1. Introduction

Prion diseases or transmissible spongiform encephalopathies are a class of lethal, transmissible neurodegenerative disorders of humans and animals.[1,2] Transmissible agents responsible for the prion diseases consist of misfolded, self-replicating states of a sialoglycoprotein referred to as PrP^{Sc}. Prions or PrP^{Sc} spread between organisms or from cell to cell by recruiting host-encoded prion protein (PrP^{C}) and replicating their disease-specific misfolded structures via a template-assisted mechanism.[3]

Prion diseases display multiple disease phenotypes characterized by diverse clinical symptoms and brain regions affected by the disease, distinct cell tropism and diverse PrP^{Sc} deposition patterns.[4] The diversity of disease phenotypes within the same host is attributed to the ability of PrP^{C} to acquire multiple, alternative, conformationally distinct, self-replicating PrP^{Sc} states referred to as prion strains or subtypes.[5–10] Structural diversity of PrP^{Sc} strains has been well documented,[11–13] yet the question how PrP^{Sc} structures elicit multiple disease phenotypes remains poorly understood. Currently, the relationship between PrP^{Sc} structure and CNS response remains empirical, whereas a mechanism that would describe their relationship in a predictable manner is lacking.

The question regarding relationship between strain-specific structures of PrP^{Sc} and disease phenotype is closely related to other set of important questions. Differences between secondary, tertiary and quaternary structures of prion strains have been well documented,[11–13] yet the extent to which strain-specific structures are differences is not known. When passaged within the same host, prion strains display different ratios of di-, mono- and unglycosylated glycoforms (reviewed in Ref. 14). While glycoform ratio is considered to be one of the features that defines strain identity, the question why different strains have different glycoform ratios has not been answered. Other important question that has not been addressed is what molecular features of PrP^{Sc} are responsible for chronic inflammation and neurodegeneration? The current article reviews emerging evidence suggesting that carbohydrates in the form of sialylated N-linked glycans, which are a constitutive part of PrP^{Sc}, are important players in defining strain-specific structures and disease phenotypes.

2. Diversity of N-glycans in PrP^{C} and PrP^{Sc}

PrP^{C} is posttranslationally modified with glycophosphatidylinositol (GPI) anchor and one or two N-linked glycans attached to the residues

Asn-181 and/or Asn-197 (mouse PrP sequence numbering).[15–18] These posttranslational modifications are carried over upon conversion of PrP^C into PrP^{Sc}.[15,17–20] PrP^C expressed in a cell exhibit astonishing heterogeneity, which is attributed to variations in the structure and composition of N-glycans with more than 400 different PrP^C glycoforms described.[20–22] Because each of the two N-glycans can carry up to five and possibly even more negatively charged sialic acid residues,[20–22] PrP^C molecules also display dramatic range of pIs or net charges at physiological pH.[23,24]

Among carbohydrate groups that compose N-glycans, sialic acid residues (Sias) play a key role in defining glycan structural and functional diversity. Sias are a family of 9-carbon containing acidic monosaccharides that are found in terminal positions of N- and O-linked glycans of glycoproteins or glycolipids.[25] Glycan sialylation is controlled by two groups of enzymes: sialyltransferases (STs) and sialidases (NEUs).[26,27] STs transfer sialic acids to the terminal positions of glycans. This process takes place in the trans-Golgi and involves 20 mammalian STs.[27] STs are divided into four families according to the type of linkages synthesized (α2–3, α2–6, α2–8 or α2–9) and the selectivity toward N- or O-linked glycans.[27,28] NEUs, on the other hand, remove Sias from glycans. Four NEUs are found in mammals, they are expressed in a tissue-specific manner and are also different with respect to cellular localization.[26,29,30]

Humans can synthesize only one type of Sias, which is *N*-acetylneuraminic acid (Neu5Ac).[31] With the exception of the ferret,[32] the rest of mammalian species produce two types of Sias. Neu5Ac is the predominant type that is synthesized in a brain, whereas Neu5Ac and N-glycolylneuraminic acid (Neu5Gc) are synthesized by peripheral organs.[25] The deficiency in synthesis of Neu5Gc in humans is linked to an irreversible mutation in the gene encoding *N*-acetylneuraminic acid hydroxylase (an enzyme that synthesize Neu5Gc from Neu5Ac) that occurred during evolution from primates to humans.[31] Like humans, ferrets can produce only Neu5Ac.[32] While humans lack the ability to synthesize Neu5Gc, this Sia can be incorporated into human tissues from diet metabolically.[33]

Sias on glycans and glycolipids of cell surface form diverse structural epitopes that are involved in a number of cellular functions. The structural diversity of Sia epitopes is produced via several mechanisms. First, Sias can be attached to galactose or *N*-acetylgalactosamine of glycans via α2–3, α2–6, α2–8 or α2–9 linkages.[25] Second, various natural substitutes including *O*-acetyl, *N*-glycolyl, *O*-lactyl, *O*-sulfate, *O*-phosphate, tauryl, hydroxyl or *O*-methyl can be synthesized on carbons of Sias at 1-, 4-, 5-, 7-, 8-, or 9-positions, where *O*-acetyl being the most common substitute.[34]

Third, in combination with Sias other groups including sulfate and fucose are involved in forming functional glycan epitopes including Sialyl Lewis^x, Sialyl Lewis^a, 6′Sulfo-Sialyl Lewis^x.[35] Fourth, complex glycans can exhibit several branching patterns that contribute to variations in density of Sia residues, a factor important for binding of multivalent ligands.

In N-linked glycans of PrP^{Sc}, Sia residues can be linked via α2–3 or α2–6 linkages with a majority being linked via α2–6-bond.[21,36] In particular, high percentages of α2–6-linked Sias were observed in PrP^{Sc} from brain, spleen or cultured cells in three prion strains.[36] Moreover, variations in the ratio of α2–3- versus α2–6-linked Sias between brain-, spleen- or cell-derived PrP^{Sc} was found to be relatively minor suggesting that the type of linkage is independent of PrP^{Sc} source.[36] Among previously described functional epitopes, Lewis^x and Sialyl Lewis^x were found in N-glycans of PrP^{Sc}.[22]

3. N-glycan are exposed on a surface of PrP^{Sc}

Each of the two N-glycans adds up to 5 kDa to PrP molecular weight. High density of packing of PrP molecules within PrP^{Sc} raises questions about spatial positioning of N-glycans within PrP^{Sc} particles and, specifically, whether PrP^{Sc} can accommodate di-glycosylated PrP. Molecular modeling of PrP^{Sc} structure suggested that the N-glycans are directed outwards exposing terminal Sia residues on the surface of PrP^{Sc} particles.[37–40] In agreement with molecular modeling, staining of brain slices from scrapie-infected animals with SNA lectin, which recognizes terminal Sia residues attached via α2–6 linkages, demonstrated very intense staining of PrP^{Sc} plaques (Fig. 1A–D).[41] In fact, PrP^{Sc} displayed much more intense staining with SNA in comparison to brain sialoglycocalix suggesting that the density of sialylation on PrP^{Sc} surface exceeds that of sialoglycocalyx (Fig. 1B). High density of sialylation on PrP^{Sc} surface is also consistent with intense staining of PrP^{Sc} plaques by Alcian Blue, a polyvalent basic dye which is generally used for staining of sialylated glycocalyx.[42]

4. Sialylation status of PrP^{Sc} and prion infectivity

High density of Sia residues on a surface of PrP^{Sc} particles raises a question whether sialylation status of PrP^{Sc} is important in determining the outcomes of prion infection. Below, we discuss specific examples illustrating involvement of carbohydrate epitopes on PrP^{Sc} surface in determining the fate of prions in an organism (Fig. 2). Our recent studies revealed

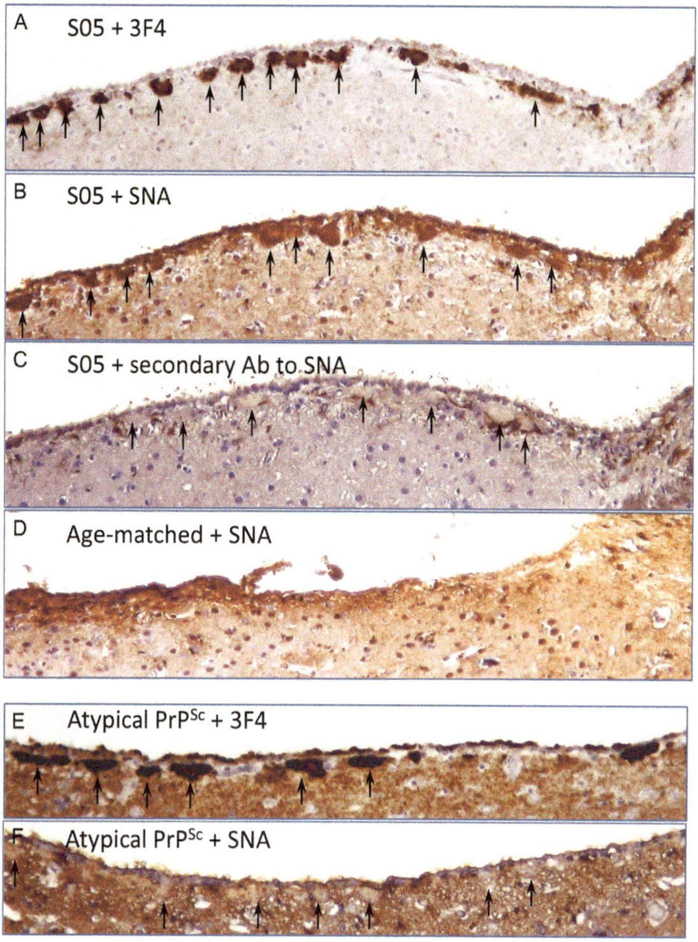

Fig. 1 Staining of PrP^{Sc} plaques with SNA lectin. Images of SO5 PrP^{Sc} plaques (A–C) or atypical PrP^{Sc} plaques (E, F) in hamster brains stained with anti-PrP 3F4 antibody (A, E), SNA lectin (B, F), or secondary antibody used for SNA staining as negative control (C). Plaques are shown by *arrows*. Staining of normal age-matched control with SNA lectin is shown in (D).

that desialylation of N-linked glycans of PrP^{Sc} impairs survival of prions in vivo.[23,43,44] On a surface of mammalian cell, sialoglycocalyx acts as a part of a "self-associated molecular pattern" helping the innate immunity to recognize "self" from "altered self" or "non-self."[45,46] Removal of Sias from cell surface glycans exposes galactose residues that generate "eat me" signals for professional and non-professional macrophages, for example, stimulating clearance of erythrocytes or platelets by Kupffer cells,[47,48] or phagocytosis of

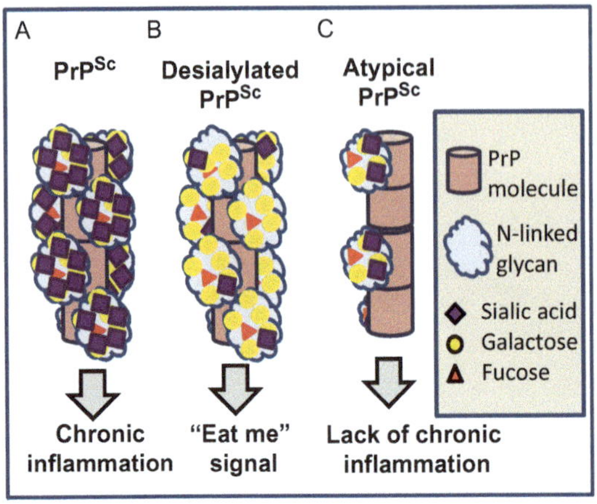

Fig. 2 Schematic representation of the hypothesis proposing that carbohydrate epitopes on PrP^{Sc} surface determine the outcomes of prion infection. (A) High density of glycans with terminal sialylation leads to chronic neuroinflammation. (B) Desialylation of PrP^{Sc} that results in a high density of exposed galactose triggers "eat me" signal in glia. (C) Atypical PrP^{Sc} has low density of glycosylation and sialylation, similar to those of sialoglycocalyx. Atypical PrP^{Sc} does not trigger "eat me" signal or chronic neuroinflammation. *Adapted from Baskakov IV, Katorcha E, Makarava N. Prion strain-specific structure and pathology: a view from the perspective of glycobiology.* Viruses. *2018;10(12):723.*

neurons by microglia.[49–51] Maintaining sufficient sialylation levels appears to be as important for prion survival, as it is for preserving healthy mammalian cells (Fig. 2A and B). Our recent studies demonstrated that donor PrP^{Sc} with reduced sialylation levels did not induce prion disease in animals after intracranial or intraperitoneal inoculations.[23,43,44] Remarkably, no PrP^{Sc} was detected in brains or spleens of animals inoculated with PrP^{Sc} with reduced sialylation levels by Western blot or by calibrated serial PMCAb that detects a single PrP^{Sc} particles, arguing that these animals were free of prions.[43,44] Moreover, other study illustrated that prion infectivity could be switched off and on in a reversible manner via removing and reinstalling sialylation of PrP^{Sc}, respectively.[43] In summary, these studies suggested that sialylation protects PrP^{Sc} against clearance and appears to be critical in defining the fate of prion infection in an organism.

5. Sialylation of PrP^{Sc} and lymphotropism

Upon prion transmission via peripheral routes, PrP^{Sc} is first sequestered by secondary lymphoid organs (SLOs), including spleen and lymph nodes, prior to invasion of the CNS.[52–55] Upon invasion of SLOs, PrP^{Sc}

replicates in spleen and lymph nodes independently of its replication in the CNS.[56–59] Prion strains show variable degrees of lymphotropism.[60] The molecular mechanism behind strain-specific lymphotropism is not known. We do not know whether limited lymphotropism of some strains is due to deficient trafficking of PrP^{Sc} to SLOs, impaired replication in SLOs, fast clearance in SLOs, or a combination of these factors.

Recent studies using hepatocarcinoma cells demonstrated that sialylation of N-linked glycans at α2–6 linkages is responsible for directed trafficking and selective adhesion of hepatocarcinoma cells to SLOs.[61,62] Another work that employed synthetic glycoclusters demonstrated that in circulation, the glycoclusters with α2–6 linked Sias were more stable and showed slower clearance rates in comparison to the glycoclusters with α2–3 linkages.[63] Notably, the two types of human CJDs, sporadic (sCJD) and variant (vCJD), show significant differences with respect to their lymphotropism, with vCJD known to be much more lymphotropic than sCJD.[64–66] PrP^{Sc} in vCJD is predominantly diglycosylated and, as a result, have a higher density of sialylation on PrP^{Sc} surface in comparison to PrP^{Sc} associated with sCJD, which is predominantly monoglycosylated.[67,68] The relative ranking of the two types of CJDs with respect to sialylation is consistent with the hypothesis that sialylation is important factor that controls trafficking of PrP^{Sc} to SLOs.

To test whether sialylation is important for prion lymphotropism, our recent studies monitored trafficking of partially desialylated PrP^{Sc} to secondary lymphoid organs upon intraperitoneal administration.[44] By 6 h post-inoculation, PrP^{Sc} with normal sialylation status were mostly found in spleen and lymph nodes, whereas partially desialylated PrP^{Sc} was targeted predominantly to the liver.[44] This work illustrated that sialylation plays a critical role in trafficking of PrP^{Sc} to SLOs and controlling the outcomes of prion transmission.

How does PrP^{Sc} survive clearance by the innate immune cells in SLOs? It has been well established that upon sequestration by SLOs, PrP^{Sc} replicates in the germinal centers of spleen and lymph nodes.[56–59] Remarkably, the spleen-derived PrP^{Sc} was found to be considerably more sialylated than brain-derived PrP^{Sc}.[69] Enhanced sialylation in SLOs was observed regardless of a prion strain, host species or inoculation route.[69] Notably, enhanced sialylation of PrP^{Sc} was found to be not due to hyper-sialylation of PrP^{C} expressed in SLOs, but attributed to post-conversion sialylation of PrP^{Sc} in SLOs by extracellular STs. While STs are traditionally believed to localize within the trans-Golgi,[70] a number of studies reported ST activity in circulation or on surfaces of the cells of the immune system including polymorphonuclear leukocytes, monocyte-derived dendritic cells, lymphocytes, and T cells.[71–76] Consistent with the hypothesis that extracellular STs are involved in enhancing sialylation of PrP^{Sc}, the sialylation status of foreign PrP^{Sc} acquired via peripheral exposure changed with colonization of SLOs.[69] Moreover, enhanced sialylation of PrP^{Sc} was recapitulated in vitro by incubating brain-derived PrP^{Sc} with primary splenocytes or cultured macrophage RAW 264.7 cells.[69] General inhibitors of STs suppressed enhanced sialylation of PrP^{Sc}.[69] One can suggests that post-conversion sialylation camouflages PrP^{Sc} in SLOs protecting it from clearance by innate immune cells. It would be interesting to test whether enhanced sialylation of PrP^{Sc} accounts for the high permissiveness of SLOs to prion infection.

6. Animal-to-human prion transmission

Humans can only synthesize Neu5Ac whether in periphery or CNS, whereas Neu5Gc is the predominant type of Sias expressed in the periphery of mammals.[31] The difference in the type of Sias expressed in the periphery of human and non-human mammals raises several important topics for discussion. First, it would be interesting to find out whether this difference contributes to the animal-to-human species barrier for prion transmission. Notably, in humans with high consumption of red meat, Neu5Gc incorporates metabolically into human cells and induces antibody responses against Neu5Gc.[33] While incorporation of Neu5Gc increases the likelihood of systemic inflammation,[33] antibodies against Neu5Gc might be beneficial for neutralizing prion infection of zoonic origin in humans. A second important aspect to consider is the functional consequences of enhanced sialylation of foreign PrP^{Sc} in SLOs.[69] Enhanced sialylation in human SLOs could "humanize" prions of animal origin transmitted to humans by decorating them with Neu5Ac and helping to deceive the human immune system.[69] A third interesting aspect is related to human-specific differences in the binding sites of human Siglecs for selective recognition of Neu5Ac over of Neu5Gc.[31] Siglecs are a family of sialic acid-binding proteins with a number of important functions (reviewed in Ref. 77). While interactions between prions and Siglecs have not yet been documented,[78] such a possibility should not be excluded considering the large number of Siglecs expressed in humans and mice. Human-specific differences in Siglecs for selective recognition of Neu5Ac over of Neu5Gc are also important for the critical assessment of the results obtained in humanized mice (mice expressing the human PrP gene). Humanized mice have often been used to assess susceptibility of humans to prion strains of animal origin or to model human-to-human transmission.[79–83] Because humanized mice express mouse but not human Siglecs, interaction between Siglecs and PrP^{Sc} might produce different outcomes in humanized mice and humans.

7. Sialylation of PrP^{Sc} and neuroinflammation

Activation and proliferation of microglia and astrocytes have been recognized as obligatory features of prion diseases.[84–86] Several protein families involved in innate immunity including siglecs, selectins, galectins, complement system components, mannose receptors and asialoglycoprotein receptors recognize carbohydrate groups.[77,87] Because the majority of carbohydrate-binding molecules have multivalent binding sites, the strength and selectivity of binding is not only dependent on the composition of functional carbohydrate epitopes but also on their density and specific configuration of carbohydrate groups. Is sialylation responsible for establishing glia reactive phenotypes? What molecular features of PrP^{Sc} glia sense and respond to? Exposure of primary microglia or BV2 cells to purified PrP^{Sc} was found to trigger pro-inflammatory responses characterized by an increase in the levels of TNFα, IL6, nitric oxide and expression of inducible Nitric Oxide Synthase.[88] Notably, partial cleavage of sialic acid residues on PrP^{Sc} surface boosted the inflammatory response of microglia to PrP^{Sc}.[88] Moreover, transient degradation of Iκβα observed upon treatment with partially desialylated PrP^{Sc} suggests that canonical NFκB activation pathway is involved in inflammatory response.[88] Thus, PrP^{Sc} can directly trigger proinflammatory response in microglia with a degree of response found to be determined by the degree of sialylation of PrP^{Sc}.

Another example that might shed light on the relationship between PrP^{Sc} molecular features and chronic neuroinflammation and neurodegeneration deals with a unique PrP^{Sc}-like state that has substantially reduced both glycosylation and sialylation levels (will be referred to as atypical PrP^{Sc}).[89–91] Atypical PrP^{Sc} is a self-replicating, transmissible state,[92–94] which accumulates in the form of small synaptic deposits and large plaques[91] (Fig. 1E and F). Yet, it is a clinically silent state that does not cause neuronal death, pathological lesions or any detectible clinical signs typically associated with prion diseases.[89–93] Remarkably, atypical PrP^{Sc} recruits mono- and un-glycosylated PrP^{C} of a host, while strongly excluding di-glycosylated

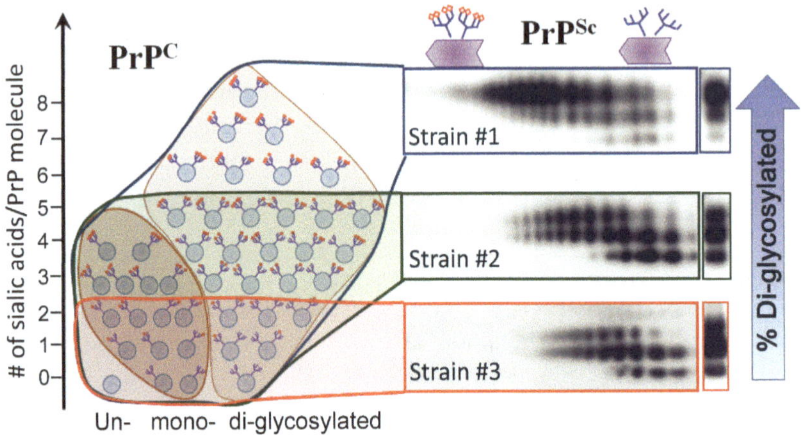

Fig. 3 Schematic diagram illustrating selective recruitment of PrP^C sialoglycoforms in a strain-specific manner according to the PrP^C sialylation status. The left panel shows distribution of PrP^C molecules according to their glycosylation status (in horizontal dimension) and sialylation status (in vertical dimension) ranging from non-sialylated to highly sialylated molecules. PrP^C molecules are shown as blue circles and sialic acid residues—as red diamonds. The panels on the right show 2D Western blots of three prion strains with different recruitment selectivity. While 263 K (strain #1) recruits PrP^C sialoglycoforms without strong preferences, hypersialylated PrP^C molecules are preferentially excluded from RML (strain #2) and excluded even stronger from atypical PrP^{Sc} (strain #3). As a result of strain-specific exclusion of highly sialylated PrP^C, ratios of glycoforms within PrP^{Sc} shift toward mono- and unglycosylated glycoform, as illustrated by corresponding 1D Western blots. *Adapted from Baskakov IV, Katorcha E. Multifaceted role of sialylation in prion diseases.* Front Neurosci. *2016;10:358*

and sialylated PrP^C molecules[92] (Fig. 3C). In fact, out of ten prion strains examined, atypical PrP^{Sc} had the lowest glycosylation and sialylation status.[92,95] Yet, unlike desialylated PrP^{Sc}, atypical PrP^{Sc} does not expose terminal galactose at high density, which is considered to be an "eat me" signal (Fig. 2B and C). As a result, it replicates and accumulates steadily in CNS. Atypical PrP^{Sc} lacks toxicity and does not case neuroinflammation,[91] a feature that appears to be attributed to its unique glycosylation and sialylation status.

8. Structural constraints imposed by N-glycans

Do N-glycans play a role in defining strain-specific PrP^{Sc} structures? Several structures that exhibit diverse PrP folding patterns have been considered in recent years to model structure of PrP^{Sc} (reviewed in Ref. 39). They include parallel in-register β-spine and two-, three- or four-rung β-solenoids.[38,96–99] To build a realistic molecular model of PrP^{Sc}, one should consider PrP^{Sc} glycosylation status, the size of the glycans and their charge distribution. Which of the proposed models can accommodate N-linked glycans? For addressing this question, a tri-antennary glycan with a size average of those found in PrP sialoglycoforms was used for modeling PrP^{Sc}.[100] In in-register parallel β-spine, the glycans of neighboring PrP molecules have to be spaced at a distance of 4.7 Å that brings them into substantial spatial interference that precludes such arrangements.[100] Considerable spatial overlap between glycans still exists in two-rung solenoid that separates glycans at a distance of 2 × 4.7 Å.[100] However, three-rung solenoids permit recruitment with minimal interference, whereas four-rung solenoid can fully accommodate diglycosylated PrP molecules without any spatial interference.[100] Modeling that takes into account glycosylation status supports the hypothesis that N-glycans limit the diversity of folding patterns accessible to glycosylated PrP^C.

Two alternative views could be taken into consideration to explain the relationship between N-glycans and strain-specific PrP^{Sc} structure. According to one view, prion strains can partially overcome constraints imposed by glycans by selectively recruiting mono- and un-glycosylated PrP^C glycoforms at the expense of diglycosylated glycoforms. Only those PrP^C glycoforms are recruited successfully that fit into a strain-specific PrP^{Sc} structure. Alternative view proposes that recruitment is not selective, and glycoforms are recruited proportionally to their relative representations in a pool of PrP^C molecules. If this is the case, the spectrum of PrP^{Sc} structures would be limited to those that can accommodate a high percentage of diglycosylated glycoforms, since diglycosylated glycoforms account for approximately 80% of all PrP^C molecules.

To test the hypothesis regarding strain-specific selectivity in recruiting PrP^C sialoglycoforms, composition of sialoglycoforms within PrP^{Sc} was analyzed using 2D gels.[69,95] Because N-linked glycans carry negatively charged Sia residues, the sialoglycoforms can be separated in horizontal dimension of 2D according to their charge.[100,101] For the 2D, PrP^{Sc} particles were denatured into monomers that were separated according to their sialylation status in horizontal dimension and their glycosylation status in vertical dimension. Analysis of multiple strains from two hosts revealed that PrP^{Sc} strains exhibits broad range of strain-specific selectivities with respect to PrP^C sialoglycoforms (Fig. 3).[95] Consistent with the first view, a group of strains shows strong preferences, as they excluded highly sialylated molecules as well as diglycosylated molecules (Fig. 3). At the same time, in support of the second view, a group of strains did not display any preferences with respect to glycosylation or sialylation status (Fig. 3). Remarkably, atypical PrP^{Sc} of synthetic origin displayed the highest selectivity,[89,92] as it excluded most of the diglycosylated and also highly and moderately sialylated PrP^{C} molecules (Fig. 3). In summary, 2D analysis revealed a broad range of selectivity displayed by prion strains in recruiting PrP^{C} sialoglycoforms, ranging from non-selective to highly selective. Notably, for the group of non-selective strains, the composition of sialoglycoforms within PrP^{Sc} was very similar to that of PrP^{C}.

9. Selective recruitment of PrP^{C} sialoglycoforms and strain-specific structure

Recently, we proposed that broad range of selectivities in recruiting PrP^{C} sialoglycoforms is attributed to strain-specific variations in quaternary structure of PrP^{Sc} (i.e. spatial orientation of PrP molecule within PrP^{Sc}) and could be achieved in the absence of significant strain-specific variations in a folding pattern.[41] Specifically, we propose that for non-selective strains, there is a considerably twist or rotation between neighboring PrP molecules assembled in PrP^{Sc} particles (Fig. 4). Owing to such rotations, glycans extend into different directions avoiding spatial interference and minimizing electrostatic repulsion between their Sia residues. In strains that select against diglycosylated and highly sialylated PrP^{C}, the twist between neighboring PrP molecules is very small or absent (Fig. 4). As a result, such mode of assembly creates spatial and electrostatic interference between glycans limiting recruitment of diglycosylated and highly sialylated PrP^{C}. The three- or four-rung solenoid models offer the best opportunity for accommodating both selective and non-selective strains (Fig. 4). The proposed arrangement of N-linked glycans and Sia residues in selective and non-selective strains could be used to distinguish these strains experimentally, based on their ultrastructure or fibril morphology. If this hypothesis is correct, non-selective strains are expected to display twisted fibril morphologies, whereas selective strains should show less twisted or non-twisted morphologies. Previous studies that employed differentially glycosylated PrP^{C} molecules support the above hypothesis.[102] Consistent with the model presented in Fig. 4, unglycosylated PrP^{C} molecules were required for propagating RML strain,[102] which belongs to the group of selective strains (Fig. 3).

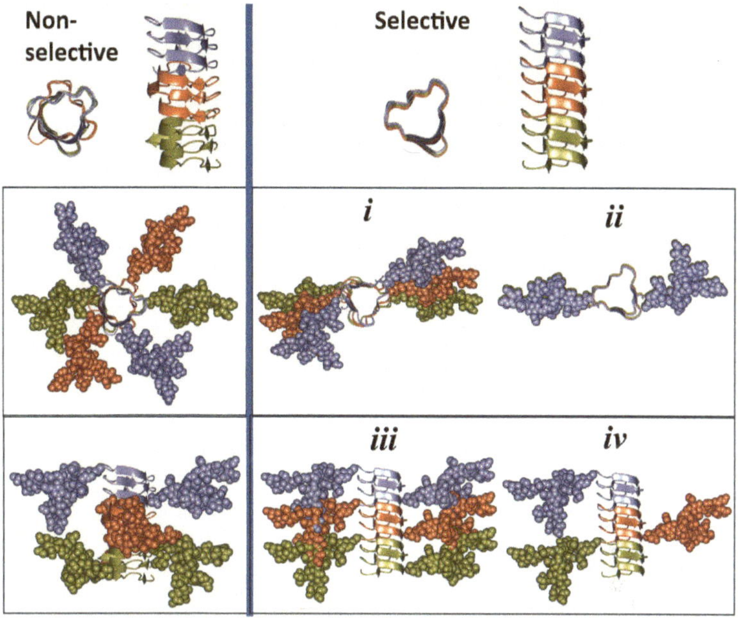

Fig. 4 Schematic diagram illustrating differences in quaternary assembly between non-selective (left panels) and selective (right panels) strains. Non-selective strains can accommodate diglycosylated sialoglycoforms because of rotation between neighboring PrP molecules, due to which glycans are spatially separated while electrostatic repulsion is reduced. In selective strains, the rotation between neighboring PrP molecules is very small (*i*) or absent (*ii*). Recruitment of diglycosylated molecules by selective strains would lead to spatial interference and electrostatic repulsion between glycans (*iii*). Negative selection of diglycosylated molecules helps to minimize spatial and electrostatic interference between glycans (*iv*). For simplicity of presentation only three-rung solenoid structures are shown here. *Adapted from Baskakov IV, Katorcha E, Makarava N. Prion strain-specific structure and pathology: a view from the perspective of glycobiology.* Viruses. *2018;10(12):723.*

10. Role of sialic acid residues and prion replication

Assuming that PrP^{Sc} adapts three- or four-rung solenoid folding patterns, sialylation of N-linked glycans is expected to result in a highly dense negative charge on PrP^{Sc} surface and impose constraints on PrP^{Sc} surfaces that might limit prion replication rate (Fig. 5). If this is true, removing Sia residues from PrP^{C} should facilitate PrP^{C} recruitment and accelerate replication of PrP^{Sc}. For testing whether this is the case, amplification rates were examined using PMCA conducted with PrP^{C} substrate desialylated by

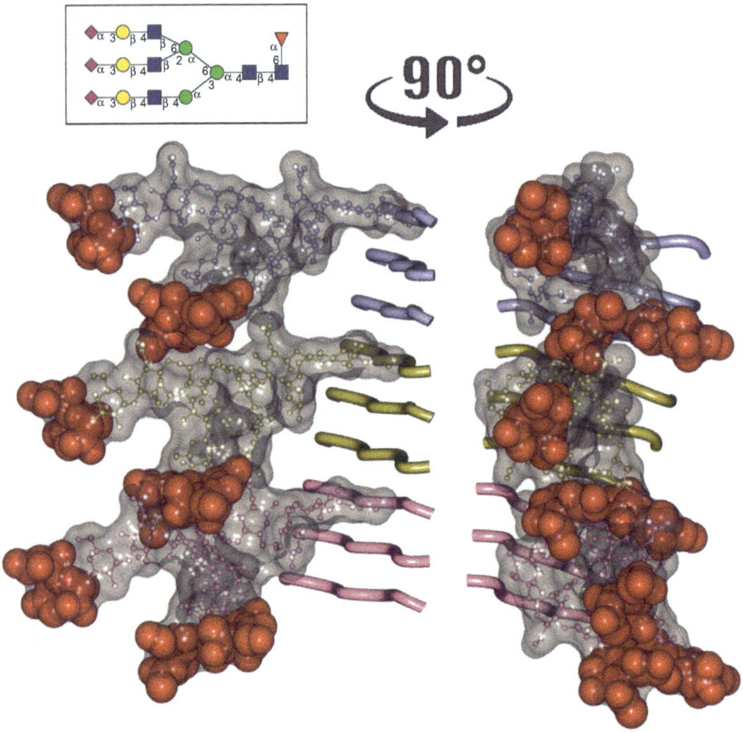

Fig. 5 Modeling of N-linked glycans in PrP^{Sc} consisting of three-rung solenoid. Two views of three-rung solenoid structures carrying tri-antennary N-glycans. Polypeptide chains are represented in the tube form, whereas N-glycans are represented in the ball-and-stick form. Each PrP molecule with corresponding N-glycan is of a different color. Sialic acid residues are colored in red. The structure of a tri-antennary N-linked glycan (shown in inset) was taken from PDB entry 3QUM, a crystal structure of human prostate specific antigen (PSA).[103] Both calculations of electrostatic surfaces and generation of images were performed with CCP4MG software. The model based on three-rung solenoid structure is shown here for simplicity of presentation and should not be considered as preferable over the four-rung solenoid model. *Adapted from Baskakov IV, Katorcha E, Makarava N. Prion strain-specific structure and pathology: a view from the perspective of glycobiology.* Viruses. *2018;10(12):723.*

neuraminidase treatment.[23,95] As predicted by the above hypothesis, significant increase in replication rates was observed for desialylated PrP^C relative to PrP^C with unaltered sialylation status for all strains tested.[23,95] An increase in replication rates was found to be strain-specific ranging from 10-fold to $\sim 10^6$-fold.[23,95] Moreover, in support of the above hypothesis, recent studies documented strong increase in efficiency of amplification and sensitivity of detections of PrP^{Sc} for several subtypes of sCJD, when reactions were conducted using unglycosylated PrP^C as a substrate.[104] The fact that removing of Sia residues or whole glycans boosts replication rate in strain-specific manner suggests that PrP^{Sc} surfaces are decorated by sialic acid in strain-specific patterns generating strain-specific density of negative charges.

As discussed above, strain-specific glycoform ratio is a result of negative selection of diglycosylated sialoglycoforms carrying bulky, highly charged glycans. Is it electrostatic repulsion between Sia residues or spatial interference between bulky glycans that imposes constrains and limits recruitment of diglycosylated glycoforms? To answer this question, glycoform ratios was analyzed in PrP^{Sc} generated from desialylated substrate in PMCA. In PMCA-derived PrP^{Sc} produced from desialylated substrate, the relative ratio of glycoforms shifted to predominantly diglycosylated regardless of the glycosylation status of strains used for seeding.[95] In fact, upon replication in desialylated substrate, prion strains lost their strain-specific glycoform ratios, which become very similar to the glycoform ratio of PrP^C.[95] In summary, these studies revealed that (i) removing Sia residues abolishes negative selection of diglycosylated glycoforms, and (ii) the sialylation status is more important than the size of glycans in defining strain-specific glycoform ratios.

11. Sialylation and strain interference

Prion strain interference occurs when a host is infected with two or more prion strains.[105–109] Strain interference refers to an extension of the incubation time to disease produced by a strain mixture relative to the incubation period produced alone by the strain with the shortest incubation time. Among factors that were previously discussed as main contributors to strain interference were competition between strains for substrate, cellular co-factors or cellular replication sites.[107,109]

Strain-specific selection of PrP^C sialoglycoforms adds an important dimension to the strain interference phenomenon.[95] Strains with substantial structural constraints rely on a narrow range of PrP^C sialoglycoforms as substrates and are unlikely to be strong competitors. In contrast, strains capable of recruiting a broad range of PrP^C sialoglycoforms have a greater chance of succeeding in competition for a substrate. Evolution of prion diseases of synthetic origin and competition between two self-propagating states, atypical PrPres and PrP^{Sc} provides remarkable illustrations of how differences in selectivity toward PrP^C sialoglycoforms determined the outcome of competition.[89,90,92–94] Only a small fraction of PrP^C sialoglycoforms that are recruited as by PrP^{Sc} was found to be suitable as a substrate to atypical PrPres.[92]

As a result, atypical PrPres replicates much slower than PrP^{Sc} and loses a competition.[89,90,92,93] In conclusion, strain-specific selection of PrP^{C} sialoglycoforms is an important factor that contributes to strain competition and interference. In a competition between selective and non-selective strains, non-selective strains are expected to win.

12. Concluding remarks

Increasing body of evidence suggest that N-linked glycans, which are a constitutive part of PrP^{Sc}, are important players in defining strain-specific structures and disease phenotypes.[110] N-glycans are exposed on PrP^{Sc} surface and impose considerable structural constrains to PrP^{Sc} assembly due to their bulky size and electrostatic repulsion between sialic acid residues. To overcome these restrains, prion strains exhibit a wide range of selectivity in recruiting PrP^{C} sialoglycoforms. While some strains recruit sialoglycoforms proportionally to their representation in PrP^{C}, others avoid diglycosylated and highly sialylated PrP^{C} glycofoms. Strain-specific selectivity in recruiting PrP^{C} sialoglycoforms reports on strain-specific differences in PrP^{Sc} quaternary structures and has to be taken into consideration when building realistic PrP^{Sc} models. Moreover, as a result of strain-specific selection, N-linked glycans appear to form individual strain-specific patterns of functional carbohydrate epitopes on PrP^{Sc} surfaces. It has been well established that the innate immune system senses terminal carbohydrate groups including galactose and sialic acid residues, which can serve as molecular cues and trigger diverse response programs by glia. Consistent with previous studies on self-associated and pathogen-associated molecular patterns, our work revealed that desialylation of PrP^{Sc} undermines survival of prions in vivo. At the same time, recruiting PrP^{C} with low sialylation levels speeds up PrP^{Sc} replication, which could be beneficial to a spread of prion under conditions with depressed innate immunity. Therefore, evolution of prion strains might represent a delicate balance between recruiting of highly sialylated glycoforms and generating sufficient density of sialylation for avoiding "eat-me" response by glia, yet limiting extremely heavily sialylated glycoforms for enabling efficient prion replication.

FUNDING

Financial support for this study was provided by National Institute of Health Grants R01 NS045585 and R01 AI128925.

References

1. Prusiner SB. Novel proteinaceous infectious particles cause scrapie. *Science*. 1982;216: 136–144.
2. Legname G, Baskakov IV, Nguyen HOB, et al. Synthetic mammalian prions. *Science*. 2004;305:673–676.
3. Cohen FE, Prusiner SB. Pathologic conformations of prion proteins. *Annu Rev Biochem*. 1998;67:793–819.
4. Collinge J, Clarke AR. A general model of prion strains and their pathogenicity. *Science*. 2007;318:930–936.
5. Bessen RA, Marsh RF. Identification of two biologically distinct strains of transmissible mink encephalopathy in hamsters. *J Gen Virol*. 1992;73:329–334.
6. Peretz D, Scott M, Groth D, et al. Strain-specified relative conformational stability of the scrapie prion protein. *Protein Sci*. 2001;10:854–863.
7. Safar J, Wille H, Itri V, et al. Eight prion strains have PrP Sc molecules with different conformations. *Nat Med*. 1998;4(10):1157–1165.
8. Ayers JL, Schutt CR, Shikiya RA, Aguzzi A, Kincaid AE, Bartz JC. The strain-encoded relationship between PrP replication, stability and processing in neurons is predictive of the incubation period of disease. *PLoS Pathog*. 2011;7(3):e1001317.
9. Gonzalez-Montalban N, Makarava N, Savtchenko R, Baskakov IV. Relationship between conformational stability and amplification efficiency of prions. *Biochemistry*. 2011;50(37):7933–7940.
10. Klimova N, Makarava N, Baskakov IV. The diversity and relationship of prion protein self-replicating states. *Virus Res*. 2015;207:113–119.
11. Caughey B, Raymond GJ, Bessen RA. Strain-dependent differences in b-sheet conformations of abnormal prion protein. *J Biol Chem*. 1998;273(48):32230–32235.
12. Spassov S, Beekes M, Naumann D. Structural differences between TSEs strains investigated by FT-IR spectroscopy. *Biochim Biophys Acta*. 2006;1760(7):1138–1149.
13. Thomzig A, Spassov S, Friedrich M, Naumann D, Beekes M. Discriminating scrapie and bovine spongiform encephalopathy isolates by infrared spectroscopy of pathological prion protein. *J Biol Chem*. 2004;279:33847–33854.
14. Lawson VA, Collins SJ, Masters CL, Hill AF. Prion protein glycosylation. *J Neurochem*. 2005;93:793–801.
15. Stahl N, Borchelt DR, Hsiao K, Prusiner SB. Scrapie prion protein contains a phosphatidylinositol glycolipid. *Cell*. 1987;51:229–240.
16. Bolton DC, Meyer RK, Prusiner SB. Scrapie PrP 27-30 is a sialoglycoprotein. *J Virol*. 1985;53:596–606.
17. Turk E, Teplow DB, Hood LE, Prusiner SB. Purification and properties of the cellular and scrapie hamster prion proteins. *Eur J Biochem*. 1988;176:21–30.
18. Stahl N, Baldwin MA, Teplow DB, et al. Structural studies of the scrapie prion protein using mass spectrometry and amino acid sequencing. *Biochemistry*. 1993;32:1991–2002.
19. Stahl N, Baldwin MA, Hecker R, Pan KM, Burlingame AL, Prusiner SB. Glycosylinositol phospholipid anchors of the scrapie and cellular prion proteins contain sialic acid. *Biochemistry*. 1992;31:5043–5053.
20. Rudd PM, Endo T, Colominas C, et al. Glycosylation differences between the normal and pathogenic prion protein isoforms. *Proc Natl Acad Sci U S A*. 1999; 96:13044–13049.
21. Endo T, Groth D, Prusiner SB, Kobata A. Diversity of oligosaccharide structures linked to asparagines of the scrapie prion protein. *Biochemistry*. 1989;28:8380–8388.
22. Stimson E, Hope J, Chong A, Burlingame AL. Site-specific characterization of the N-linked glycans of murine prion protein by high-performance liquid chromatography/electrospray mass spectrometry and exoglycosidase digestions. *Biochemistry*. 1999;38:4885–4895.

23. Katorcha E, Makarava N, Savtchenko R, D'Azzo A, Baskakov IV. Sialylation of prion protein controls the rate of prion amplification, the cross-species barrier, the ratio of PrPSc glycoform and prion infectivity. *PLoS Pathog*. 2014;10(9):e1004366.
24. Katorcha E, Klimova N, Makarava N, et al. Knocking out of cellular neuraminidases Neu1, Neu3 or Neu4 does not affect sialylation status of the prion protein. *PLoS One*. 2015;10(11):e0143218.
25. Varki A. Sialic acids. In: Varki A, Cummings R, Esko J, Freeze H, Hart G, Marth J, eds. *Essentials of Glycobiology*. Cold Spring Harbor, NY: Cold Spring Harbor Laboratory Press; 1999:195–210.
26. Miyagi T, Yamaguchi K. Mammalian sialidases: physiological and pathological roles in cellular functions. *Glycobiology*. 2012;22(7):880–896.
27. Audry M, Jeanneau C, Imberty A, Harduin-Lepers A, Delannoy P, Breton C. Current trend in the structure-activity relationships of sialylatransferases. *Glycobiology*. 2011; 21(6):716–726.
28. Takashima S. Characterization of mouse sialylatransferase gene: their evolution and diversity. *Biosci Biotechnol Biochem*. 2008;72:1155–1167.
29. Pshezhetsky AV, Ashmarina LI. Desialylation of surface receptors as a new dimension in cell signaling. *Biochemistry (Mosc)*. 2013;78(7):736–745.
30. Monti E, Bonten E, D'Azzo A, et al. Sialidases in vertebrates: a family of enzymes tailored for several cell functions. *Adv Carbohydr Chem Biochem*. 2010;64:403–479.
31. Varki A. Uniquely human evolution of sialic acid genetics and biology. *Proc Natl Acad Sci U S A*. 2010;107:8939–8946.
32. Ng PS, Bohm R, Hartley-Tassell LE, et al. Ferrets exclusively synthesize Neu5Ac and express naturally humanized influenza A virus receptors. *Nat Commun*. 2014;5:5750.
33. Samraj AN, Pearce OMT, Laubli H, et al. A red meat-derived glycan promotes inflammation and cancer progression. *Proc Natl Acad Sci U S A*. 2015;112(2):542–547.
34. Schauer R, Srinivasan GV, Wipfer D, Kniep B, Schwartz-Albiez R. O-acetylated sialic acids and their role in immune defence. In: Wu AM, ed. *The Molecular Immunology of Complex Carbohydrates-3*. New York, Dordrecht, Heidelberg, London: Sprinenger Science; 2011:525–548.
35. Fukuda M, Hiraoka N, Yeh JC. C-type lectins and sialyl lewis X oligosaccharides versatile roles in cell–cell interaction. *J Cell Biol*. 1999;147(3):467–470.
36. Katorcha E, Baskakov IV. Analyses of N-linked glycans of PrPSc revealed predominantly 2,6-linked sialic acid residues. *FEBS J*. 2017;284(21):3727–3738.
37. Wille H, Michelitsch MD, Guenebaut V, et al. Structural studies of the scrapie prion protein by electron crystallography. *Proc Natl Acad Sci U S A*. 2002;99:3563–3568.
38. Govaerts C, Wille H, Prusiner SB, Cohen FE. Evidance for assembly of prions with left-handed b-helices into trimers. *Proc Natl Acad Sci U S A*. 2004;101:8342–8347.
39. Requena JR, Wille H. The Structure of the infectious prion protein: experimental data and molecular models. *Prion*. 2014;8(1):60–66.
40. Spagnolli G, Rigoli M, Orioli S, et al. Full atomistic model of prion structure and conversion. *PLoS Pathog*. 2019;15(7):e1007864.
41. Baskakov IV, Katorcha E, Makarava N. Prion strain-specific structure and pathology: a view from the perspective of glycobiology. *Viruses*. 2018;10(12):723.
42. Makarava N, Kovacs GG, Bocharova OV, et al. Recombinant prion protein induces a new transmissible prion disease in wild type animals. *Acta Neuropathol*. 2010; 119(2):177–187.
43. Katorcha E, Daus ML, Gonzalez-Montalban N, et al. Reversible off and on switching of prion infectivity via removing and reinstalling prion sialylation. *Sci Rep*. 2016; 6:33119.
44. Srivastava S, Katorcha E, Daus ML, Lasch P, Beekes M, Baskakov IV. Sialylation controls prion fate in vivo. *J Biol Chem*. 2017;292(6):2359–2368.

45. Varki A. Sialic acids in human health and disease. *Trends Mol Med.* 2008;14(8):351–360.
46. Brown GC, Neher JJ. Microglial phagocytosis of live neurons. *Nat Rev Neurosci.* 2014;15:209–216.
47. Aminoff D, Bruegge WF, Bell WC, Sarpolis K, Williams R. Role of sialic acid in survival of erythrocytes in the circulation: interaction of neuraminidase-treated and untreated erythrocytes with spleen and liver at the cellular level. *Proc Natl Acad Sci U S A.* 1977;74(4):1521–1524.
48. Jansen AJG, Josefsson EC, Rumjantseva V, et al. Desialylation accelerates platelet clearance after refrigeration and initiates GPIba metalloproteinase-mediated cleavage in mice. *Blood.* 2012;119(5):1263–1273.
49. Linnartz B, Kopatz J, Tenner AJ, Neumann H. Sialic acid on the neuronal glycocalyx prevents complement C1 binding and complement receptor-3-mediated removal by microglia. *J Neurosci.* 2012;32(3):946–952.
50. Linnartz-Gerlach B, Mathews M, Neumann H. Sensing the neuronal glycocalyx by glial sialic acid binding immunoglobulin-like lectins. *Neuroscience.* 2014;275:113–124.
51. Linnartz-Gerlach B, Schuy C, Shahraz A, Tenner AJ, Neumann H. Sialylation of neurites inhibits complement-mediated macrophage removal in a human macrophage-neuron Co-Culture System. *Glia.* 2016;64(1):35–47.
52. Huang FP, Farquhar CF, Mabbott NA, Bruce ME, MacPherson GG. Migrating intestinal dendritic cells transport PrPSc from the gut. *J Gen Virol.* 2002;83:267–271.
53. Castro-Seoane R, Hummerich H, Sweeting T, et al. Plasmacytoid dendritic cells sequester high prion titres at early stages of prion infection. *PLoS Pathog.* 2012;8(2): e1002538.
54. Michel B, Meyerett-Reid C, Johnson T, et al. Incunabular imminological events in prion trafficing. *Sci Rep.* 2012;2:440.
55. Takakura I, Muiyazawa K, Kanaya T, et al. Orally administered prion protein is incorporated by m cells and spreads into lymphoid tissues with macrophages in prion protein knockout mice. *Am J Pathol.* 2011;179(3):1301–1309.
56. McCulloch L, Brown KL, Bradford BM, et al. Follicular dendritic cell-specific prion protein (PrP) expression alone is sufficient to sustain prion infection in the spleen. *PLoS Pathog.* 2011;7(12):e1002402.
57. Kujala P, Raymond CR, Romeijn M, et al. Prion uptake in the gut: identification of the first uptake and replication sites. *PLoS Pathog.* 2011;7(12):e1002449.
58. Brown KL, Stewart K, Ritchie DL, et al. Scrapie replication in lyphoid tissues depends on prion protein-expressing follicular dendritic cells. *Nat Med.* 1999;5(11): 1308–1312.
59. Montrasio F, Frigg R, Glatzel M, et al. Impaired prion replication in spleens of mice lacking functional follicular dendritic cells. *Science.* 2000;288:1257–1259.
60. Aguzzi A, Nuvolone M, Zhu C. The immunology of prion diseases. *Nat Rev Immunol.* 2013;13(12):888–902.
61. Wang S, Chen X, Wei A, Yu X, Niang B, Zhang J. α2,6-linked sialic acids on N-glycans modulate the adhesion of hepatocarcinoma cells to lymph nodes. *Tumour Biol.* 2015;36(2):885–892.
62. Zhang Z, Sun J, Hao L, Liu C, Ma H, Jia L. Modification of glycosylation mediates the invasive properties of murine hepatocarcinoma cell lines to lymph nodes. *PLoS One.* 2013;8(6):e65218.
63. Tanaka K, Siwu ERO, Minami K, et al. Noninvasive imaging of dendrimer-type N-glycan clusters: in vivo dynamics dependence on oligosaccharide structure. *Angew Chem Int Ed.* 2010;49:8195–8200.
64. Hill AF, Butterworth RJ, Joiner S, et al. Investigation of variant Creutzfeldt-Jakob disease and other human prion diseases with tonsil biopsy samples. *Lancet.* 1999;353: 183–189.

65. Wadsworth JD, Joiner S, Hill AF, et al. Tissues distribution of protease resistant prion protein in variant Creutzfeldt-Jakob disease using a highly sensitive immunobloting assay. *Lancet*. 2001;358:171–180.
66. Halliez S, Reine F, Herzog L, et al. Accelerated, spleen-based titration of variant Creutzfeldt-Jakob disease infectivity in transgenic mice expressing human prion protein with sensitivity comparable to that of survival time bioassay. *J Virol*. 2014;88(15): 8678–8686.
67. Pan T, Li R, Kang SC, et al. Biochemical fingerprints of prion diseases: scrapie prion protein in human prion diseases that share prion genotype and type. *J Neurochem*. 2005;92(1):132–142.
68. Zanusso G, Farinazzo A, Prelli F, et al. Identification of distinct N-terminal truncated forms of prion protein in different Creutzfeldt-Jakob disease subtypes. *J Biol Chem*. 2004;279:38936–38942.
69. Srivastava S, Makarava N, Katorcha E, Savtchenko R, Brossmer R, Baskakov IV. Post-conversion sialylation of prions in lymphoid tissues. *Proc Natl Acad Sci U S A*. 2015;112(48):E6654–E6662.
70. Harduin-Lepers A, Vallejo-Ruiz V, Krzewinski-Recchi MA, Samyn-Petit B, Julien S, Delannoy P. The human sialyltransferase family. *Biochimie*. 2001;83(8):727–737.
71. Rifat S, Kang TJ, Mann D, et al. Expression of sialyltransferase activity on intact human neutrophils. *J Leukoc Biol*. 2008;84:1075–1081.
72. Gross HJ, Merling A, Moldenhauer G, Schwartz-Albiez R. Ecto-sialyltransferase of human B lymphocytes reconstitutes differentiation markers in the presence of exogenous CMP-N-acetyl neuraminic acid. *Blood*. 1996;87(12):5113–5126.
73. Cabral MG, Piteira AR, Silva Z, Ligeiro D, Brossmer R, Videira PA. Human dendritic cells contain cell surface sialyltransferase activity. *Immunol Lett*. 2010;131(1):89–96.
74. Kaufmann M, Blaser C, Takashima S, Schwartz-Albiez R, Tsuji S, Pircher H. Identification of an alpha2,6-sialyltransferase induced early after lymphocyte activation. *Int Immunol*. 1999;11(5):731–738.
75. Schwartz-Albiez R, Merling A, Martin S, Haas R, Gross HJ. Cell surface sialylation and ecto-sialyltransferase activity of human CD34 progenitors from peripheral blood and bone marrow. *Glycoconj J*. 2004;21(8–9):451–459.
76. Nasirikenari M, Veillon L, Collins CC, Azadi P, Lau JT. Remodeling of marrow hematopoietic stem and progenitor cells by non-self ST6Gal-1 sialyltransferase. *J Biol Chem*. 2014;289(20):7178–7189.
77. Rabinovich GA, Croci DO. Regulatori circuits mediated by lectin-glycan interaction in autoimmunity and cancer. *Immunity*. 2012;36:322–335.
78. Bradford BM, Crocker PR, Mabbott NA. Peripheral prion disease pathogenesis is unaltered in the absence of sialoadhesin (Siglec-1/CD169). *Immunology*. 2014;143(1): 120–129.
79. Collinge J, Palmer MS, Sidle KC, et al. Unaltered susceptibility to BSE in transgenic mice expressing human prion protein. *Nature*. 1995;378:779–783.
80. Asante EA, Linehan JM, Desbruslais M, et al. BSE prions propagate as either variant CJD-like or sporadic CJD-like prion stains in transgenic mice expressing human prion protein. *EMBO J*. 2002;21(23):6358–6366.
81. Wadsworth JD, Asante EA, Desbruslais M, et al. Human prion protein with valine 129 prevents expression of variant CJD phenotype. *Science*. 2004;306(5702):1793–1796.
82. Bishop MT, Hart P, Aitchison L, et al. Predicting susceptibility and incubation time of human-to-human transmission of vCJD. *Lancet Neurol*. 2006;5:393–398.
83. Cassard H, Torres JM, Lacroux C, et al. Evidence for zoonotic potential of ovine scrapie prions. *Nat Commun*. 2014;5:5821.
84. Lu ZY, Baker CA, Manuelidis L. New molecular markers of early and progressive CJD brain infection. *J Cell Biochem*. 2004;93:644–652.

85. Vincenti JE, Murphy L, Grabert K, et al. Defining the microglia response during the time course of chronic neurodegeneration. *J Virol.* 2016;90(6):3003–3017.
86. Gomez-Nicola D, Fransen NL, Suzzi S, Perry VH. Regulation of microglial proliferation during chronic neurodegeneration. *J Neurosci.* 2013;33(6):2481–2493.
87. Linnartz B, Bodea L-G, Neumann H. Microglia carbohydrate-binding receptors for neural repair. *Cell Tissue Res.* 2012;349:215–227.
88. Srivastava S, Katorcha E, Makarava N, Barrett JP, Loane DJ, Baskakov IV. Inflammatory response of microglia to prions is controlled by sialylation of PrP^{Sc}. *Sci Rep.* 2018;8(1):e11326.
89. Makarava N, Kovacs GG, Savtchenko R, et al. Genesis of mammalian prions: from non-infectious amyloid fibrils to a transmissible prion disease. *PLoS Pathog.* 2011; 7(12):e1002419.
90. Makarava N, Kovacs GG, Savtchenko R, et al. A new mechanism for transmissible prion diseases. *J Neurosci.* 2012;32(21):7345–7355.
91. Kovacs GG, Makarava N, Savtchenko R, Baskakov IV. Atypical and classical forms of the disease-associated state of the prion protein exhibit distinct neuronal tropism, deposition patterns, and lesion profiles. *Am J Pathol.* 2013;183(5):1539–1547.
92. Makarava N, Savtchenko R, Baskakov IV. Two alternative pathways for generating transmissible prion disease de novo. *Acta Neuropathol Commun.* 2015;3:69.
93. Makarava N, Savtchenko R, Alexeeva I, Rohwer RG, Baskakov IV. New molecular insight into mechanism of evolution of mammalian synthetic prions. *Am J Pathol.* 2016;186(4):1006–1014.
94. Makarava N, Savtchenko R, Baskakov IV. Selective amplification of classical and atypical prions using modified protein misfolding cyclic amplification. *J Biol Chem.* 2013;288(1):33–41.
95. Katorcha E, Makarava N, Savtchenko R, Baskakov IV. Sialylation of the prion protein glycans controls prion replication rate and glycoform ratio. *Sci Rep.* 2015; 5:16912.
96. Tycko R, Savtchenko R, Ostapchenko VG, Makarava N, Baskakov IV. The a-helical C-terminal domain of full-length recombinant PrP converts to an in-register parallel á-sheet structure in PrP fibrils: evidence from solid state nuclear magnetic resonance. *Biochemistry.* 2010;49(44):9488–9497.
97. Cobb NJ, Sonnichsen FD, McHaourab H, Surewicz W. Molecular architecture of human prion protein amyloid: a parallel, in-register b-structure. *Proc Natl Acad Sci U S A.* 2007;104:18946–18951.
98. Groveman BR, Dolan MA, Taubner LM, Kraus A, Wickner RB, Caughey B. Parallel in-register intermolecular β-sheet architectures for prion-seeded prion protein (PrP) amyloids. *J Biol Chem.* 2014;289(35):24129–24142.
99. Varquez-Fernandez E, Vos MR, Afanasyev P, et al. The structural architecture of an infectious mammalian prion using electron cryomicroscopy. *PLoS Pathog.* 2016;12(9): e1005835.
100. Baskakov IV, Katorcha E. Multifaceted role of sialylation in prion diseases. *Front Neurosci.* 2016;10:358.
101. Katorcha E, Baskakov IV. Analysis of covalent modifications of amyloidogenic proteins using two-dimensional electrophoresis: prion protein and its sialylation. *Methods Mol Biol.* 2018;1779:241–255. 241–255.
102. Nishina K, Deleault NR, Mahal S, et al. The stoichiometry of host PrPC glycoforms modulates the efficiency of PrPSc formation in vitro. *Biochemistry.* 2006;45: 14129–14139.
103. Stura EA, Muller BH, Bossus M, Michel S, Jolivet-Reynaud C, Ducancel F. Crystal structure of human prostate-specific antigen in a sandwich antibody complex. *J Mol Biol.* 2011;414:530–544.

104. Camacho MV, Telling G, Kong Q, Gambetti P, Notari S. Role of prion protein glycosylation in replication of human prions by protein misfolding cyclic amplification. *Lab Invest*. 2019;99:1741–1748.
105. Dickinson AG, Fraser H, Meikle VMH, Outram GW. Competition between different scrapie agents in mice. *Nat New Biol*. 1972;237:244–245.
106. Kimberlin RH, Walker CA. Competition between strains of scrapie depends on the blocking agent being infectious. *Intervirology*. 1985;23:74–81.
107. Bartz JC, Kramer ML, Sheehan MH, et al. Prion interference is due to a reduction in strain-specific PrPSc levels. *J Virol*. 2007;81(2):689–697.
108. Schutt CR, Bartz JC. Prion interference with multiple prion isolates. *Prion*. 2008; 2(2):61–63.
109. Shikiya RA, Ayers JI, Schutt CR, Kincaid AE, Bartz JC. Coinfecting prion strains compete for a limiting cellular resource. *J Virol*. 2010;84(11):5706–5714.
110. Makarava N, Chang JC-Y, Molesworth K, Baskakov IV. Posttranslational modifications define course of prion strain adaptation and disease phenotype. *J Clin Invest*. 2020;130:4382–4395. https://doi.org/10.1172/JCI138677.

CHAPTER FOUR

Cofactor molecules: Essential partners for infectious prions

Surachai Supattapone*
Department of Biochemistry and Cell Biology and Department of Medicine, Geisel School of Medicine at Dartmouth College, Hanover, NH, United States
*Corresponding author: e-mail address: supattapone@dartmouth.edu

Contents

Abstract

The protein-only hypothesis predicts that infectious mammalian prions are composed solely of PrP^{Sc}, a misfolded conformer of the normal prion protein, PrP^{C}. However, to date, all wild type protein-only PrP^{Sc} preparations lack significant levels of prion infectivity. Using a systemic biochemical approach, our laboratory isolated and identified two different endogenous cofactor molecules, RNA (Deleault et al., 2003 [50]; Deleault et al., 2007 [59]) and phosphatidylethanolamine (Deleault et al., 2012 [61]; Deleault et al., 2012 [18]), which facilitate the formation of prions with high levels of specific infectivity, leading us to propose to the alternative hypothesis that cofactor molecules are required to form wild type infectious prions (Deleault et al., 2007 [59]; Deleault et al., 2012 [18]; Geoghegan et al., 2007 [57]). In addition, we found that purified cofactor molecules restrict the strain properties of chemically defined infectious prions (Deleault et al., 2012 [18]), suggesting a "cofactor selection" model in which natural variation in the distribution of strain-specific cofactor molecules in different parts of the brain may be responsible for strain-dependent patterns of neurotropism (Deleault et al., 2012 [18]; Geoghegan et al., 2007 [57]).

Progress in Molecular Biology and Translational Science, Volume 175
ISSN 1877-1173
https://doi.org/10.1016/bs.pmbts.2020.07.009

1. Introduction

Prion diseases are a group of invariably fatal neurodegenerative disorders that occur in inherited, sporadic, and infectious forms.[1] Specific examples include familial, sporadic, and iatrogenic Creutzfeldt-Jakob disease (CJD) in humans, bovine spongiform encephalopathy (BSE) in cattle, scrapie in sheep and goats, and chronic wasting disease (CWD) in cervid species. Prion diseases with either sporadic or infectious etiologies as well as many types of inherited prion disease are subsequently transmissible to normal, wild type hosts. When inoculated directly into the central nervous system (CNS), wild type prions appear to be very potent infectious agents; there are typically $>10^7$ infectious units (LD_{50}) per gram of brain tissue in animals with terminal prion disease.[2] Only mammalian species are naturally susceptible to prion infection, but individual species can exhibit markedly different degrees of susceptibility. Whereas most species, including humans, are susceptible to a subset of prion isolates derived from other isolates, rabbits appear to be resistant to prion infection and bank voles are almost universally susceptible.[3–7] Presently there are no clinically effective vaccines or therapies to combat these devastating disorders.[8]

The molecular mechanisms responsible for the pathogenesis prion disease are fascinating, unusual, and largely unknown. Although wild type prions are potent infectious agents, they are unorthodox because they lack a nucleic acid genome, and therefore are unrelated to all other known, conventional infectious agents, such as viruses, bacteria, fungi, and parasites.[9,10] The fundamental event underlying all forms of prion disease appears to be the autocatalytic misfolding of a host-encoded protein, PrP^C, into various non-native conformers collectively termed PrP^{Sc}.[11,12] PrP^C is a glycophosphatidylinositol (GPI)-anchored glycoprotein which is attached to the outer leaflet of the plasma membrane in many different cell types throughout the body, but it is most highly expressed in CNS neurons.[13–15] According to the widely accepted "protein-only hypothesis," infectious prions are simply comprised of PrP^{Sc} without any other informational molecules. Indeed, much evidence shows that PrP^{Sc} is required for prion infectivity,[16,17] but it remains unclear whether it is sufficient.[18–22] Whereas PrP^{Sc} misfolding in inherited forms of prion disease is caused the presence of specific pathogenic mutations in the PrP sequence,[23,24] the molecular basis for wild type PrP conformational change in the sporadic and infectious forms of prion disease remain obscure. Sporadic CJD is the most common form of prion disease in humans and is not associated with any particular epidemiological risk factors other than age.[25,26] It has been hypothesized that sporadic prion disease may be triggered either by the somatic mutation or stochastic conformational change of PrP, but there is currently no direct evidence showing that either of these processes occurs in patients.[27,28] The pathogenic mechanisms responsible for neuronal death downstream of PrP conformational change are also unknown. All forms of prion diseases are characterized by progressive spongiform degeneration in the CNS, but surprisingly these neuropathological changes are not caused either by loss of PrP^{C} or accumulation of PrP^{Sc} in the absence of PrP^{C}, and must therefore be triggered by some other consequence of PrP conversion.[29,30]

Mammalian prions appear to be mechanistically unique. Although recent work suggests that other proteins such as amyloid beta, tau, and synuclein can also propagate as misfolded conformers in various experimental systems,[31] it is not likely that any human neurodegenerative diseases other than prion diseases can be transmitted between individual patients. In a large study conducted by the National Institutes of Health, various forms of prion disease were found to be the only neurodegenerative diseases tested that could be transmitted to non-human primates.[32] Also, whereas infectious prion diseases typically occur in clusters, outbreaks, or epidemics (e.g., kuru, iatrogenic CJD, and variant CJD), there is no evidence that more prevalent neurodegenerative diseases such as Alzheimer's disease have ever displayed a similar pattern of disease transmission.[26] Epidemiologic investigation of rare Parkinson's disease and amyotrophic lateral sclerosis (ALS) clusters have found evidence for common exposure to specific chemical compounds, such as the neurotoxic by-product MPTP and pesticides that trigger reactive oxygen species, rather than infected CNS tissues.[33] Mammalian prions are also quite dissimilar to fungal amyloids. Unlike mammalian prions, fungal amyloids appear to have relatively low levels of specific "infectivity,"[34] and PrP lacks Q/N-rich domains that are essential for propagation of yeast prions.[35] Mechanistically, fungal amyloids appear to be more closely related to mutant proteins with expanded polyglutamine tracts such as Huntingtin and ataxin, which cause dominant genetic diseases such as Huntington's disease and spinocerebellar ataxia.[36]

Interestingly, it has been observed that multiple prion "strains" can be propagated indefinitely in a single animal species.[37] Each prion strain appears to be associated with a unique PrP^{Sc} conformation, clinical presentation, and pattern of regional pathology within the CNS (i.e., strain-specific neurotropism).[37]

Historically, the paradoxical existence of strains presents a significant theoretical challenge for the protein-only hypothesis. In particular, it has been difficult to understand why different PrP^{Sc} conformers are associated with different patterns of neurotropism since there do not appear to be receptors that preferentially bind specific prion strains on the surface on neurons.

2. Prion formation *in vitro*

It has been difficult to study the pathogenic mechanisms underlying prion disease experimentally. A number of different factors complicate using a forward genetics approach. Low incidence rates hinder clinical studies in humans, and long incubation times limit the number of experiments that can be performed in experimental animals.[38,39] A variety of cultured mammalian cell lines can be chronically infected with prions, but these cell lines generally display no obvious phenotype.[40,41] Many simple organisms such as yeast and *Caenorhabditis elegans* are not susceptible to prion infection even when PrP^{C} is deliberately over-expressed. A notable exception is *Drosophila melanogaster*, which become susceptible to prion infection upon transgenic expression of PrP^{C} in brain neurons.[42]

On the other hand, a number of significant mechanistic insights have been gained using biochemical approaches. Before discussing this work, it is important to highlight two significant limitations with biochemical studies of prion disease. First, biochemical studies can only be used to study the mechanism of PrP^{Sc} formation, not the downstream mechanism(s) responsible for neuronal cell death. Second, although *in vitro* experiments provide a powerful approach to rigorously determine the minimal chemical requirements for specific process (such as prion infectivity), they cannot by themselves prove that a specific molecule participates in a given process *in vivo*.

Early attempts to produce PrP^{Sc} in cell-free systems were unsuccessful.[43] In retrospect, these experiments probably failed because of the lack physical agitation, which is necessary to break apart large PrP^{Sc} aggregates. The first successful cell-free conversion assay was developed by Kocisko et al.[44] This system used [^{35}S]-labeled PrP^{C} substrate whose conversion into protease-resistant PrP^{Sc} (PrP-res) could be detected with great sensitivity by autoradiography. Critically, this assay was the first to show that prion-seeded *in vitro* conversion of PrP^{C} to PrP^{Sc} is a specific process, which faithfully propagates strain-dependent PrP^{Sc} conformations and correlates with known species barriers.[44–46] The major limitation of this assay is inefficiency:

a large molar excess of non-radioactive PrP^{Sc} seed is required to drive formation of radioactive PrP^{Sc} product.[47,48] In a major advance, Soto and colleagues developed protein misfolding cyclic amplification (PMCA), a cell-free method that efficiently amplifies PrP^{Sc} levels through the use of intermittent sonication.[49] Although PrP^{Sc} levels can also be amplified *in vitro* by continuous shaking instead of intermittent sonication,[50–54] the kinetics of PrP^{Sc} formation are significantly faster with PMCA than with shaking. Most importantly, Castilla et al. showed that PrP^{Sc} molecules produced *in vitro* from crude brain homogenate substrate are infectious and retain the strain phenotype of the prion seed, including neurotropism.[55,56] These results provided critical evidence that the mechanism of infectious prion replication could be studied by using a reductionist biochemical approach.

3. Isolation and identification of endogenous prion cofactors

The existence of prion cofactors was first detected by Deleault et al. by applying specific enzymatic treatments to crude brain homogenate substrates in single-round shaking PrP^{Sc} amplification assays.[50] Unexpectedly, *in vitro* formation of Sc237 PrP^{Sc} molecules from crude Syrian hamster brain homogenate was inhibited by enzymatic depletion of single-stranded RNA.[50] Moreover, PrP^{Sc} amplification activity could be rescued in RNase-treated by total RNA preparations isolated from mammalian tissues but not yeast or bacterial cells.[50] Further characterization of the cofactor requirements for Sc237-seeded PrP^{Sc} formation and serial propagation *in vitro* showed that some single-stranded homopolymeric RNA and DNA molecules (such as poly(A) RNA and oligo(dT) DNA) could serve as potent cofactors, whereas glycosaminoglycans and heparan sulfate proteoglycans were significantly less potent, and double-stranded nucleic acids were inactive.[53] A systematic test of increasing oligo(dT) lengths revealed that the minimum length necessary to facilitate hamster PrP^{Sc} formation successful is ~40 bases.[57] Collectively, these experiments suggest that unstructured, linear polyanionic molecules represent a major class of endogenous prion cofactors.

When used as prion conversion cofactors, single-stranded nucleic acids appear to be selectively incorporated into nuclease-resistant physical complexes with hamster PrP^{Sc} molecules.[57] Interestingly, synthetic prions formed with a photocleavable oligo(dT) 100-mer maintained their infectivity and strain properties after ultraviolet irradiation-triggered degradation of the incorporated oligonucleotide into five base fragments.[58] There are three potential explanations for these results: (1) once formed, prions no longer require cofactor molecules to maintain their infectious properties, consistent with the protein-only hypothesis; (2) maintenance of infectious properties may have been mediated by endogenous lipid molecules which co-purify with the native PrP^C substrate used in these experiments[59]; or (3) five base oligonucleotides are sufficiently large to maintain the structure of hamster prions, even though longer oligonucleotides are initially required to facilitate their formation.

Unexpectedly, it was later discovered that polyanionic molecules could not serve as cofactors for RML-seeded formation of native PrP^{Sc} molecules in reconstituted serial PMCA reactions.[60] Moreover, unlike the situation with Sc237-seeded PMCA reactions using crude hamster brain homogenate, PMCA reactions using crude mouse or vole brain homogenates seeded with RML, 22L, or 301C prions were not inhibited by nuclease treatment.[60] These observations suggested the existence of a novel class of endogenous cofactor molecules that are not nucleic acids and prompted a concentrated effort to purify and identify these alternative cofactor molecules. Using a systematic purification and reconstitution approach to isolate the alternative cofactor activity from $PrP^{0/0}$ (knockout) mouse brains, Deleault et al. obtained a preparation containing a mixture of five types of phospholipids (phosphatidylethanolamine (PE), phosphatidylcholine (PC), phosphatidylinositol (PI), lyso PE, and lyso PC) that fully reconstituted the ability of mouse PrP^C to convert into PrP^{Sc}.[61] Further testing with purified and synthetic preparations of individual phospholipids demonstrated that PE was the only component required for cofactor activity.[61] PE is a zwitterionic phospholipid that is enriched on neuronal plasma membranes where PrP^C molecules are normally localized. Interestingly, stoichiometric amounts of lipids co-purify with native PrP^C isolated by a combination of immunoaffinity and conventional chromatography techniques from brain tissue,[50] supporting a role for endogenous lipids in prion formation *in vivo*. Similar to single-stranded nucleic acids, PE appears to be incorporated into physical complexes with PrP^{Sc} during the *in vitro* conversion process.[18] Unlike RNA, PE appears to be a universal cofactor for prion formation: reconstitution experiments show that PE can be used as a cofactor for serial PMCA (sPMCA) reactions using mouse, hamster, vole, a deer PrP^C as well as mouse and vole bacterially expressed recombinant (rec)PrP substrates seeded with a wide variety of prion strains.[61]

4. Cofactor molecules are required for the formation of infectious prions

Due to their unorthodox replication mechanism, there has been longstanding interest in the chemical composition of infectious mammalian prions. If the protein-only hypothesis is correct, it should be possible to produce fully infectious prions by chemically refolding pure wild type PrP molecules, and many groups have attempted to accomplish this goal. Logically, one might anticipate that the process of PrP^{Sc} formation would be thermodynamically favored since infectious prions are notoriously difficult to inactivate, and much harsher conditions are required to denature PrP^{Sc} than PrP^{C}.[28] However, so far it has only been possible to produce pure misfolded PrP conformers with extremely low levels of specific infectivity, often requiring mutation of PrP or bioassays in transgenic hosts expressing mutant PrP molecules at non-physiological levels.[18–22] Authentic brain-derived prions typically have high levels of specific infectivity, i.e., $>10^6$ LD_{50} units/μg PrP^{Sc}, but samples containing only pure misfolded PrP conformers all display much lower levels of specific infectivity, i.e., <1 LD_{50}/μg PrP^{Sc}.[62] These observations suggest that either cofactor molecules are required to form infectious prions, or the protein-only hypothesis is incorrect (i.e., cofactor molecules are essential components of infectious prions). Either way, we can infer that prions formed with cofactor molecules must have different molecular structures than misfolded PrP conformers formed without cofactors.

The essential role of cofactor molecules for infectious prion formation was initially discovered by Deleault et al. using immunopurified native hamster PrP^{C} as a substrate in chemically defined sPMCA reactions.[59] These experiments revealed that PrP^{C} alone could not produce protease-resistant PrP^{Sc} in Sc237- or 139H-seeded reactions, but a combination of PrP^{C} and synthetic poly(A) RNA produced autocatalytic PrP^{Sc} with high conversion efficiency.[59] Strikingly, the minimal cocktail of PrP^{C} and synthetic poly(A) RNA was able to produce PrP^{Sc} spontaneously even without seeding, showing for the first time that infectious wild type prions could be produced from non-infectious reactants.[59] Critically, end-point titration bioassays in hamsters confirmed that PrP^{Sc} produced in both seeded or unseeded sPMCA reactions had moderately high levels of specific infectivity ($\sim10^4$–10^5 LD_{50}/μg PrP^{Sc}).[59] The ability to form infectious prions spontaneously in unseeded sPMCA reactions containing PrP and RNA was later confirmed by Wang and Ma using recPrP and cellular RNA purified from normal mouse liver (rather than immunopurified PrP^{C} and synthetic poly(A) RNA) as substrates.[63]

The first PrP^{Sc} molecules composed entirely from non-mammalian sources were produced by Deleault et al. using bacterially expressed recPrP and synthetic PE substrates in sPMCA reactions.[18] The combination of recPrP and PE facilitated the formation of a PrP^{Sc} conformer with an ~18 kDa proteinase K (PK)-resistant core, which was designated "cofactor PrP^{Sc}" (Fig. 1). Interestingly, propagation of cofactor PrP^{Sc}-seeded sPMCA reactions in recPrP substrate without cofactor resulted in two different outcomes: (1) in ~40% of the reactions, no PrP^{Sc} formation was observed (i.e., similar to the negative outcome of all reactions using pure native PrP^{C} substrate), and (2) in ~60% of the reactions, there was successful propagation of alternative PrP^{Sc} conformer with an ~16 kDa PK-resistant core, termed "protein-only PrP^{Sc}" (Fig. 1). The observation that recPrP but not PrP^{C} is able to form protein-only PrP^{Sc} suggests that post-translational modifications such as N-linked glycans and the GPI-anchor may normally prevent native PrP^{C} from entering folding pathways that produce protein-only PrP^{Sc} conformers. Regardless, the availability of two chemically defined wild type $recPrP^{Sc}$ conformers derived from a common seed and differing

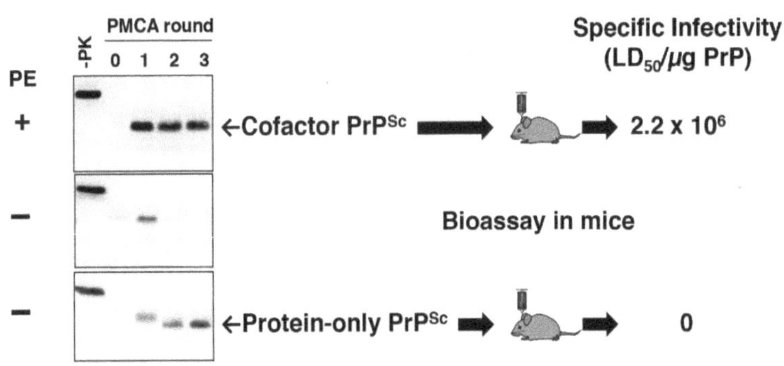

Fig. 1 Cofactor molecules are required to form infectious prions. Western blots of reconstituted sPMCA reactions show adaptation of autocatalytic PrP^{Sc} molecules in the presence and absence of PE cofactor, as indicated. (−PK) = samples not subjected to proteinase K digestion; all other samples were proteolyzed. All reactions were initially seeded with cofactor PrP^{Sc} molecules. Reactions containing PE continued to propagate an ~18 kDa PK-resistant core (top blot). Approximately 40% of reactions lacking PE failed to propagate (middle blot), while ~60% adapted into a protein-only PrP^{Sc} conformer with an ~16 kDa PK-resistant core (bottom blot). End-point bioassays were performed in normal C57BL mice to determine the specific infectivity of *in vitro*-generated recombinant PrP^{Sc} molecules.

only by the presence of cofactor has provided a powerful system to study the role of cofactor molecules in forming and maintaining prion infectivity. End-point titration bioassays in wild type mice indicate that cofactor PrP^{Sc} molecules have a level of specific infectivity at least 10^5-fold greater than protein-only PrP^{Sc} molecules (Fig. 1).[18] These results are consistent with previous experiments using native PrP^C and RNA substrate[59] in showing that wild type prions with significant levels of specific infectivity cannot be formed by PrP alone.

It is important to note that cofactor molecules may not be required for the formation of mutant prions. Noble et al. discovered that recPrP harboring either the D177N or E199K mutations (linked to the inherited prion diseases fatal familial insomnia and familial CJD, respectively) spontaneously misfold into autocatalytic PK-resistant PrP^{Sc} conformers in the absence of cofactor molecules.[64] Interestingly, the mutant conformations could be faithfully propagated using wild type recPrP substrate, in reactions that did require cofactor molecules.[64] In addition, a PrP stop mutant comprised of residues 23–144 (associated with a Gerstmann-Sträussler-Scheinker-like prion disease) folded into amyloid fibrils without cofactors was shown be infectious for wild type mice (albeit with a relatively low level of specific infectivity).[65] Similarly, artificial mutants in which the four lysine residues within central polybasic domain (i.e., residues 101–110) were substituted with either alanine (K ➔ A) or asparagine (K ➔ N) residues could be folded into protein-only amyloids that display a low but detectable level of infectivity for wild type mice.[66]

5. Cofactor molecules restrict the strain properties of purified prions

The observation that both single-stranded nucleic acids and PE become incorporated into physical complexes with PrP^{Sc} molecules during the *in vitro* conversion process suggests that cofactor molecules might play a critical role in forming or maintaining strain-dependent PrP^{Sc} conformations. To test this hypothesis, Deleault et al. produced PrP^{Sc} molecules from only recPrP and synthetic PE substrate in sPMCA experiments originally seeded with three different prion strains.[18] Castilla et al. had previously shown that sPMCA reactions using crude brain homogenate faithfully preserves the PrP^{Sc} conformations and biological strain properties of multiple prion strains.[56] In contrast, Deleault et al. found that the three samples formed with a single cofactor (i.e., PE) had similar PrP^{Sc} conformations and

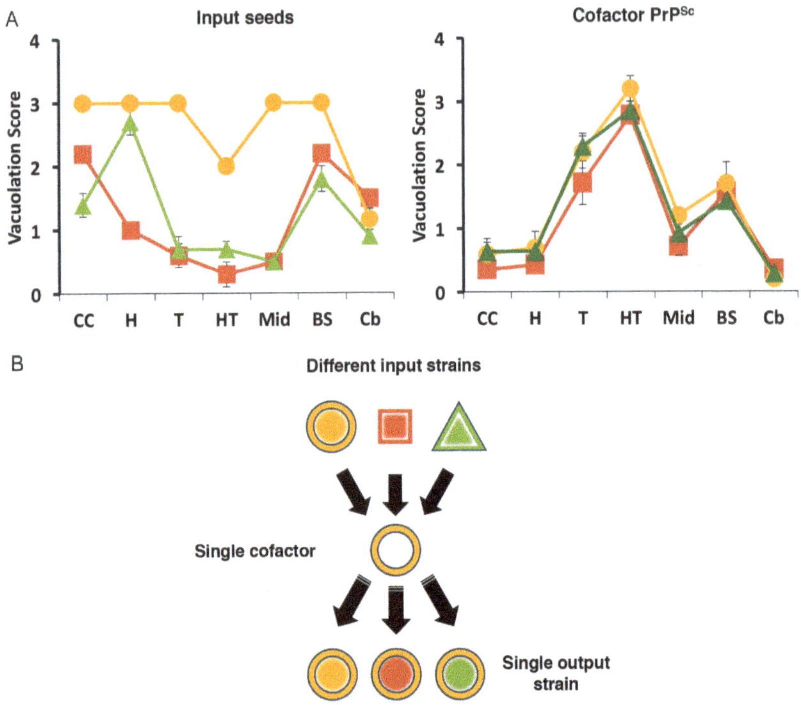

Fig. 2 Propagation with a single cofactor restricts prion strain diversity. (A) Regional neuropathology of infected mice. Profiles of vacuolation scores of animals inoculated with samples containing either input prions or cofactor PrP^{Sc} molecules, as indicated. Prion strains: OSU, red squares; Me7, orange circles; 301C, green triangles. Brain regions: CC, cerebral cortex (all layers); H, hippocampus; T, thalamus; HT, hypothalamus; Mid, midbrain; BS, brain stem; Cb, cerebellum. The mean values ($n = 5–8$) are shown ± SEM. (B) Cofactor restriction *in vitro*. Schematic diagram illustrating the principle of "cofactor restriction." *In vitro* propagation with a single cofactor forces multiple prion strains to adapt into a single PrP^{Sc} conformer with identical strain properties.

biological strain phenotypes, including neurotropism (Fig. 2A).[18] These results showed for the first time that providing PrP only a single cofactor molecule during the conversion process restricts and determines its folding pathway, regardless of input seed (Fig. 2B). Similar results were subsequently obtained with other classes of cofactor molecules, including GM1 ganglioside[67] and various polyanions.[68] Interestingly, each type of cofactor molecule appears to induce the formation of a distinctive PrP^{Sc} structure and biological prion strain.

6. Restoration of specific infectivity and strain properties from protein-only PrP^{Sc}

The lack of infectivity exhibited by protein-only PrP^{Sc} in wild type mice correlates well with the inability of this conformer to seed sPMCA reactions with crude mouse brain homogenate substrate.[18] Since bank voles are near universal prion recipients, Burke et al. investigated whether bank vole brain homogenate might be a more permissive substrate than mouse brain homogenate in sPMCA reactions seeded by protein-only PrP^{Sc}.[69] Indeed, this turned out to be the case, and serial dilution sPMCA experiments showed that crude bank vole brain homogenate could be potently seeded by protein-only PrP^{Sc} (the minimum amount protein-only PrP^{Sc} needed to seed 3-round sPMCA reaction = 6 pg PrP^{Sc}). The susceptibility of bank vole PrP^C to seeding by protein-only PrP^{Sc} provided a unique opportunity to test the protein-only hypothesis in an internally controlled experiment. Accordingly, Burke et al. measured the specific infectivity of cofactor PrP^{Sc}, protein-only PrP^{Sc}, and protein-only PrP^{Sc}-seeded bank vole brain homogenate (designated [protein-only ➔ BH PrP^{Sc}]) by performing end-point titration bioassays in wild type bank voles (Fig. 3).[69] As expected, the cofactor PrP^{Sc} and [protein-only ➔ BH PrP^{Sc}] samples both displayed high levels of specific infectivity (Fig. 3, top and bottom rows).

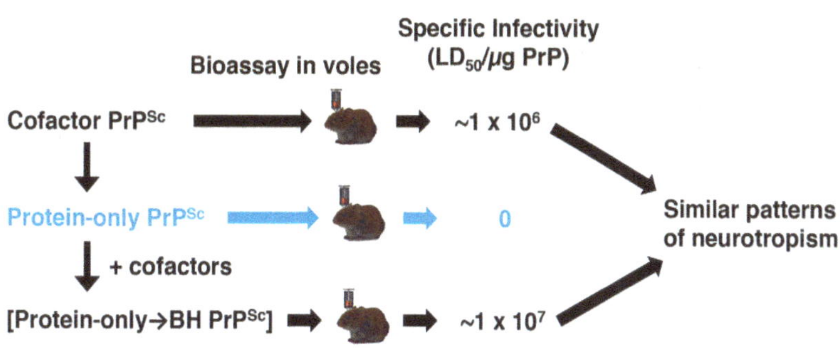

Fig. 3 Infectivity and strain information can be fully restored from protein-only PrP^{Sc}. End-point bioassays were performed in normal bank voles to determine the specific infectivity of *in vitro*-generated samples. Cofactor PrP^{Sc} was used as a seed to produce protein-only PrP^{Sc} by sPMCA propagation in the absence of cofactors. Protein-only PrP^{Sc} was subsequently used as a seed to produce [protein-only ➔ BH PrP^{Sc}] by sPMCA propagation in crude bank vole brain homogenate in a cofactor-dependent reaction. Bioassay results show restoration of both high specific infectivity and original strain properties.

However, protein-only PrP^{Sc} was not directly infectious to bank voles (Fig. 3, middle row). This result was particularly surprising because (1) protein-only PrP^{Sc} can seed conversion of bank vole PrP^{C} potently, and (2) protein-only PrP^{Sc} is positioned in the middle of an *in vitro* propagation sequence cofactor PrP^{Sc} → protein-only PrP^{Sc} → [protein-only → BH PrP^{Sc}] in which both its progenitor and product are highly infectious (Fig. 3).

To determine the factors needed to produce infectious [protein-only → BH PrP^{Sc}], Burke et al. used immunopurified bank vole PrP^{C} and $PrP^{0/0}$ brain homogenate (as a source of endogenous cofactors) as substrates in reconstitution assays.[69] These experiments confirmed that formation of [protein-only → BH PrP^{Sc}] requires cofactor molecules (Fig. 3). Taken together, these experiments show that cofactor molecules are required either to form or restore prion infectivity. The observation that protein-only PrP^{Sc} is infectious even though both its direct progenitor and product (both of which were formed with cofactor molecules) are both fully infectious represents internally controlled evidence against the protein-only hypothesis. The ability of protein-only PrP^{Sc} to acquire biological infectivity in a cofactor-dependent process is reminiscent of the ability of pure PrP amyloid fibrils to stimulate the formation of infectious prions *in vivo*[70] and *in vitro*.[71] Consistent with this explanation, serial propagation of the infectious SSLOW strain (produced by intracerebral inoculation of amyloid fibrils followed by blind passage from aged asymptomatic recipients) appears to be dependent on RNA as a cofactor.

In the work of Burke et al., a particularly striking result emerged during neuropathological analysis of bank voles inoculated with the various *in vitro*-generated PrP^{Sc} samples.[69] Specifically, it was found that voles inoculated with cofactor PrP^{Sc} displayed the same pattern of neurotropism as voles inoculated with [protein-only → BH PrP^{Sc}] (Figs. 3 and 4A). This remarkable convergence of strain characteristics was very surprising because the two samples contain different types of PrP substrates (i.e., $recPrP^{Sc}$ vs native PrP^{Sc}) and cofactors (i.e., purified phospholipid vs crude $PrP^{0/0}$ brain homogenate). Therefore, we can infer that: (1) PrP post-translational modifications do not influence the biological properties of this prion strain; (2) despite being non-infectious and lacking a cofactor, protein-only PrP^{Sc} is somehow able to maintain the original strain information of cofactor PrP^{Sc} and faithfully transmit this information to [protein-only → BH PrP^{Sc}]; and (3) during protein-only PrP^{Sc}-seeded conversion of crude brain homogenate, bank vole PrP^{C} appears to have selected and used the endogenous cofactor needed to maintain the original strain properties of cofactor PrP^{Sc} (presumably PE) out of a diverse mixture containing a variety of potential cofactors (including RNA).

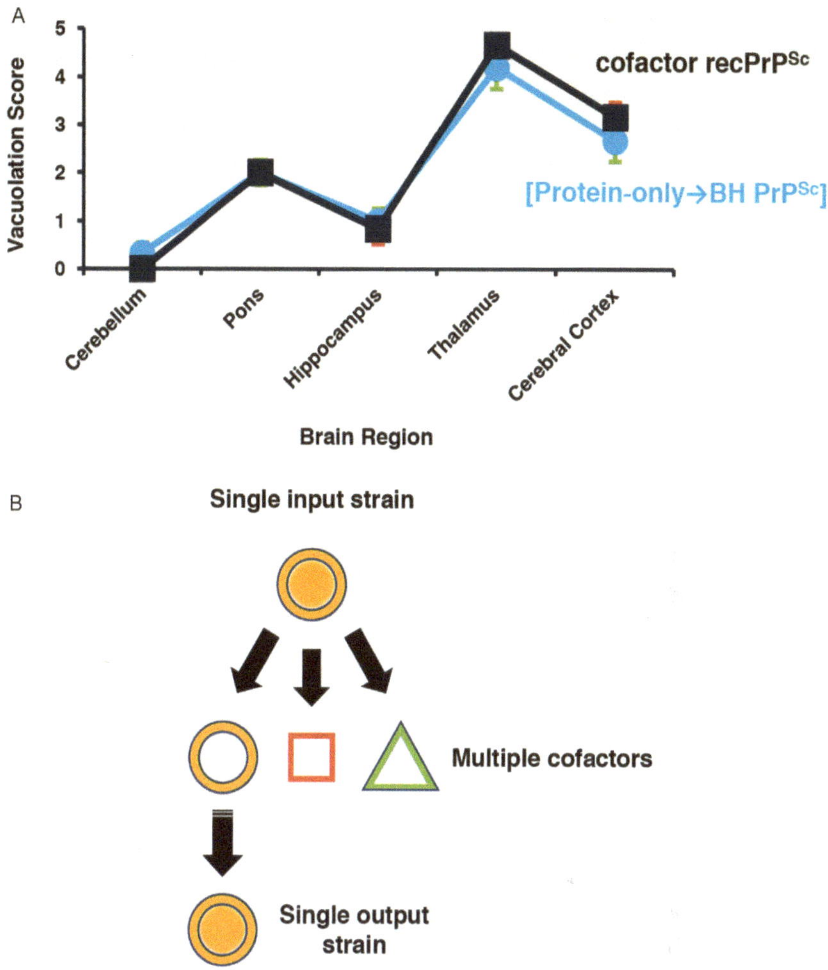

Fig. 4 Restored PrPSc preserves original strain information. (A) Regional neuropathology bank voles with cofactor recPrP^{Sc} or [protein-only ➔ BH PrP^{Sc}]. Profiles of vacuolation scores of animals inoculated with either cofactor recPrP^{Sc} (black squares) or [protein-only ➔ BH PrP^{Sc}] (blue circles) show nearly identical patterns of neurotropism. Mean values ± SEM are shown. $N = 6$ for all measurements except for [protein-only ➔ BH PrP^{Sc}] cerebellum and pons, where $N = 3$. (B) Cofactor selection *in vitro*. Schematic diagram illustrating the principle of "cofactor selection." *In vitro* propagation of single prion strain in a diverse mixture of potential cofactor molecules (e.g., crude brain homogenate) faithfully maintains the original PrP^{Sc} conformation and strain properties of the seed.

7. Cofactor selection model of neurotropism

The ability of protein-only PrP^{Sc} to maintain and faithfully propagate the original strain properties of cofactor PrP^{Sc} during propagation in crude brain homogenate (Fig. 4B) provides strong additional support for the cofactor selection model of prion neurotropism (Fig. 5), which my colleagues and I originally proposed on the basis that cofactor molecules are selectively incorporated into physical complexes with PrP^{Sc} molecules[57] and can restrict PrP^{Sc} conformation and strain properties.[18]

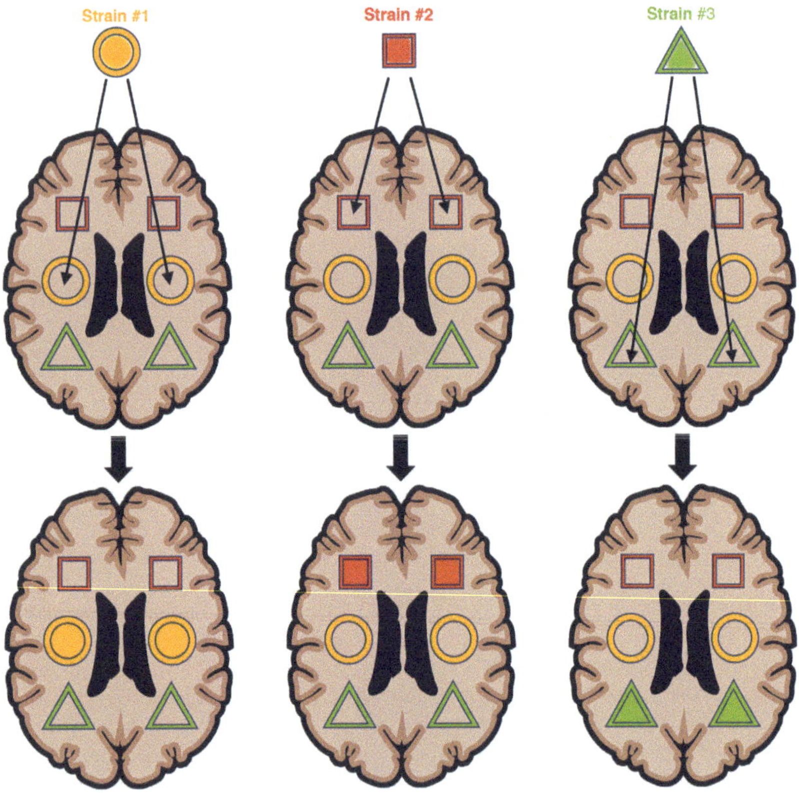

Fig. 5 Cofactor selection model explains neurotropism. Schematic diagram shows hypothetical model to explain why strain-specific PrP^{Sc} conformers have characteristic patterns of neurotropism. The model proposes that each prion strain has a PrP^{Sc} conformation that can only be formed by a specific set of endogenous cofactors and that different endogenous cofactors are proposed to be naturally enriched in different regions of the brain (top of diagram). Therefore, each strain-specific PrP^{Sc} conformer would be expected replicate and accumulate preferentially in the brain regions that have higher levels of the specific cofactor molecules that are able to facilitate its formation (bottom of diagram).

In this model, we propose that each prion strain has a PrP^{Sc} conformation that can only be formed by a specific set of endogenous cofactors (Fig. 5, top and middle). Furthermore, we hypothesize that different endogenous cofactors may be enriched in different regions of the brain (Fig. 5, middle). Therefore, for any particular prion strain, PrP^{Sc} would replicate and accumulate preferentially in the brain regions that have higher levels of the specific cofactor molecules that are able to facilitate its formation (Fig. 5, bottom). This model is compatible with all of the experimental observations currently known, and provides a logical explanation as to why strain-specific PrP^{Sc} conformers produce characteristic patterns of neurotropism. Alternative hypotheses that do not involve cofactor molecules include selective targeting, in which neurons in different parts of the brain might express different strain-specific receptors, and selective degradation, in which cells in different brain regions employ degradation mechanisms that can distinguish between different strains. However, there is currently no evidence for the existence of either neuronal cell surface receptors or degradation mechanisms that can specifically distinguish between different PrP^{Sc} conformers.

Because purified prions made with either RNA or PE molecules display biological strain properties that differ from any naturally occurring prion strain,[18,59,69] we can infer that there are probably multiple endogenous cofactors that have not yet been identified. We predict that there will be one or more specific endogenous cofactors that are needed to form each prion strain. It is difficult to identify strain-specific cofactor molecules by using a reconstitution assay because strain properties can generally only be determined *in vivo* by time consuming bioassays. There may be a few exceptions in which a strain-specific PrP^{Sc} conformer can be identified more quickly by surrogate biochemical methods (e.g., 301C PrP^{Sc} is characterized by having a relatively small PK-resistant core and high ratio of diglycosylated to unglycosylated PrP[18,72]). Alternatively, if cofactor molecules are indeed complexed to PrP^{Sc} molecules *in situ*,[57] it may be possible to identify strain-specific endogenous cofactor molecules analytically in highly purified preparations of brain-derived prions.

8. Are prion cofactors essential components or catalysts?

A key mechanistic question raised by the discovery of prion conversion cofactor molecules is whether cofactors are essential structural components of prions or simply disposable catalysts for prion formation.

Several observations argue that cofactor molecules are likely to be necessary structural and informational components of infectious prions: (1) cofactor molecules are selectively incorporated into degradation-resistant complexes with PrP^{Sc} in infectious prions[18,57]; (2) cofactors are invariably required to form wild type prions with significant levels of specific infectivity[18,59,69]; (3) cofactors are required to restore infectivity from protein-only PrP^{Sc} molecules[69]; and (4) purified cofactor molecules restrict PrP^{Sc} conformation and strain properties.[18,68] On the other hand, the argument that cofactor molecules are essential components of infectious prions has not yet been confirmed because: (1) there is no genetic evidence to show that cofactor molecules are necessary for PrP^{Sc} formation in living cells; and (2) there is no evidence that cofactor molecules bind to specific biding sites within the PrP^{Sc} structure. To resolve this issue, it will be necessary to determine the detailed structure of infectious cofactor PrP^{Sc} to determine whether the incorporated cofactor molecule(s) occupy a specific binding site. Interestingly, Yamasaki et al. recently used cryo electron microscopy (EM) to show that tau filaments isolated from patients with chronic traumatic encephalopathy (CTE) contain a hydrophobic cofactor which appears to bind to a specific binding pocket in the tau structure.[73] A similar approach using cofactor PrP^{Sc} may be the best way to determine the mechanistic and structural role of cofactor molecules for infectious prions.

9. Cofactors are critical tools for determining the structural mechanism of prion infectivity

One of the most significant gaps in our knowledge about mammalian prions is that we do not know the molecular structural changes responsible for PrP^{Sc} infectivity. Part of the difficulty in closing this gap is technical. Because prions are insoluble, they are not amenable to several well-established methods for structural determination, such as solution nuclear magnetic resonance (NMR) spectroscopy or X-ray crystallography. However, recent improvements in alternative structure determination methods such as solid-state (ss)NMR and cryo EM which can be used on insoluble proteins has helped to partially mitigate this technical obstacle. A second problem is the intrinsic morphological heterogeneity of mammalian prions and their natural tendency to aggregate into large clusters. This poor behavior is a hindrance for data acquisition in cryo EM, but can be somewhat mitigated by the use of various dispersion techniques. A final problem is that it may be difficult to infer the prion replication mechanism from a single static structural model.

Our work with prion cofactors is providing an important new approach to deal with all of these problems. Specifically, PE can be used to convert recPrP into chemically defined prions with very high levels of specific infectivity (i.e., $>10^6$ $LD_{50}/\mu g$ PrP^{Sc}) comparable to brain-derived prions.[18] The ability to use recPrP instead of native PrP substrate has several critical advantages: (1) recPrP can be isotopically labeled, which is a requirement for ssNMR; (2) recPrP can be produced in milligram quantities, also required for ssNMR; and (3) the lack of variable post-translational modifications reduces heterogeneity and simplifies structural analysis for both ssNMR and cryo EM. The ability to perform ssNMR also helps circumvent the obstacle of morphological heterogeneity because ssNMR signals are mostly derived from relatively short-range interactions. Finally, the availability of cofactor PrP^{Sc} and protein-only PrP^{Sc} as a pair of closely related conformers with a stark difference in specific infectivity will help us to identify the specific structural elements that are responsible for the difference in prion infectivity between these two conformers. Although there is a dramatic difference in their levels of biological infectivity, these two conformers are derived from the same original seed[18] and display similar strain properties following restoration of infectivity from protein-only PrP^{Sc}.[69] These observations suggest that cofactor PrP^{Sc} and protein-only PrP^{Sc} must have different conformations to account for their difference in infectivity, but likely also share many structural features due to their common provenance and their similar strain properties (following restoration of protein-only PrP^{Sc} infectivity). This is supported by the results of Noble et al., who used deuterium exchange mass spectrometry (DXMS) to compare the solvent accessibility profile of the two conformers.[74] This comparison found that cofactor PrP^{Sc} and protein-only PrP^{Sc} generally have similar levels of solvent accessibility throughout most of their structures. However, two specific regions of PrP^{Sc} (residues 91–115, 144–163) were significantly more solvent accessible in protein-only PrP^{Sc} than cofactor PrP^{Sc}.[74] This observation suggests that these two regions may play significant roles in the mechanism of cofactor-mediated prion infectivity, but a higher resolution structural comparison will be needed to confirm that hypothesis.

10. Conclusion

Much progress has been made over the past 15 years in our understanding about the essential role of cofactor molecules for the formation of infectious wild type prions. Cofactor molecules allow us to produce both native and recombinant prions with high levels of specific infectivity, and withdrawal of cofactor molecules during propagation also enables us to produce non-infectious protein-only PrP^{Sc} molecules that serve as well-matched controls for a wide variety of studies.[18,59] Several landmark studies have used cofactor molecules to rigorously test major hypotheses in prion research.

For instance, the chemical nature of infectious mammalian prions has long been controversial. It has been difficult to study this question directly by purifying PrP^{Sc} from prion-infected mammalian tissues because existing purification protocols are not fully efficient and therefore potentially subject to contamination. This has led to the proposal of many different competing hypotheses, some of which have now been rigorously refuted by various experiments using chemically defined cofactor and PrP substrates to form prions with high levels of specific infectivity. Most notably, Deleault et al. used synthetic PE and recPrP to produce, for the first time, fully infectious mammalian prions without nucleic acids or any components derived from a mammalian source.[18] This result provided the first definitive evidence that mammalian prions are neither viruses or virinos (hypothetical particles composed of a host protein which protects a small nucleic acid genome[75]). Burke et al. demonstrated that protein-only PrP^{Sc} lacks a significant level of specific infectivity, but nonetheless could faithfully maintain strain properties.[69] This result directly refuted the "co-prion" hypothesis, which posits that protein-only PrP^{Sc} is infectious and that small informational nucleic acid "co-prions" encode strain properties.[76] At this time, there appear to be only two feasible models for the chemical nature of infectious prions. The protein-only hypothesis may yet be correct: although cofactor molecules appear to be necessary to form infectious mammalian prions, it is possible that the continued presence of cofactor molecules may not be necessary to maintain the infectious conformation of PrP^{Sc}.[58] Alternatively, and perhaps more likely, cofactor molecules may be essential structural components of mammalian prions, which interact with specific PrP^{Sc} residues to maintain its infectious conformation.[18]

The etiology of sporadic prion diseases has also been the subject of much speculation. Several hypotheses have been proposed, including somatic mutation, spontaneous PrP conversion, and various environmental exposures, but until recently there has been no experimental support for any of these hypotheses.[25,26,77–79] Deleault et al. showed that a cocktail of purified PrP^C and poly(A) RNA molecules could spontaneously and stochastically generate infectious prions in the absence of any infectious seed or starting materials.[57,59] This result provided the first and only experimental support for the concept of rare spontaneous PrP conversion events as the etiology of sporadic prion disease, and suggests that interaction between PrP^C and cofactor molecules may be the critical event of that stochastic pathogenic process.

Several areas for future research are suggested by our current understanding of prion cofactors. The cofactor selection model of neurotropism[18,57,69] predicts that each prion strain is requires its own unique set of specific endogenous cofactor molecules. Currently, two classes of endogenous cofactor molecules have been identified (i.e., polyanions and lipids), and thus far these cofactors have produced artificial rather than natural prion strains.[50,53,61,67,80] It will be important to identify the endogenous cofactor molecules that are required to form specific, naturally occurring prion strains.

Thus far, all experimental support for involvement of cofactor molecules in mammalian prion formation has relied on reductionist biochemical experiments coupled with bioassays. It will be eventually important to complement these studies with genetic studies in cells or *in vivo*. A technical limitation in this regard is that the currently known prion cofactors are essential molecules whose levels and proper trafficking are carefully regulated by healthy cells.

Finally, prospects for structural determination of cofactor PrP^{Sc} and protein-only PrP^{Sc} appear promising. These samples are good candidates for using a combination of complementary experimental approaches, including cryo EM, ssNMR, and mass spectrometry fingerprinting techniques. Moreover, the availability of paired samples derived from the same seed but differing in biological infectivity will help us decipher the structural elements that are most vital for encoding infectivity. These structural studies should also reveal the specific mechanism by which cofactor molecules influence PrP^{Sc} structure, and determine whether cofactor molecules are essential and informational components of infectious prions.

References

1. Prusiner SB. Prions. *Proc Natl Acad Sci U S A*. 1998;95(23):13363–13383.
2. Bessen RA, Marsh RF. Identification of two biologically distinct strains of transmissible mink encephalopathy in hamsters. *J Gen Virol*. 1992;73(pt 2):329–334.
3. Loftus B, Rogers M. Characterization of a prion protein (PrP) gene from rabbit; a species with apparent resistance to infection by prions. *Gene*. 1997;184(2):215–219.
4. Agrimi U, Nonno R, Dell'Omo G, et al. Prion protein amino acid determinants of differential susceptibility and molecular feature of prion strains in mice and voles. *PLoS Pathog*. 2008;4(7):e1000113.

5. Espinosa JC, Nonno R, Di Bari M, et al. PrPC governs prion strain susceptibility in bank vole while others host factors modulate strain features. *J Virol.* 2016;90:10660–10669.
6. Nonno R, Angelo Di Bari M, Agrimi U, Pirisinu L. Transmissibility of Gerstmann-Straussler-Scheinker syndrome in rodent models: new insights into the molecular underpinnings of prion infectivity. *Prion.* 2016;10:421–433.
7. Nonno R, Di Bari MA, Cardone F, et al. Efficient transmission and characterization of Creutzfeldt-Jakob disease strains in bank voles. *PLoS Pathog.* 2006;2(2):e12.
8. Aguzzi A, Lakkaraju AKK, Frontzek K. Toward therapy of human prion diseases. *Annu Rev Pharmacol Toxicol.* 2018;58:331–351.
9. Alper T. Scrapie agent unlike viruses in size and susceptibility to inactivation by ionizing or ultraviolet radiation. *Nature.* 1985;317(6039):750.
10. Prusiner SB. Novel proteinaceous infectious particles cause scrapie. *Science.* 1982; 216(4542):136–144.
11. Caughey BW, Dong A, Bhat KS, Ernst D, Hayes SF, Caughey WS. Secondary structure analysis of the scrapie-associated protein PrP 27-30 in water by infrared spectroscopy. *Biochemistry.* 1991;30(31):7672–7680.
12. Pan KM, Baldwin M, Nguyen J, et al. Conversion of alpha-helices into beta-sheets features in the formation of the scrapie prion proteins. *Proc Natl Acad Sci U S A.* 1993;90(23):10962–10966.
13. Stahl N, Baldwin MA, Burlingame AL, Prusiner SB. Identification of glycoinositol phospholipid linked and truncated forms of the scrapie prion protein. *Biochemistry.* 1990;29(38):8879–8884.
14. Harris DA. Cellular biology of prion diseases. *Clin Microbiol Rev.* 1999;12(3):429–444.
15. Saeki K, Matsumoto Y, Hirota Y, Onodera T. Three-exon structure of the gene encoding the rat prion protein and its expression in tissues. *Virus Genes.* 1996;12(1):15–20.
16. Bueler H, Aguzzi A, Sailer A, et al. Mice devoid of PrP are resistant to scrapie. *Cell.* 1993;73(7):1339–1347.
17. Prusiner SB, Groth D, Serban A, et al. Ablation of the prion protein (PrP) gene in mice prevents scrapie and facilitates production of anti-PrP antibodies. *Proc Natl Acad Sci U S A.* 1993;90(22):10608–10612.
18. Deleault NR, Walsh DJ, Piro JR, et al. Cofactor molecules maintain infectious conformation and restrict strain properties in purified prions. *Proc Natl Acad Sci U S A.* 2012;109(28):E1938–E1946.
19. Legname G, Baskakov IV, Nguyen HO, et al. Synthetic mammalian prions. *Science.* 2004;305(5684):673–676.
20. Colby DW, Wain R, Baskakov IV, et al. Protease-sensitive synthetic prions. *PLoS Pathog.* 2010;6(1):e1000736.
21. Makarava N, Kovacs GG, Bocharova O, et al. Recombinant prion protein induces a new transmissible prion disease in wild-type animals. *Acta Neuropathol.* 2010;119(2):177–187.
22. Kim JI, Cali I, Surewicz K, et al. Mammalian prions generated from bacterially expressed prion protein in the absence of any mammalian cofactors. *J Biol Chem.* 2010;285(19): 14083–14087.
23. Hsiao K, Prusiner SB. Inherited human prion diseases. *Neurology.* 1990; 40(12):1820–1827.
24. Prusiner SB. Inherited prion diseases. *Proc Natl Acad Sci U S A.* 1994;91(11):4611–4614.
25. Brown P, Cathala F, Raubertas RF, Gajdusek DC, Castaigne P. The epidemiology of Creutzfeldt-Jakob disease: conclusion of a 15-year investigation in France and review of the world literature. *Neurology.* 1987;37(6):895–904.
26. Pedersen NS, Smith E. Prion diseases: epidemiology in man. *APMIS.* 2002; 110(1):14–22.
27. Prusiner SB. Molecular biology and transgenetics of prion diseases. *Crit Rev Biochem Mol Biol.* 1991;26(5–6):397–438.

28. Cohen FE. Protein misfolding and prion diseases. *J Mol Biol.* 1999;293(2):313–320.
29. Bueler H, Fischer M, Lang Y, et al. Normal development and behaviour of mice lacking the neuronal cell-surface PrP protein. *Nature.* 1992;356(6370):577–582.
30. Brandner S, Isenmann S, Raeber A, et al. Normal host prion protein necessary for scrapie-induced neurotoxicity. *Nature.* 1996;379(6563):339–343.
31. Scialo C, De Cecco E, Manganotti P, Legname G. Prion and prion-like protein strains: deciphering the molecular basis of heterogeneity in neurodegeneration. *Viruses.* 2019; 11(3):261.
32. Brown P, Gibbs Jr. CJ, Rodgers-Johnson P, et al. Human spongiform encephalopathy: the National Institutes of Health series of 300 cases of experimentally transmitted disease. *Ann Neurol.* 1994;35(5):513–529.
33. Khandhar SM, Marks WJ. Epidemiology of Parkinson's disease. *Dis Mon.* 2007;53(4):200–205.
34. Tanaka M, Chien P, Naber N, Cooke R, Weissman JS. Conformational variations in an infectious protein determine prion strain differences. [See comment] *Nature.* 2004; 428(6980):323–328.
35. Michelitsch MD, Weissman JS. A census of glutamine/asparagine-rich regions: implications for their conserved function and the prediction of novel prions. *Proc Natl Acad Sci U S A.* 2000;97(22):11910–11915.
36. La Spada AR, Taylor JP. Repeat expansion disease: progress and puzzles in disease pathogenesis. *Nat Rev Genet.* 2010;11(4):247–258.
37. Bartz JC. Prion strain diversity. *Cold Spring Harb Perspect Med.* 2016;6(12):a024349.
38. Mead S, Poulter M, Uphill J, et al. Genetic risk factors for variant Creutzfeldt-Jakob disease: a genome-wide association study. *Lancet Neurol.* 2009;8(1):57–66.
39. Lloyd SE, Onwuazor ON, Beck JA, et al. Identification of multiple quantitative trait loci linked to prion disease incubation period in mice. *Proc Natl Acad Sci U S A.* 2001;98(11):6279–6283.
40. Beranger F, Mange A, Solassol J, Lehmann S. Cell culture models of transmissible spongiform encephalopathies. *Biochem Biophys Res Commun.* 2001;289(2):311–316.
41. Lehmann S, Beranger F, Solassol J, et al. Cell culture models of transmissible spongiform encephalopathies. *C R Acad Sci III.* 2002;325(1):59–65.
42. Thackray AM, Andreoletti O, Bujdoso R. Mammalian prion propagation in PrP transgenic Drosophila. *Brain.* 2018;141(9):2700–2710.
43. Raeber AJ, Borchelt DR, Scott M, Prusiner SB. Attempts to convert the cellular prion protein into the scrapie isoform in cell-free systems. *J Virol.* 1992;66(10):6155–6163.
44. Kocisko DA, Come JH, Priola SA, et al. Cell-free formation of protease-resistant prion protein. *Nature.* 1994;370(6489):471–474.
45. Bessen RA, Kocisko DA, Raymond GJ, Nandan S, Lansbury PT, Caughey B. Non-genetic propagation of strain-specific properties of scrapie prion protein. *Nature.* 1995;375(6533):698–700.
46. Kocisko DA, Priola SA, Raymond GJ, Chesebro B, Lansbury Jr PT, Caughey B. Species specificity in the cell-free conversion of prion protein to protease-resistant forms: a model for the scrapie species barrier. *Proc Natl Acad Sci U S A.* 1995;92(9):3923–3927.
47. Caughey B, Kocisko DA, Raymond GJ, Lansbury Jr. PT. Aggregates of scrapie-associated prion protein induce the cell-free conversion of protease-sensitive prion protein to the protease-resistant state. *Chem Biol.* 1995;2(12):807–817.
48. Caughey B, Raymond GJ, Priola SA, et al. Methods for studying prion protein (PrP) metabolism and the formation of protease-resistant PrP in cell culture and cell-free systems. An update. *Mol Biotechnol.* 1999;13(1):45–55.
49. Saborio GP, Soto C, Kascsak RJ, et al. Cell-lysate conversion of prion protein into its protease-resistant isoform suggests the participation of a cellular chaperone. *Biochem Biophys Res Commun.* 1999;258(2):470–475.

50. Deleault NR, Lucassen RW, Supattapone S. RNA molecules stimulate prion protein conversion. *Nature*. 2003;425(6959):717–720.
51. Lucassen R, Nishina K, Supattapone S. In vitro amplification of protease-resistant prion protein requires free sulfhydryl groups. *Biochemistry*. 2003;42(14):4127–4135.
52. Nishina K, Deleault NR, Lucassen RW, Supattapone S. In vitro prion protein conversion in detergent-solubilized membranes. *Biochemistry*. 2004;43(9):2613–2621.
53. Deleault NR, Geoghegan JC, Nishina K, Kascsak R, Williamson RA, Supattapone S. Protease-resistant prion protein amplification reconstituted with partially purified substrates and synthetic polyanions. *J Biol Chem*. 2005;280(29):26873–26879.
54. Orem NR, Geoghegan JC, Deleault NR, Kascsak R, Supattapone S. Copper (II) ions potently inhibit purified PrPres amplification. *J Neurochem*. 2006;96(5):1409–1415.
55. Castilla J, Saa P, Hetz C, Soto C. In vitro generation of infectious scrapie prions. *Cell*. 2005;121(2):195–206.
56. Castilla J, Morales R, Saa P, Barria M, Gambetti P, Soto C. Cell-free propagation of prion strains. *EMBO J*. 2008;27(19):2557–2566.
57. Geoghegan JC, Valdes PA, Orem NR, et al. Selective incorporation of polyanionic molecules into hamster prions. *J Biol Chem*. 2007;282(50):36341–36353.
58. Piro JR, Harris BT, Supattapone S. In situ photodegradation of incorporated polyanion does not alter prion infectivity. *PLoS Pathog*. 2011;7(2):e1002001.
59. Deleault NR, Harris BT, Rees JR, Supattapone S. Formation of native prions from minimal components in vitro. *Proc Natl Acad Sci U S A*. 2007;104(23):9741–9746.
60. Deleault NR, Kascsak R, Geoghegan JC, Supattapone S. Species-dependent differences in cofactor utilization for formation of the protease-resistant prion protein in vitro. *Biochemistry*. 2010;49(18):3928–3934.
61. Deleault NR, Piro JR, Walsh DJ, et al. Isolation of phosphatidylethanolamine as a solitary cofactor for prion formation in the absence of nucleic acids. *Proc Natl Acad Sci U S A*. 2012;109(22):8546–8551.
62. Supattapone S. Synthesis of high titer infectious prions with cofactor molecules. *J Biol Chem*. 2014;289(29):19850–19854.
63. Wang F, Wang X, Yuan CG, Ma J. Generating a prion with bacterially expressed recombinant prion protein. *Science*. 2010;327(5969):1132–1135.
64. Noble GP, Walsh DJ, Miller MB, Jackson WS, Supattapone S. Requirements for mutant and wild-type prion protein misfolding in vitro. *Biochemistry*. 2015;54(5):1180–1187.
65. Choi JK, Cali I, Surewicz K, Kong Q, Gambetti P, Surewicz WK. Amyloid fibrils from the N-terminal prion protein fragment are infectious. *Proc Natl Acad Sci U S A*. 2016;113:13851–13856.
66. Groveman BR, Raymond GJ, Campbell KJ, et al. Role of the central lysine cluster and scrapie templating in the transmissibility of synthetic prion protein aggregates. *PLoS Pathog*. 2017;13(9):e1006623.
67. Kim C, Xiao X, Chen S, et al. Artificial strain of human prions created in vitro. *Nat Commun*. 2018;9(1):2166.
68. Fernandez-Borges N, Di Bari MA, Erana H, et al. Cofactors influence the biological properties of infectious recombinant prions. *Acta Neuropathol*. 2018;135(2):179–199.
69. Burke CM, Walsh DJ, Steele AD, et al. Full restoration of specific infectivity and strain properties from pure mammalian prion protein. *PLoS Pathog*. 2019;15(3):e1007662.
70. Makarava N, Kovacs GG, Savtchenko R, et al. Stabilization of a prion strain of synthetic origin requires multiple serial passages. *J Biol Chem*. 2012;287(36):30205–30214.
71. Moda F, Le TN, Aulic S, et al. Synthetic prions with novel strain-specified properties. *PLoS Pathog*. 2015;11(12):e1005354.
72. Piro JR, Harris BT, Nishina K, et al. Prion protein glycosylation is not required for strain-specific neurotropism. *J Virol*. 2009;83(11):5321–5328.

73. Yamasaki TR, Holmes BB, Furman JL, et al. Parkinson's disease and multiple system atrophy have distinct alpha-synuclein seed characteristics. *J Biol Chem*. 2019;294(3): 1045–1058.
74. Noble GP, Wang DW, Walsh DJ, et al. A structural and functional comparison between infectious and non-infectious autocatalytic recombinant PrP conformers. *PLoS Pathog*. 2015;11(6):e1005017.
75. Kimberlin RH. Scrapie agent: prions or virinos? *Nature*. 1982;297(5862):107–108.
76. Weissmann C. A 'unified theory' of prion propagation. *Nature*. 1991;352(6337): 679–683.
77. Prusiner SB. Molecular biology of prions causing infectious and genetic encephalopathies of humans as well as scrapie of sheep and BSE of cattle. *Dev Biol Stand*. 1991;75:55–74.
78. Cohen FE, Prusiner SB. Pathologic conformations of prion proteins. *Annu Rev Biochem*. 1998;67:793–819.
79. Chen C, Dong XP. Epidemiological characteristics of human prion diseases. *Infect Dis Poverty*. 2016;5(1):47.
80. Wong C, Xiong LW, Horiuchi M, et al. Sulfated glycans and elevated temperature stimulate PrP(Sc)-dependent cell-free formation of protease-resistant prion protein. *EMBO J*. 2001;20(3):377–386.

CHAPTER FIVE

The role of prion strain diversity in the development of successful therapeutic treatments

Sara A.M. Holec[a,b], Alyssa J. Block[b], and Jason C. Bartz[b,*]
[a]Institute for Applied Life Sciences and Department of Biology, University of Massachusetts Amherst, Amherst, MA, United States
[b]Department of Medical Microbiology and Immunology, School of Medicine, Creighton University, Omaha, NE, United States
*Corresponding author: e-mail address: jbartz@creighton.edu

Contents

Abstract

Prions are a self-propagating misfolded conformation of a cellular protein. Prions are found in several eukaryotic organisms with mammalian prion diseases encompassing a wide range of disorders. The first recognized prion disease, the transmissible spongiform encephalopathies (TSEs), affect several species including humans. Alzheimer's disease, synucleinopathies, and tauopathies share a similar mechanism of self-propagation of the prion form of the disease-specific protein reminiscent of the infection process of TSEs. Strain diversity in prion disease is characterized by differences in the phenotype of disease that is hypothesized to be encoded by strain-specific conformations of the prion form of the disease-specific protein. Prion therapeutics that target the prion form of the disease-specific protein can lead to the emergence of drug-resistant strains of prions, consistent with the hypothesis that prion strains exist as a dynamic mixture of

Progress in Molecular Biology and Translational Science, Volume 175
ISSN 1877-1173
https://doi.org/10.1016/bs.pmbts.2020.07.001

a dominant strain in combination with minor substrains. To overcome this obstacle, therapies that reduce or eliminate the template of conversion are efficacious, may reverse neuropathology, and do not result in the emergence of drug resistance. Recent advancements in preclinical diagnosis of prion infection may allow for a combinational approach that treats the prion form and the precursor protein to effectively treat prion diseases.

1. Prion diseases

Prions are a self-propagating misfolded conformation of a cellular protein. The conversion of the normal isoform to the misfolded prion form occurs in a stepwise manner, but the exact mechanism is unknown.[1–4] Prions are found in several eukaryotic organisms. Yeast prions are a dominant, non-Mendelian form of epigenetic inheritance that causes a detectable phenotype. For example, the [URE3], [SWI+], and [GAR+] prions in *Saccharomyces cerevisiae* enhance growth on nutrient-poor sources, while the [PSI+] prion enables translational readthrough.[5–9] The prion state in yeast is dependent on a chromosomal gene and transient overproduction of the normal cellular protein promotes the prion state which can be reversibly cured.[8,10,11] The [Het-s] prion of *Podospora anserina* is infectious and has a necessary role in heterokaryon incompatibility, a process that restricts fusion of hyphae to genetically similar partners.[12–14] Prions found in yeast, fungi, and mammals can have advantageous roles. Functional prion-like proteins involved in memory, RNA metabolism, and immunity have been identified.[15–19] However, prions in mammals are most notably associated with neurodegenerative diseases.

Mammalian prion diseases encompass a wide range of disorders. The first recognized prion disease, the transmissible spongiform encephalopathies (TSEs), are characterized by long incubation periods followed by a relatively short duration of clinical signs that ultimately leads to death of the host.[1,20] TSEs occur in both domesticated populations of sheep and goats (scrapie), cattle (bovine spongiform encephalopathy, BSE), and mink (transmissible mink encephalopathy, TME). Chronic wasting disease (CWD) is unique among TSEs as it can affect both wild and domestic populations of cervids.[21–23] Camel prion disease (CPD) in domesticated dromedary camels was recently identified, but the geographic distribution and potential of transmission to wild populations is unknown.[24] Human TSEs can have infectious (Kuru), inherited (Gerstmann-Sträussler-Scheinker, GSS), or sporadic (Creutzfeldt-Jakob disease, CJD) etiologies, with sporadic being the most prevalent.[25] Iatrogenic transmission of PrP prions can occur via cadaveric human growth hormone (c-hGH), dura mater grafts, or corneal transplants.[26–29] TSE agent replication involves the misfolded infectious prion conformation of the prion protein, PrP^{Sc}, binding to the normal, cellular conformation, PrP^{C}, initiating further seeded conversion.

Alzheimer's disease, synucleinopathies, and tauopathies are characterized by the misfolding and aggregation of amyloid-β (Aβ; a peptide derived from the host cellular amyloid precursor protein (APP)), α-synuclein protein (α-syn), and tau protein, respectively. Similar to TSEs, these protein misfolding diseases share a similar mechanism of self-propagation of the prion form of the disease-specific protein and cell to cell spreading of the host protein in the prion state.[30–39] Experimental injection of the prion form of the protein into animals that express the respective precursor protein can accelerate the pathogenesis of disease, reminiscent of the infection process of TSEs (or PrP prions).[34,40–46] Although these non-prion protein disorders have both inherited and sporadic etiologies, conflicting evidence for infectious transmission has been reported. Epidemiological studies concluded that a history of one or more blood transfusions was not associated with a higher risk of developing AD compared to age- and gender-matched controls for each patient case.[47,48] Injection of nonhuman primates with over 600 patient samples from individuals diagnosed with nonspongiform diseases (i.e., AD, PD, etc.) failed to transmit disease.[49] No cases of AD or PD were identified among recipients of c-hGH in the United States despite finding mild amounts of pathological Aβ, tau, and α-synuclein by immunohistochemical analysis of pituitary glands from neurodegenerative disease and control patients.[50] However, a recent study reported that recipients of c-hGH contaminated with PrP prions subsequently developed CJD with concurrent Aβ deposition in the brain, suggesting iatrogenic transmission of Aβ.[51] Retrospective studies on the archived c-hGH samples identified significant amounts of tau, $A\beta_{42}$ and $A\beta_{40}$.[52] Intracerebral injection of these c-hGH samples to transgenic mice expressing a human mutant of the amyloid precursor protein resulted in the development of brain pathology similar to the patients injected with the same material.[52] This finding suggests, under certain conditions, iatrogenic transmission of AD is possible. Overall, the aforementioned neurodegenerative disorders share profound similarities at the biochemical and cellular level.

2. Prion strain diversity

Prion strains are operationally defined by heritable differences in the biological properties of disease under controlled agent and host parameters.[53–56] In yeast, prion strains are classified based on several traits that differ depending on the prion state (i.e., [PSI+] or [URE3]) but were initially distinguished by strength of phenotype (i.e., "strong" versus "weak").[57–59] TSE strains are classified by differences in incubation period and neuropathology but can differ in their clinical signs and biochemical features of PrP^{Sc}.[60,61] Similar to TSEs, Aβ, tau, and α-synuclein strains are defined by differences in abnormal protein deposition and pathology in the brain as well as protein conformation or seeding activity of the prion form of the respective protein.[38,44,62–64] The relationship between the strain-specific biochemical features of the prion form of the protein and the phenotype of disease is poorly understood.

2.1 Prion strain diversity in yeast

Studies of yeast prion strain diversity have added much to the knowledge of prion strain diversity. Yeast prion strains differ in strength of phenotype, stability of propagation, toxicity/lethality, ability to overcome inter- or intraspecies transmission barriers, and biochemical and physical properties.[57–59,65–67] Nuclear magnetic resonance (NMR) spectroscopy identified conformational differences between [*PSI+*] yeast prion strains, and atomic force microscopy (AFM) reveals strain-specific differences in yeast prion fibril morphology.[68] There are strong correlations between biochemical properties of the prion form of the protein and strain phenotype in yeast. The strength of the yeast prion phenotype corresponds with an increased fragility of protein aggregates.[69] Mechanistically, increased fragility of prion aggregates results in rapid generation of new free ends, accelerating prion formation (Fig. 1, step 5).[69]

2.2 TSE strain diversity

Strain diversity is hypothesized to be encoded by strain-specific conformations of PrP^{Sc}.[63,64,68,70–72] Strain-specific differences in the biochemical properties of PrP^{Sc} were first identified in mice.[73] These differences were later observed in hamsters infected with the hyper (HY) or drowsy (DY) strains of hamster-adapted transmissible mink encephalopathy (TME).[74] Detergent extraction and protease digestion of HY and DY PrP^{Sc} revealed

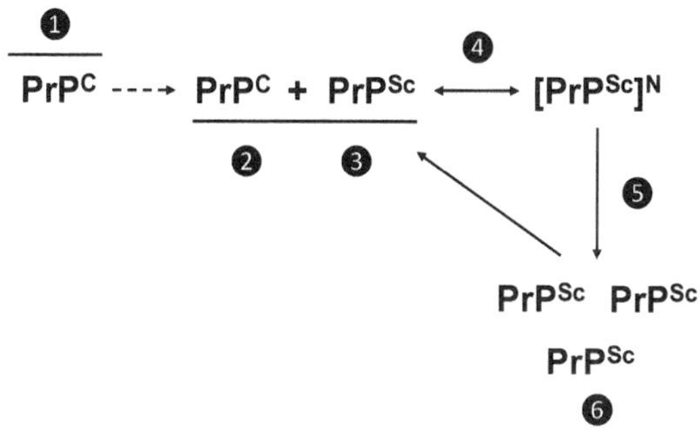

Fig. 1 Mechanistic location of therapeutic targets for PrP prion diseases. Therapeutic targets: ❶ PrP^{C} post-translational processing, metabolism, cellular trafficking, or localization; ❷ PrP^{C} by binding or inducing conformational change that prevents PrP^{C} interaction; ❸ PrP^{Sc} by stabilization, redistribution; ❹ conversion process by blocking binding site on PrP^{C}, capping growth, inhibition of cofactor interactions, or ❺ promoting fragmentation or ❻ clearance from the host.

strain-specific Western blot migration profiles suggesting strain-specific differences in the PK digestion site on PrP^{Sc}.[75] Edman protein sequencing of PrP^{Sc} from HY or DY TME-infected hamsters confirmed that PrP^{Sc} from these two strains had different PK cleavage sites.[70] Based on these observations, the authors proposed the hypothesis that strain-specific conformations of PrP^{Sc} encoded prion strain diversity.[70,75,76] Consistent with this hypothesis, the molecular weight of the proteinase K-resistant fragment of PrP^{Sc} after deglycosylation in fatal familial insomnia is 19 kDa while the same fragment was 21 kDa in sporadic and familial Creutzfeldt Jakob disease.[72] When inoculated into mice, the strain-specific migration pattern of PrP^{Sc} from human prion strains was preserved supporting the hypothesis that strain-specific conformations of PrP^{Sc} is the basis of strain diversity.[72]

The presence of cellular cofactors during prion conversion may be responsible for the maintenance of strain-specific, infectious conformations of PrP^{Sc}. Development of synthetic, *in vitro*-generated prions able to successfully infect wild-type rodents relied on cofactors such as polyanions and lipids.[77,78] Propagation of a recombinant prion strain in PMCA in the presence of both recombinant murine PrP and the lipid phosphatidylethanolamine (PE) retained infectivity.[79] Withdrawal of the cofactor (PE) led to the emergence of a protein-only strain, named OSU, that did not cause disease in mice suggesting cellular cofactors are required for infectivity. Additionally, *in vitro* PMCA propagation of three distinct prion strains with PE as the sole cofactor resulted in the three strains evolving into a single strain.[79] This work on the directed evolution of prions suggests that cellular cofactors are involved in prion strain diversity. Recent work has clarified the role of prion cofactors. Using a novel *in vitro* system where protein only PrP^{Sc} preparations, in the absence of PMCA adaptation, are directly infectious to animals, but only in the presence of co-factors.[80] Importantly, protein-only and cofactor generated PrP^{Sc} had identical strain properties indicating that the conformation of PrP^{Sc} is sufficient to encode prion strain information and that cellular co-factors are needed for prion propagation.[80] Overall, these data provide a unified model of prion infectivity where strain information is encoded by the structure of PrP^{Sc} and host cellular cofactors are necessary for prion conversion and subsequent infectivity.

Lesion profiles, cellular assays, and the biochemical properties of PrP^{Sc} are used to differentiate prion strains. The scrapie cell assay (SCA) and the cell panel assay (CPA) assess the relative susceptibility or resistance of murine neuroblastoma cell lines to infection by different prion strains.[81–83] The Western blot migration profile, conformational stability, and sensitivity to proteolytic digestion of PrP^{Sc} can be strain specific. It is unclear which, if any, of the strain-specific biochemical differences of PrP^{Sc} directly contributes the strain-specific phenotype of disease. For example, in murine PrP prions, decreased PrP^{Sc} conformational stability is positively correlated with a shorter incubation period, however, in humans and hamsters the inverse relationship between PrP^{Sc} conformational stability and incubation period was identified.[71,84–87]

2.3 Aβ strain diversity

The heterogeneous clinical presentation of AD may be attributed to strain diversity.[88] The Arctic and Swedish APP mutations in humans result in distinct AD pathologies.[89–92] It was hypothesized that the two AD clinical disease outcomes were a result of two distinct Aβ strains. Conformational stability studies of Aβ from patient brains with Arctic AD and Swedish AD mutations established that aggregates of Aβ found in the Arctic AD patients were significantly more sensitive to denaturation with guanidine hydrochloride compared to Swedish AD.[44] Further, injecting transgenic mice expressing human APP containing the Swedish mutation with the Arctic or Swedish AD brain material resulted in two distinct phenotypes of disease that were maintained upon serial passage consistent with the predicted behavior of prion strains.[44] Luminescent conjugated oligothiophenes, a class of amyloid binding dyes, to investigate how the Aβ amyloid structure in familial AD differs from sporadic AD.[93] The emission spectra of luminescent conjugated oligothiophenes changes when bound to amyloid of differing structure. The emission spectra of luminescent conjugated oligothiophenes bound to Aβ some of the subtypes of familial and sporadic AD varied significantly, suggesting a variety of Aβ structures exist in different AD cases.[93] Solid state nuclear magnetic resonance imaging analysis of Aβ from patient tissue samples from PCA-AD or rp-AD indicated that Aβ from PCA-AD resembled typical AD with predominant Aβ40 fibril structure while rp-AD samples were composed of a cloud of variant Aβ40 structures.[94] Overall, this data is consistent with the hypothesis that Aβ structural variance contribute to the clinical and pathological heterogeneity of AD.

2.4 Tau strain diversity

Alzheimer's disease (AD), corticobasal degeneration (CBD), progressive supranuclear palsy (PSP), Pick's disease (PiD), and argyrophilic grain disease (AGD) are unique tauopathies defined by the pattern of tau deposition in the brain along with distinct clinical phenotypes.[95] Transmission studies in wild-type and transgenic mouse models have established that tau extracts from various tauopathies transmit distinct tau neuropathology.[38,64,96–98] For example, AD-tau variants can differentially seed tau pathology after intracranial injection of wild-type mice; the tau variants were determined to be structurally different by trypsin digests that resulted in three distinct trypsin resistant fragments.[38] Injecting wild-type mice with tau from AD, CBD, or PSP resulted in differences in tau conversion efficiency and cellular tropism in the CNS suggesting the three tau isolates are distinct prion strains.[64] Consistent with this hypothesis, tau fibrils from the tau strains showed varying resistance to Gdn-HCl denaturation with CBD-tau, AD-tau, and PSP-tau having the lowest to highest stability, respectively.[64] Two recombinant tau strains in HEK293 cells expressing the tau repeat domain that propagated conformationally distinct forms of tau that could be serially passaged in PS19 transgenic tau mice that retained the strain-specific conformation of tau.[99] Further, the HEK293 cell model was used to examine AD, CBD, and AGD patient samples, and each tauopathy induced unique tau inclusion morphology.[99] Consistent with this finding, 18 tau strains were identified based on differences in limited proteolysis and *in vitro* seeding activity of tau in combination with hippocampal tau pathology when transmitted to PS19 transgenic tau mice.[100] Taken together, these data support that tau structural variance may explain the distinct pathologies and clinical presentation of the tauopathies.

2.5 α-synuclein strain diversity

Strain-specific prion forms of α-synuclein have been identified in synucleinopathies including Parkinson's disease (PD), dementia with Lewy bodies, and multiple system atrophy (MSA). Synthetic α-synuclein fibrils generated under varying salt concentrations form distinct conformations with differences in seeding ability and toxicity *in vivo* and *in vitro*.[62,101,102] Recombinant fibrillar α-synuclein created under different laboratory conditions varied in ability to cross-seed tau aggregation in cell culture and *in vivo*.[63] Consistent with this observation, proteinase K digestion revealed differences in protein cleavage between the synthetic α-synuclein as well as two distinct α-synuclein digestion products isolated from Parkinson's disease with dementia (PDD) patient brains.[63] Distinct digestion products are suggestive of two α-synuclein conformations reminiscent of prior observations in PrP prions.[72,75,103–105] *In vitro* seeding assays utilizing a cell line that readily propagates MSA α-synuclein was not susceptible to α-synuclein from PD patient brains suggesting that PD contains a different strain of α-synuclein.[106] Brain material from PD and MSA patients has distinct seeding activity and cellular inclusion pathology in HEK293T cells, suggestive of distinct strain of α-synuclein.[107] Overall, α-synuclein strains share many features of PrP prion strains that include differences in biochemical features, neuropathology, incubation period, and clinical manifestation of disease.[38,62,63]

3. Prion strains exist as a mixture

Multiple prion strains can co-exist in an individual host. In cases of sporadic Creutzfeldt-Jakob disease (sCJD), the polymorphism at codon 129 and different proteinase K cleavage patterns via Western blot classify disease as Type 1 (21 kDa) or 2 sCJD (19 kDa).[103] Type 1 and Type 2 sCJD PrP^{Sc} can co-exist in the CNS of the same patient.[108–116] It is suggested that up to 40% of sCJD cases contain more than one strain, however, the prevalence rate is controversial since methodological differences in strain typing and the region of the brain that is evaluated can both affect the outcome.[108–116] It is not well understood if strains of Aβ can co-exist in a single host, however, distinct strains of Aβ have been identified raising the possibility that sporadic AD may be a mixture of strains.[44] Mapping phospho-tau patterns of deposition in the hippocampal regions of individuals with various tauopathies established that the pattern of pathology is disease-specific with regional selectivity reminiscent of strain targeting in TSE's.[117] This same study determined that AD-related tau hippocampal deposition is greatly influenced by simultaneous argyrophilic grain disease (AGD), another tauopathy, suggesting that coexisting tau strains can create heterogeneous brain pathologies and overlapping clinical features highlighting the need for discriminatory methods of post-mortem diagnosis.[117] Antibodies that distinguish AD tau from non-AD tau deposits of CBD, PSP, and Pick's Disease (PiD) can discern the differential deposition in co-occurring pathologies.[118] While ample experimental evidence suggests that different α-synucleinopathy strains exist, the coexistence of α-synucleinopathy strains has not been described.

Coexisting prion strains can interfere with each other. Strain interference is a phenomenon in which a slowly-converting blocking strain can delay or prevent the emergence of a quickly-converting strain. A number of parameters govern prion strain interference. Increasing the titer of the blocking strain results in a corresponding increase in the interference effect.[119–122] Importantly, prion conversion in a common population of cells is required for interference to occur as the strains compete for a limiting cellular resource, PrP^C.[123,124] Consistent with this hypothesis, co-infection of two low conversion efficiency prion strains results in each strain converting independently without interference that is due to PrP^C not becoming limiting.[125] Overall, strain interference is an important parameter in the emergence of a dominant strain from a mixture.

Currently, strain interference has not been reported for mammalian non-PrP prions, and further investigation is required to determine if this phenomenon is a common property of mammalian prions. In yeast, there is evidence that strain variants compete within a cell during and after mating, with "strong" variants outcompeting "weak" variants for the template of conversion.[126] Mechanistically, "weak" variants are characterized by quick growth but low frangibility, whereas "strong" variants grow slowly but are more easily fragmented, leading to increased free ends for growth and therefore a competitive advantage over "weak" variants.[69]

4. Prion therapeutics

Therapeutics that target prion conversion can work via several mechanisms. Drugs can target PrP^C prior to its interaction with PrP^{Sc} by affecting the post-translational processing, metabolism, cellular trafficking, or localization of PrP^C (Fig. 1, step 1). Drugs that bind to or induce a conformational change in PrP^C can hinder the interaction between PrP^C and PrP^{Sc} (Fig. 1, step 2).

Drugs can target PrP^{Sc} by stabilizing or redistributing PrP^{Sc}, interfering with the interaction between PrP^{C} and PrP^{Sc} (Fig. 1, step 3). The conversion of PrP^{C} to PrP^{Sc} can be targeted by drugs that block the PrP^{Sc} binding site on PrP^{C}, cap PrP^{Sc} aggregate growth, or inhibit cofactors that enable conversion (Fig. 1, step 4). Therapeutics that target clearance of PrP^{Sc} either increase fragmentation of PrP^{Sc} aggregates (Fig. 1, step 5) or promote clearance of PrP^{Sc} from cells (Fig. 1, step 6).

4.1 Treatments that target PrP^{Sc}

Several drugs that directly interact with PrP^{Sc} have anti-prion effects (Table 1 and Fig. 1). Congo red, an amyloid binding dye, was one of the first compounds identified to extend the incubation period of prion-infected animals and can reduce the formation of PrP^{Sc} in scrapie-infected mouse neuroblastoma cells (Sc^{+} MNB).[127] Congo red is hypothesized to directly bind and over-stabilize PrP^{Sc}, preventing further prion formation.[127–129] Oral administration of Compound B, an amyloidophilic compound, extends the incubation period of TSEs and is hypothesized to inhibit formation of PrP^{Sc} by directly binding PrP^{Sc} or by interacting with cofactors/chaperones necessary for prion conversion.[134,135] The diphenyl-pyrazole anle138b inhibits PrP^{Sc} amplification and reduces neurotoxicity by shifting PrP^{Sc} oligomers to a smaller size by reducing intermolecular hydrogen bonding and obstructing the formation of higher order oligomers, modulating the formation of toxic PrP^{Sc} aggregates.[136,137] Anle138b has anti-prion activity both *in vitro* and *in vivo* and substantially prolonged survival even when treatment began after onset of signs of prion disease.[137]

Cellular redistribution of PrP^{Sc} can inhibit prion formation. Chlorpromazine (CPZ), an antipsychotic, redistributes PrP^{Sc} from the early endosomal/endocytic recycling pathway to the late endosomal/lysosomal pathway resulting in an inhibition of prion conversion in prion-infected N2a cells; however, CPZ was ineffective *in vivo*.[139] A cholesterol synthesis inhibitor, U18666A, can redistribute PrP^{Sc} from the early endosomal/endocytic recycling pathway to the late endosomal/lysosomal pathway, causing an increase in PrP^{Sc} degradation that corresponds to a decrease in total PrP^{Sc} in prion-infected N2a cells.[139]

Increasing degradation of PrP^{Sc} can ameliorate prion infection (Fig. 1, step 6). The tyrosine kinase inhibitor STI571 decreases the half-life of PrP^{Sc} in prion-infected ScN2a cells via interaction with the tyrosine kinase c-Abl without affecting the cellular location or trafficking of PrP^{Sc}.[208]

Table 1 Anti-prion therapeutics.

Drug	Type	Model	Efficacy	Suggested mechanism	Strain Specificity	References
PrP^{Sc} target						
Congo Red	Sulfonated amyloid dye	Neuronal cell culture, hamster	Decreases PrP^{Sc} accumulation and inhibits propagation in culture, prolongs mean incubation period *in vivo*	Stabilizes PrP^{Sc}, inhibits GAG-PrP binding and potentially PrP^{C}-PrP^{Sc} binding	127S, Sc237, PrP (27–30), 263K, 139H	127–131
LIN5001, LIN7002, LIN5044	Luminescent conjugated polythiophenes (LCPs)	Mouse	IV infusion prolongs survival, reduces infectivity	Aggregate hyperstabilization	RML6, 263K	132,133
Compound B	4-pyridinecarboxaldehyde, 2-[4-(5-oxazolyl)phenyl] hydrazone	Neuronal cell culture, mouse	Eliminates PrP^{Sc} in culture; prolongs incubation period/life, reduces PrP deposition, large quantity needed for efficacy	Binds to abnormal PrP, inhibits new formation of abnormal PrP	RML, 22L (marginally), Fukuoka-1 GSS, 263K (barely)	134,135
anle138b	3,5-diphenyl-pyrazole (DPP)	Neuronal cell culture, mouse	Antiprion activity *in vitro* and *in vivo*, substantial prolongation of survival (txt after clinical onset), good bioavailability, BBB penetration	Directly blocks PrP^{Sc} amplification, causes shift towards smaller PrP^{Sc} oligomer size, inhibits pathological aggregation (PrP^{Sc} specific)	RML, ME7, 301c, sCJD, vCJD, a-syn	136–138

Continued

Table 1 Anti-prion therapeutics.—cont'd

Drug	Type	Model	Efficacy	Suggested mechanism	Strain Specificity	References
U18666A	Cholesterol synthesis inhibitor	Neuronal cell culture	Significantly decreased PrP-res levels	Redistributes PrP^{Sc} to late endosome-lysosome, degradation of PrP^{Sc} in secondary lysosomes	22L	139
Pentosan Polysulfate (PPS)	Cysteine protease inhibitor	Neuronal cell culture, mouse, human	IV infusion prolonged survival even when administered after abnormal PrP deposition, rapidly and significantly decreased PrP^{Sc} levels, various efficacy in humans	May act as a competitive coreceptor (with endogenous GAG or other proteoglycans) for PrP at cell surface or causes fragmentation of PrP^{Sc} at the cell surface	263K, RML, Fukuoka-1 GSS, 22L, iCJD, vCJD	139–147
Heterologous protein	Hamster PrP	Neuronal cell culture, mouse	Decreases PrP-res in culture, slows disease progression, increases survival, significantly delays onset of clinical signs, decreases PrP^{Sc} accumulation in brain/spleen	Either binds to PrP^{Sc} creating a functionally impotent aggregate unable to produce additional PrP^{Sc} or binds and blocks the conversion site from PrP^{C}	RML-Chandler	148–151

IND24, IND81, IND114338, IND125, IND126461	2-aminothiazoles	Neuronal cell culture, mouse	Strain-specific and administration time-dependent extension in survival, decreased & altered distribution of PrP^{Sc} in brain, can lead to drug resistance (e.g. IND24-resistant RML)	Inhibits formation of new PrP^{Sc}, reduces PrP^{Sc} load via drug-like mechanism, might promote clearance of PrP^{Sc}	RML, ME7, 22L, CWD	138,152–156
Amphotericin B	Fungizone (polyene antibiotic)	Neuronal cell culture, mouse, hamster, human	Administration time-dependent prolongation of incubation period, delays onset of clinical signs without improvement of neurological symptoms or prolongation of clinical disease, delays accumulation of PrP^{Sc} in the brain	Potentially could: interact with PrP^{C}/PrP^{Sc}, interfere with uptake of PrP^{Sc} into cells, target detergent-resistant microdomains (modify cell surface distribution and/or trafficking of PrP), regulate microglial/glial activation, directly bind and either "cap" amyloid growth or overstabilize PrP^{Sc}	263K, CJD	140,157–164

Continued

Table 1 Anti-prion therapeutics.—cont'd

Drug	Type	Model	Efficacy	Suggested mechanism	Strain Specificity	References
PrP^C Target						
771, active ASO 1, active ASO 2	Antisense oligonucleotide	Neuronal cell culture, mouse	Reduces PrP^C/PrP^{Sc} in cell culture and *in vivo*, delays onset of clinical signs, slows disease progression, toxicity issues when administered after prion inoculation	Decreased PrP^C levels slow accumulation of PrP^{Sc}	RML, potentially more since PrP^C is the target	165,166
J1, J8, J20, J35	Chalcones	Neuronal cell culture	Decreases PrP^{Sc} levels, completely blocks PrP109–149 aggregation	Directly binds to PrP^C (C-terminal), decreases amount of PrP^C on cell surface (arrest in ER), may stabilize PrP^C or bind seeding surface	RML, 22L, CWD, 263K, CJD	167,168
Y13, Y17	Oxadiazole	Neuronal cell culture	Could not totally block conversion but Increased lag phase for conversion; minor decrease in PrP-res formed in RT-QuIC	Direct binding to PrP^C (stabilize PrP^C or bind at seeding surface), potentially interacts with PrP^C/PrP^{Sc} N-terminal region, decrease amount of PrP^C on cell surface causing arrest in ER	RML, 263K, CJD	167,168

Fe(III)-TMPyP	Cationic tetrapyrrole	Neuronal cell culture, mouse	Inhibits replication *in vitro*/cell culture, prolongs survival, inhibits amplification of PrP^{Sc}, inhibits cytotoxic effects of delta CR PrP	Interacts with C-terminal domain of PrP^{C} (Helix-3), acts as pharmacological chaperone for PrP^{C} (decreases prion-induced misfolding)	RML, 22L, bank vole strains (Italian, UK (SCR1), and CH1641)	169,170
D18	Anti-recombinant PrP (29–231) antibody (residues 132–156; helix A) Fabs	Neuronal cell culture, mouse	Abolishes prion replication and clears existing PrP^{Sc} in cell culture (2 week txt), prolongs incubation period and decreased titer in mice	Binds to PrP^{C} on cell surface and hinders docking of PrP^{Sc}/a cofactor needed for conversion	Scrapie	171
6H4	Anti-murine PrP^{C} antibody (Mab and expressed as transgene)	Neuronal cell culture, mouse	Prevents infection of susceptible cells, cures chronically infected cultures, conferred anti-PrP titers ($Prnp^{0/0}$ mice) w/o induction of autoimmune disease, no infectivity detected ($Prnp^{+/0}$-6H4u mice) when challenged, prevents or drastically delays pathogenesis	Occludes PrP^{C} at prion replication sites, captures and degrades (immune-mediated) incoming PrP^{Sc} inoculum, steric competition with template-directed refolding, interference with seeded PrP^{Sc} nucleation reaction	RML	172,173

Continued

Table 1 Anti-prion therapeutics.—cont'd

Drug	Type	Model	Efficacy	Suggested mechanism	Strain Specificity	References
ICSM 18	Isotype IgG1, anti-murine alpha-PrP antibody (residues 146–159)	Neuronal cell culture, mouse	Reduces splenic PrPSc levels, increases survival, passive transfer of Abs had no effect late in incubation period	Direct inhibition of PrP^{Sc} production, binding may bury "active residues" and stabilize helix 1	RML	174,175
110	Monoclonal anti-PrP antibody (PHGGGWG at aa 59–65 and aa 83–89; octarepeat region)	Neuronal cell culture	Reduced PrP^{Sc} accumulation (dose-dependent) in cell culture, long term txt increased PrP^{C} levels	Retains PrP^{C} on cell surface (may interfere with PrP^{C} metabolism), mAb-PrP^{C} binding inhibits PrP^{C}-PrP^{Sc} interaction by occupying binding domain or through steric interference	Chandler (139A), RML	176
6D11	Monoclonal antibody, recognizes PrP^{C}, PrP^{Sc}, and recPrP (epitope between 93–109)	Neuronal cell culture	Complete abrogation of PrP^{Sc}, no re-emergence for at least 14 days, (following removal of txt), pre-incubation prevents infection	May prevent PrP^{C}-PrP^{Sc} interaction or interfere with binding auxiliary molecules needed for prion propagation	22L	177
7H6	Monoclonal Ab, recognizes residues 130–140	Neuronal cell culture	Complete abrogation of PrP^{Sc}, no re-emergence for at least 14 days, (following removal of txt)	Similar to 6D11?	22L	177

7A12	Monoclonal Ab, recognizes residues 143–155 (alpha-helix A)	Neuronal cell culture	Complete abrogation of PrP^{Sc}, no re-emergence for at least 14 days, (following removal of txt), weakest preventative effect	Similar to 6D11?	22L	177
SAF34	Monoclonal Ab, recognizes residues 59–89 (HuPrP)	Neuronal cell culture	Dose-dependent decrease of PrP^{Sc} levels, increased inhibition when used in conjunction with SAF61, re-emergence of PrP^{Sc} following removal of txt	Decreases half-life of PrP^{C}, antibodies act on PrP^{C} or PrP^{Sc} isoforms, antibodies prevent PrP^{C}-PrP^{Sc} interaction	22L	178
SAF61	Monoclonal Ab, recognizes residues 144–152 (HuPrP)	Neuronal cell culture	Dose-dependent decrease of PrP^{Sc} levels, increased inhibition of PrP^{Sc} when used in conjunction with SAF34, no re-emergence of PrP^{Sc} following removal of txt	Decreases half-life of PrP^{C}, antibodies act on PrP^{C} or PrP^{Sc} isoforms, antibodies prevent PrP^{C}-PrP^{Sc} interaction, increase clearance/degradation of PrP^{C}, modulate cellular trafficking of PrP, induce conformational change exposing PrP to protease attacks	22L	178

Continued

Table 1 Anti-prion therapeutics.—cont'd

Drug	Type	Model	Efficacy	Suggested mechanism	Strain Specificity	References
PrioV3	Camelid anti-PrP antibody (residues between 171 and 190, YYR motif)	Neuronal cell culture, *endo*-thelial cells, mouse	Crosses BBB *in vitro* and *in vivo*, abrogates PrP^{Sc} replication in cells, permanent depletion of PrP^{Sc} in cells, marked inhibition of PrP^{Sc} accumulation in spleen	Alters PrP^{C} expression (direct neutralizing effect on PrP^{C}/PrP^{Sc}), could block PrP^{C} incorporation into infectious PrP^{Sc}	RML	179,180
44B1	Monoclonal antibody	Neuronal cell culture	Rapidly and significantly decreased PrP^{Sc} levels	Retains PrP^{C} on cell surface as an antigen–antibody complex, interferes with internalization and trafficking of PrP^{C} to endocytic compartments	22L	139
POM2	Anti-PrP antibody to flexible tail of PrP^{C}	Mouse COCS	Prevented prion-mediated neurodegeneration but prion titer was not decreased	Induces shift in distribution of PrP^{Sc} moieties without affecting overall quantity	RML	181
LD7, JZ107	Phenethyl piperidine	Neuronal cell culture	Reduced PrP^{Sc} by 50%, permanently cured RML-infected cells (1 month exposure), protects against dendritic spine loss in presence of PrP^{Sc}	Might interact with PrP^{C} in a cellular context, perhaps in conjunction with other cell-surface receptors, may interact with PrP^{C} substrate	RML, 22L	182

Alprenolol hydrochloride	B-adrenergic blocker	Neuronal cell culture, mouse	Reduced levels of PrP^{Sc} in cells and brains of infected mice, inhibited PrP^{Sc} accumulation and spongiform changes but did not prolong survival	May inhibit via interaction with PrP^{C}	Fukuoka-1	183
R12, R24, R12-A-R12	RNA aptamer	Neuronal cell culture	Significantly reduces PrP^{Sc} levels in cell culture	Tightly binds to and stabilizes PrP^{C}, blocking conversion to PrP^{Sc}	Fukuoka-1	184,185
DE10, DC2, EB8, EF2	Monoclonal antibodies to N-terminal region of PrP^{Sc}	Neuronal cell culture	Promoted complete clearance of PrP^{Sc} in cell culture	Binds to a region on PrP^{C} that PrP^{Sc} also binds, blocking PrP^{C}-PrP^{Sc} interaction	RML	186
PrP^{C}/PrP^{Sc}						
Chlorpromazine (CPZ)	Phenothiazine	Neuronal cell culture	No to modest effect on PrP^{Sc} accumulation, inhibits prion replication in cells but not *in vitro*	Decreases PrP^{C} at cell surface by inhibiting clathrin-mediated endocytosis, redistributes PrP^{Sc} to late endosome-lysosome	sheep scrapie 127S, Sc237, Type 1 CJD, RML, 22L	129,139,187
Quinacrine	Lysosomotropic agent	Neuronal cell culture, mouse, humans	Inhibits cell growth at concentrations greater than 2.0 μM, no prolongation effect in mice and humans, may lead to drug resistance	Competitive inhibition of GAG-PrP interaction, unfolding of PrP-res, destablilization of PrP-res conformation, induces conformational change that disfavors PrP^{Sc} conformation	Scrapie, 263K, vCJD, sCJD, iatrogenic CJD, genetic CJD	140,188–193

Continued

Table 1 Anti-prion therapeutics.—cont'd

Drug	Type	Model	Efficacy	Suggested mechanism	Strain Specificity	References
E-64d	Cysteine protease inhibitor	Neuronal cell culture, mouse	No toxicity to cell growth at [ ] up to 100 μM, no prolongation effect	Competitive inhibition of GAG-PrP interaction, unfolding of PrP-res, destabilization of PrP-res conformation	Scrapie, 263K	140,190
MS-8209	Amphotericin B derivative	Neuronal cell culture, mouse, hamster	Prolongs incubation period (dose- and timing dependent), delays onset of PrP-res/GFAP accumulation, vacuolation, spongiosis, and astrogliosis in brain, variable efficacy across prion species	Potentially affects conversion, may interact with astrocyte lysosomal system and limit propagation of PrP at inoculation site	Sheep scrapie 127S, 139A, Sc237 (weaker with 139H), type 1 CJD, C506M3 (similar to ME7), 263K	129,194–200
31C6	Monoclonal anti-PrP antibody (residues 143–149), IgG1	Neuronal cell culture, mouse	Reduces PrP^{Sc} in cell culture (dose-dependent), no re-emergence following txt, mAb txt at 120 dpi increased survival (not statistically significant), slowed weight loss, disease progression, and accumulation of PrP^{Sc} in brain	Direct inhibition of PrP^{C}-PrP^{Sc} interaction by occupying binding domains, could interfere with PrP^{C} metabolism by retaining PrP^{C} on cell surface	Chandler (139A), RML	176,201

Doxycycline	Antibiotic	Neuronal cell culture, human	Variably affected PrP-res accumulation in culture, variable prolongation of survival, reduction in widespread & severe lesions (early txt)	Destabilize abnormal PrP, could operate at cell level by modulating formation of PrP aggregates	CJD (sporadic and genetic–E200K or V210I), sCJDMM1, sCJDVV2a, vCJDMM2b, iCJDMM1	202–206
Other						
GSK2606414	Protein kinase RNA-like ER kinase (PERK) inhibitor	Mouse	Oral treatment reversed cognitive deficits and prevented clinical disease, effective pre- and post-symptomatic, halted progression in spongiform degeneration, protected from neuronal loss	Prevents activation of UPR branch that mediates prion neurotoxicity by inhibiting PERK	RML	207

Several drugs increase degradation of PrP^{Sc} by inducing autophagy. A-12 and A-14, derivatives of the antitumor drug celecoxib, can reduce or eliminate PrP^{Sc} in prion-infected neuronal cells lines by stimulating autophagy.[209] Lithium, which is used to treat depression, reduces PrP^{Sc} in prion-infected neuronal and non-neuronal cells cultures by inducing autophagy.[210]

Pentosan polysulfate (PPS), a cysteine protease inhibitor, induces a reduction in prion conversion by causing fragmentation of PrP^{Sc} at the cell surface or utilizing endogenous glycosaminoglycans (GAGs) to competitively interfere with the binding of PrP^{C} to PrP^{Sc}.[139,140] Treatment of prion-infected mice with PPS decreased PrP^{Sc} accumulation and prolonged survival compared to untreated controls even when administered after the onset of PrP^{Sc} deposition in the CNS.[139,140] While effective in rodents, treatment of human prion diseases with PPS has varied results. In some patients treated with intravenous PPS, there is a significant increase in survival compared with untreated patients[141–143]; however, there are several cases indicating treatment is not effective at extending survival or ameliorating clinical features of disease.[144–146]

Treatment of prion disease with heterologous prion proteins (HetPrP) can interfere with the formation of PrP^{Sc} in both Sc^{+}MNB cells and *in vivo*.[148–151] Treatment of RML-infected mice with hamster HetPrP significantly delayed onset of clinical symptoms and significantly decreased PrP^{Sc} accumulation in brain and spleen.[150] Mechanistically, it is hypothesized that HetPrP either incorporates itself into a growing PrP^{Sc} aggregate and, unable to be converted to PrP^{Sc} by the host species PrP^{Sc}, hinders conversion or binds directly to the conversion site on PrP^{Sc} blocking the PrP^{C} conversion site.[148] Overall, drugs that target PrP^{Sc} that are effective in prion-infected cell cultures and animals are ineffective in human clinical trials. The lack of efficacy in human clinical trials could be partly due to timing of treatment initiation. In prion-infected rodents, anti-prion drugs administered prior to or shortly after prion infection have the highest efficacy, with administration at the onset of clinical disease being less effective. In human prion disease, treatment is typically not begun until after onset of clinical signs, decreasing treatment efficacy.[211] In an attempt to overcome this, an ongoing study is treating subjects at risk for developing fatal familial insomnia with doxycycline.[212]

4.2 Strain-specific efficacy of anti-prion treatments

Treatment of prion-infected animals or cell cultures with anti-prion drugs can lead to the emergence of drug resistant prion strains. Swainsonine inhibits the processing of asparagine-linked glycoproteins by impeding the action of Golgi α-mannosidase II as well as lysosomal mannosidase.[213–215] Treatment of prion-infected neuroblastoma-derived R33 cells with swainsonine can result in the emergence of swainsonine-resistant prions.[216] Importantly, PrP^{Sc} from swainsonine-resistant and swainsonine-sensitive prions from cell culture have different PrP^{Sc} conformational stabilities suggesting they are distinct prion strains.[216,217] Treatment of prion-infected cells with swainsonine selected for a drug resistant substrain, but when the inhibitor is removed, the susceptible substrain reverted to a drug sensitive population in cell culture.[218] Additionally, passage of swainsonine-resistant RML prions from AMO10 to PK1 cells in the presence of swainsonine resulted in the selection of swainsonine-dependent prion variants with an increased efficiency to propagate in the presence of the drug.[219] However, passaging the swainsonine-dependent prions in the absence of drug resulted in reversion to a swainsonine-sensitive prion population in cells.[219]

RML-infected ScN2a cells treated with the 2-aminothiazole IND24 resulted in the emergence of an IND24-resistant prion strain (RML [IND24]) that had a different host cell range, biochemical and neurological features compared to RML-infected mice treated with vehicle (RML [V]).[152] In addition to IND24, this drug-resistant prion strain was also resistant to the anti-prion drugs Compound B and quinacrine.[152] This suggests that the mechanism of resistance between the three anti-prion compounds is similar. Susceptibility of the RML[IND24] strain to IND24 could be restored when passaged in the absence of IND24 resulting in the reversion of the RML[IND24] phenotype to the RML[V] phenotype in mice.[152]

The mechanism underlying the emergence of drug-resistant prion strains is not known. Anti-prion treatments can be strain specific.[152,157] One hypothesis is that compounds effective against the most predominant, but not all, strains in a mixture can allow for the emergence of a drug resistant strain. Prion strains are thought to exist as a dynamic mixture of substrains, or mutant spectra.[220,221] In this scenario, the dominant strain suppresses the emergence of substrains and, once the dominant strain is removed by the anti-prion drug, the suppressive effect is diminished allowing for emergence of a preexisting drug resistant strain. Studies of strain interference have established the conditions and mechanism of how a dominant strain can suppress the emergence of a minor strain consistent with this hypothesis.[121,123,221–223] Selective removal of a dominant strain from a mixture using strain-specific differences in the stability of PrP^{Sc} can allow for the emergence of a highly pathogenic substrain from a mixture, providing additional support for this hypothesis.[224] Alternatively, it is possible that anti-prion compounds directly interact with PrP^{Sc} altering its conformation (i.e., mutation) and therefore strain properties. Anti-prion compounds can bind to PrP^{Sc} consistent with this hypothesis, however, RML[IND24] remained resistant to IND24 for up to 20 passages in CAD5 cells in the absence of IND24 suggesting that IND24 interaction with PrP^{Sc} is not required for the resistant phenotype.[152] Overall, it is unclear if the emergence of drug resistant prion strains is due to selection of a preexisting mutant strain from the mutant spectra and/or if the anti-prion drug acts as a prion mutagen to generate the drug resistant strain.

Maintenance of a drug resistant phenotype may have a high fitness cost. Passage of drug resistant prion strains in the absence of the anti-prion drug can result in rapid reversion to the original drug sensitive strain. This suggests that the drug resistant prion strain has relatively poor fitness for the host. In the absence of drug, the mutant spectra that arises from prion conversion produces mutants that have increased fitness for the host. This observation raises several questions. First, reversion of the drug resistant strain in the drug-free environment leads to the emergence of the original drug sensitive strain. It is unclear why the drug resistant strain would revert to the parent strain as opposed to another strain if generation of the mutant spectra is random. This suggests that the drug resistant strain may contain a "memory" that favors reversion to the original strain. Second, treatment of a host with an anti-prion drug can extend the incubation period of disease without the emergence of a drug resistant strain. The mechanism behind this observation is unknown. It is possible that certain strains can still retain sufficient suppressive effect of substrains during anti-prion treatment. The relative distribution of the mutant spectra in any given isolate may differ depending on its passage history. For example, biologically cloned strains may have a more limited mutant spectra compared to uncloned or field isolates. Finally, highly stable class I strains may have a more limited mutant spectra compared to unstable class III strains.[225] Overall, very little is known about how the interaction between the host and the strain influences prion strain fitness.

The emergence of drug resistant prions is a significant barrier to the development of effective treatments of prion diseases that target the prion form of the protein. Treatments that reduce or eliminate the abundance or the ability of the conversion template to form the prion form of the protein may offer an alternative strategy for anti-prion therapies.

4.3 Treatments that target PrP^C

Prion conversion is dependent on the presence of the template of conversion (Fig. 1, step 2). Genetic ablation of PrP in mice results in complete resistance to prion infection with little known detriment to development or function.[226,227] Similarly, ablation of PrP^C in cattle does not result in major functional or immunological deficits and tissue from these animals does not support prion conversion.[228] In a conditional PrP^C knockout model, transgenic mice express PrP^C until approximately 12 weeks of age before undergoing gene-directed depletion of PrP^C.[229] When these mice are challenged with RML prions at 3–4 weeks of age, prion infection was established, but the depletion of PrP^C at 12 weeks halted progression to the clinical phase of prion disease and reversed spongiosis.[229] Overall, these studies provide evidence that elimination of PrP^C, even after the establishment of prion infection, has potential as a therapeutic strategy.

Antisense oligonucleotides (ASOs) can modulate expression of PrP^C. Sequence-specific ASOs lower levels of PrP^C and PrP^{Sc} in ScN2a cells.[165] When sequence specific ASOs were administered to mice via either i.p. or intracerebroventricular (ICV) routes, ASOs were well tolerated and resulted in a decrease in *Prnp* mRNA levels in the brain compared to PBS negative controls.[165,166] When ASOs were administered prophylactically or at time of inoculation with the RML prion strain, there was diminished PrP^C expression and a corresponding reduction in PrP^{Sc} in the brains of ASO-treated mice compared to PBS controls. Reduction of PrP^{Sc} levels correlated to a prolongation of the incubation period of disease, but animals ultimately succumbed to prion disease.[165,166] When ASOs are administered just prior to the onset of clinical signs of disease, this resulted in many of the animals rapidly succumbing to the toxic effect of ASO administration.[166] Encouragingly, the mice that survived the ASO treatment had a delayed onset of clinical disease as well as an extended clinical duration of disease that was three times longer than saline treated controls.[166] Overall, ASOs can delay the onset of prion disease whether given prophylactically or after the establishment of infection and, one issues of toxicity are addressed, may be an attractive method of depressing PrP^C expression to treat prion diseases.

Treatments that alter the expression of PrP^C on the cell surface can reduce prion conversion. For example, a group of chalcones and oxadiazoles can directly bind to PrP^C and decrease its cell surface expression arresting PrP^C trafficking to the endoplasmic reticulum resulting in a decrease in

PrP^{Sc} abundance and aggregation in prion-infected N2a cells.[167,168] Alternatively, some of these compounds prevent conversion of PrP^C to PrP^{Sc} by binding to and stabilizing PrP^C or directly binding to PrP^C at the seeding surface.[168,184,185]

Alteration of post translational processing of PrP^C can inhibit prion formation. PrP^C can undergo either α- or β-endoproteolytic cleavage resulting in C1 and C2 fragments, respectively.[230] The C2 amino acid sequence approximately aligns with that of the PrP^{Sc} protease resistant core, and has been found in abundance in prion infected brains.[231–237] The production of proteolytic cleavage products with varying pathogenic propensities is similar to what is observed of amyloid precursor protein, where APP cleavage is central to the development of AD.[238] Processing of APP by the enzymes β-secretase and γ-secretase leads to production of the peptides Aβ-40 and Aβ-42. While Aβ-40 has neuroprotective qualities,[239,240] Aβ-42 has been implicated in AD pathogenesis.[241,242] Endopeptidase cleavage of α-synuclein results in aggregation and toxicity that ultimately causes PD.[243] Pharmacological alteration of cellular proteolytic cleavage of the precursor proteins is an attractive druggable target. For example, a drug that preferentially inhibits the production of the C2 fragment would be a viable anti-prion treatment as the unfavorable PrP^C cleavage product (C2) would be sequestered while the C1 fragment would be preserved and able to continue to perform a protective function.[244] The metalloprotease, ADAM10, plays a role PrP^C shedding from the cell surface, but has also been suggested to contribute the α-cleavage of PrP^C.[232,245–249] Inhibition of ADAM10 resulted in an earlier onset of prion disease and slower spread to other brain regions in mice as a result of PrP^C sequestration, suggesting ADAM10 may cleave newly formed PrP^{Sc} from the surface of infected cells.[247,250] These data suggest that a therapeutic that increases PrP^C shedding, instead of inhibiting post-translational processing of PrP^C, may be more effective at dampening disease progression.

Monoclonal antibodies (mAb) directed against PrP^C can inhibit prion conversion. Several mAbs (D18, 6H4, ICSM 18, 31C6, etc.) are able to abrogate prion conversion and cure prion-infected N2a cells.[171,172,176–180,251] In prion-infected N2a cells, the inhibition of prion conversion depends on the binding affinity between the mAb and PrP^C at the cell surface, with the most potent inhibitors often binding the α-helical domain of PrP.[174,176,177] Inhibition of PrP^{Sc} formation by antibodies in cell culture has been proposed to occur via several mechanisms. The antibodies D18 and 6D11 can abolish prion conversion by binding to PrP^C at the cell surface and hindering

PrP^{Sc}-templated conversion.[171,177] Similarly, treatment of prion-infected N2a cells with mAb 44B1 retains PrP^C at the cell surface in an antigen–antibody complex rendering it unavailable for conversion resulting in a significant decrease of PrP^{Sc} levels.[139] A group of N-terminal monoclonal antibodies (DE10, DC2, EB8, and EF2) are hypothesized to directly bind to a region on PrP^C that has been shown to tightly bind PrP^{Sc}, effectively blocking the PrP^C-PrP^{Sc} interaction.[186] Antibodies that disrupt PrP^C metabolism or trafficking either by preventing internalization of PrP^C or reducing the half-life of PrP^C, result in a decrease in PrP^{Sc} in cell culture models.[176,178,251] Although the aforementioned antibodies were reported to have no effect on cell growth or induce cellular toxicity, there are reports of anti-PrP antibodies (e.g., POM1) that are neurotoxic.[181,252–255] Anti-PrP antibodies that are toxic tend to bind epitopes in the globular domain of PrP and may trigger similar pathogenic cell signaling pathways as prion disease.[181,254]

Monoclonal antibodies directed against PrP^C can alter prion pathogenesis and delay the onset of clinical signs of prion infection. ICSM 18, when administered to mice either intraperitoneal (i.p.) or i.c. 7 or 30 days after i.p. inoculation with RML was found to decrease PrP^{Sc} levels in the spleen and delay the onset of clinical signs of prion infection compared to untreated controls.[175] However, treatment was ineffective when administered at the onset of clinical signs or if mice were i.c. inoculated with RML.[175] Treatment of mice infected with the Chandler strain of murine-adapted scrapie with mAb 31C6 120 dpi led to an increase in the incubation period, a decrease in PrP^{Sc} levels, and delayed progression of neuropathological lesions in the cerebellum.[201] Creating $PrP^{0/0}$ mice expressing the 6H4 antibody as a transgene (*Prnp*$^{0/0}$-6H4μ) led to these mice producing anti-PrP titers in sera.[173] Crossbreeding of *Prnp*$^{0/0}$-6H4μ mice with C57BL/6 mice led to re-introduction of one or two PrP alleles, generating *Prnp*$^{+/0}$-6H4μ and *Prnp*$^{+/+}$-6H4μ mice, respectively, which developed anti-PrP serum levels.[173] Following intraperitoneal inoculation of the RML strain of scrapie, *Prnp*$^{+/0}$ 6H4μ mice lacked deposition of PrP^{Sc} in the spleen or the brain, whereas *Prnp*$^{+/0}$ mice had deposition in both tissues.[173] Expression of the 6H4 antibody as a transgene drastically delayed prion pathogenesis in mice.[173] Overall, anti-PrP antibodies can reduce or eliminate PrP^{Sc} formation in cells and animals and provides proof of principle of the efficacy of this approach.

Vaccines that result in host induced production of antibodies that recognize PrP^{Sc} or PrP^C have been attempted. Vaccination of mice or white-tailed deer with attenuated *Salmonella* expressing PrP resulted in significant prolongation of onset of clinical disease in the vaccinated group with complete protection in a portion of the animals as determined by the absence of PrP^{Sc} deposition; however, these studies involved a relatively small sample size.[256,257] Vaccines consisting of peptides derived from the primary amino acid sequence of the prion protein resulted in a humoral immune response to all peptides and delayed the onset of clinical signs of prion infection compared to control prion-infected mice.[258] Other studies have focused on protein epitopes specific to PrP^{Sc} with strong induction of antibody responses after vaccination of mice, sheep, and white-tailed deer, strengthening the possibility of an effective PrP^{Sc}-specific vaccine.[259–262] In contrast to the previous studies, vaccinating elk to induce a antibody response against PrP^{Sc} resulted in a significant shortening of incubation period of CWD.[263] The acceleration in disease onset could be attributed to antibody-mediated misfolding of PrP^{C} or by promoting the uptake of PrP^{Sc} in the gut.[263] Since PrP^{C} is a host-encoded protein, development of an effective vaccine directed immune response is impeded by host tolerance. While many approaches have aimed to overcome tolerance, there are negative consequences of inducing an antibody response to a host-encoded cellular protein. The effects of antibodies directed at the globular domain of PrP^{C} results in varying degrees of neurotoxicity while other PrP^{C}-specific antibodies lead to apoptosis and improper signaling cascades.[181,253,264] Vaccination strategies have successfully induced strong humoral and mucosal immune responses in multiple animal models; however, more work is required to evaluate the most effective vaccination strategy to prevent disease onset while avoiding toxicity.

5. Future prospects

Development of ultra-sensitive methods of PrP^{Sc} detection have greatly improved and may lead to routine detection of prion infection in asymptomatic individuals. Protein misfolding cyclic amplification (PMCA) and real time quaking induced conversion (RT-QuIC) can detect a single infectious prion particle.[265,266] Both *in vitro* methods can detect PrP^{Sc} in urine, feces, blood, saliva, and cerebrospinal fluid (CSF).[267–272] PMCA detects PrP^{Sc} from vCJD patient blood and urine samples, while RT-QuIC is currently being used for the diagnosis of sCJD from CSF and nasal mucosa swab samples.[267,269–271,273] Advancements in *in vitro* techniques have enabled reliable preclinical detection in experimental prion disease. PMCA can detect prions in the blood or skin from pre-clinical prion-infected mice and hamsters.[274–276] The sensitivity of RT-QuIC can detect prions in the blood of prion-infected preclinical hamsters and deer at timepoints as early as 5 days and 1 month post inoculation, respectively.[277,278] Like PrP^{Sc}, Aβ, tau, and α-synuclein are present in biological fluids making these *in vitro* techniques promising for preclinical diagnosis.[279–283] Overall, preclinical diagnosis of prion diseases will provide a larger window for therapeutic treatments and consequently a greater likelihood of success. Challenging this is the presence of multiple prion strains. Strain-specific efficacy of anti-prion drugs or vaccines can result in emergence of resistant strains that may be more pathogenic or have a different host range compared to the original dominant strain. These challenges are reminiscent of those faced by bacterial, fungal, and viral drug development and a successful anti-prion therapy may require a multidrug approach. The unique nature of prions, however, provides an additional therapeutic target, namely the precursor protein. Treatments that target the prion precursor protein can inhibit prion formation and may reverse pathological changes. A combinational approach that treats the prion form and the precursor protein will likely be needed to effectively treat the broad spectrum of prion diseases.

References

1. Prusiner SB. Novel proteinaceous infectious particles cause scrapie. *Science*. 1982;216 (4542):136–144.
2. Horiuchi M, Caughey B. Specific binding of normal prion protein to the scrapie form via a localized domain initiates its conversion to the protease-resistant state. *EMBO J*. 1999;18(12):3193–3203.
3. Bessen RA, Kocisko DA, Raymond GJ, Nandan S, Lansbury PT, Caughey B. Non-genetic propagation of strain-specific properties of scrapie prion protein. *Nature*. 1995;375(6533):698–700.
4. Kocisko DA, Come JH, Priola SA, et al. Cell-free formation of protease-resistant prion protein. *Nature*. 1994;370(6489):471–474.
5. Du Z, Park KW, Yu H, Fan Q, Li L. Newly identified prion linked to the chromatin-remodeling factor Swi1 in Saccharomyces cerevisiae. *Nat Genet*. 2008;40(4):460–465.
6. Jarosz DF, Lancaster AK, Brown JCS, Lindquist S. An evolutionarily conserved prion-like element converts wild fungi from metabolic specialists to generalists. *Cell*. 2014;158(5):1072–1082.
7. Masison DC, Wickner RB. Prion-inducing domain of yeast Ure2p and protease resistance of Ure2p in prion-containing cells. *Science*. 1995;270(5233):93–95.
8. Wickner RB. [URE3] as an altered URE2 protein: evidence for a prion analog in Saccharomyces cerevisiae. *Science*. 1994;264(5158):566–569.
9. Uptain SM, Sawicki GJ, Caughey B, Lindquist S. Strains of [PSI(+)] are distinguished by their efficiencies of prion-mediated conformational conversion. *EMBO J*. 2001;20(22):6236–6245.
10. Chernoff Y, Lindquist S, Ono B, Inge-Vechtomov S, Liebman S. Role of the chaperone protein Hsp104 in propagation of the yeast prion-like factor [psi+]. *Science*. 1995;268(5212):880–884.

11. Wickner RB, Bezsonov E, Bateman DA. Normal levels of the antiprion proteins Btn2 and Cur1 cure most newly formed [URE3] prion variants. *Proc Natl Acad Sci U S A*. 2014;111(26):E2711–E2720.
12. Coustou V, Deleu C, Saupe S, Begueret J. The protein product of the het-s heterokaryon incompatibility gene of the fungus Podospora anserina behaves as a prion analog. *Proc Natl Acad Sci U S A*. 1997;94(18):9773–9778.
13. Maddelein ML, Dos Reis S, Duvezin-Caubet S, Coulary-Salin B, Saupe SJ. Amyloid aggregates of the HET-s prion protein are infectious. *Proc Natl Acad Sci U S A*. 2002;99(11):7402–7407.
14. Saupe SJ. The [Het-s] prion of Podospora anserina and its role in heterokaryon incompatibility. *Semin Cell Dev Biol*. 2011;22(5):460–468.
15. Fioriti L, Myers C, Huang YY, et al. The persistence of hippocampal-based memory requires protein synthesis mediated by the prion-like protein CPEB3. *Neuron*. 2015;86(6):1433–1448.
16. Gilks N, Kedersha N, Ayodele M, et al. Stress granule assembly is mediated by prion-like aggregation of TIA-1. *Mol Biol Cell*. 2004;15(12):5383–5398.
17. Hou F, Sun L, Zheng H, Skaug B, Jiang QX, Chen ZJ. MAVS forms functional prion-like aggregates to activate and propagate antiviral innate immune response. *Cell*. 2011;146(3):448–461.
18. Li X, Rayman JB, Kandel ER, Derkatch IL. Functional role of Tia1/Pub1 and Sup35 prion domains: directing protein synthesis machinery to the tubulin cytoskeleton. *Mol Cell*. 2014;55(2):305–318.
19. Stephan JS, Fioriti L, Lamba N, et al. The CPEB3 protein is a functional prion that interacts with the actin cytoskeleton. *Cell Rep*. 2015;11(11):1772–1785.
20. Bolton DC, Mckinley MP, Prusiner SB. Identification of a protein that purifies with the scrapie prion. *Science*. 1982;218(4579):1309–1311.
21. Baeten LA, Powers BE, Jewell JE, Spraker TR, Miller MW. A natural case of chronic wasting disease in a free-ranging moose (Alces alces shirasi). *J Wildl Dis*. 2007; 43(2):309–314.
22. Benestad SL, Mitchell G, Simmons M, Ytrehus B, Vikoren T. First case of chronic wasting disease in Europe in a Norwegian free-ranging reindeer. *Vet Res*. 2016; 47(1):88.
23. Williams ES, Young S. Chronic wasting disease of captive mule deer: a spongiform encephalopathy. *J Wildl Dis*. 1980;16(1):89–98.
24. Babelhadj B, Di Bari MA, Pirisinu L, et al. Prion disease in dromedary camels Algeria. *Emerg Infect Dis*. 2018;24(6):1029–1036.
25. Imran M, Mahmood S. An overview of human prion diseases. *Virol J*. 2011;8:559.
26. Duffy P, Wolf J, Collins G, DeVoe AG, Streeten B, Cowen D. Letter: possible person-to-person transmission of Creutzfeldt-Jakob disease. *N Engl J Med*. 1974;290(12):692–693.
27. Heckmann JG, Lang CJ, Petruch F, et al. Transmission of Creutzfeldt-Jakob disease via a corneal transplant. *J Neurol Neurosurg Psychiatry*. 1997;63(3):388–390.
28. Koch TK, Berg BO, De Armond SJ, Gravina RF. Creutzfeldt-Jakob disease in a young adult with idiopathic hypopituitarism. Possible relation to the administration of cadaveric human growth hormone. *N Engl J Med*. 1985;313(12):731–733.
29. Thadani V, Penar PL, Partington J, et al. Creutzfeldt-Jakob disease probably acquired from a cadaveric dura mater graft. Case report. *J Neurosurg*. 1988;69(5):766–769.
30. Frost B, Jacks RL, Diamond MI. Propagation of tau misfolding from the outside to the inside of a cell. *J Biol Chem*. 2009;284(19):12845–12852.
31. Kfoury N, Holmes BB, Jiang H, Holtzman DM, Diamond MI. Trans-cellular propagation of Tau aggregation by fibrillar species. *J Biol Chem*. 2012;287(23):19440–19451.
32. Langer F, Eisele YS, Fritschi SK, Staufenbiel M, Walker LC, Jucker M. Soluble Aβ seeds are potent inducers of cerebral β-amyloid deposition. *J Neurosci*. 2011;31:14488–14495.

33. Eisele YS, Obermuller U, Heilbronner G, et al. Peripherally applied Abeta-containing inoculates induce cerebral beta-amyloidosis. *Science*. 2010;330(6006):980–982.
34. Clavaguera F, Bolmont T, Crowther RA, et al. Transmission and spreading of tauopathy in transgenic mouse brain. *Nat Cell Biol*. 2009;11(7):909–913.
35. Desplats P, Lee HJ, Bae EJ, et al. Inclusion formation and neuronal cell death through neuron-to-neuron transmission of alpha-synuclein. *Proc Natl Acad Sci U S A*. 2009; 106(31):13010–13015.
36. Fehlinger A, Wolf H, Hossinger A, et al. Prion strains depend on different endocytic routes for productive infection. *Sci Rep*. 2017;7(1):6923.
37. Freundt EC, Maynard N, Clancy EK, et al. Neuron-to-neuron transmission of alpha-synuclein fibrils through axonal transport. *Ann Neurol*. 2012;72(4):517–524.
38. Guo JL, Narasimhan S, Changolkar L, et al. Unique pathological tau conformers from Alzheimer's brains transmit tau pathology in nontransgenic mice. *J Exp Med*. 2016;213(12):2635–2654.
39. Holmes BB, DeVos SL, Kfoury N, et al. Heparan sulfate proteoglycans mediate internalization and propagation of specific proteopathic seeds. *Proc Natl Acad Sci U S A*. 2013;110(33):E3138–E3147.
40. Chandler RL. Encephalopathy in mice produced by inoculation with scrapie brain material. *Lancet*. 1961;1(7191):1378–1379.
41. Bolmont T, Clavaguera F, Meyer-Luehmann M, et al. Induction of tau pathology by intracerebral infusion of amyloid-beta -containing brain extract and by amyloid-beta deposition in APP x Tau transgenic mice. *Am J Pathol*. 2007;171(6):2012–2020.
42. Luk KC, Kehm VM, Zhang B, O'Brien P, Trojanowski JQ, Lee VM. Intracerebral inoculation of pathological alpha-synuclein initiates a rapidly progressive neurodegenerative alpha-synucleinopathy in mice. *J Exp Med*. 2012;209(5):975–986.
43. Watts JC, Giles K, Oehler A, et al. Transmission of multiple system atrophy prions to transgenic mice. *Proc Natl Acad Sci U S A*. 2013;110(48):19555–19560.
44. Watts JC, Condello C, Stöhr J, et al. Serial propagation of distinct strains of Aβ prions from Alzheimer's disease patients. *Proc Natl Acad Sci U S A*. 2014;111: 10323–10328.
45. He Z, Guo JL, McBride JD, et al. Amyloid-β plaques enhance Alzheimer's brain tau-seeded pathologies by facilitating neuritic plaque tau aggregation. *Nat Med*. 2018;24:29–38.
46. Woerman AL, Oehler A, Kazmi SA, et al. Multiple system atrophy prions retain strain specificity after serial propagation in two different Tg(SNCA*A53T) mouse lines. *Acta Neuropathol*. 2019;137(3):437–454.
47. O'Meara ES, Kukull WA, Schellenberg GD, et al. Alzheimer's disease and history of blood transfusion by apolipoprotein-E genotype. *Neuroepidemiology*. 1997;16(2):86–93.
48. Bohnen NI, Warner MA, Kokmen E, Beard CM, Kurland LT. Prior blood transfusions and Alzheimer's disease. *Neurology*. 1994;44(6):1159.
49. Brown P, Gibbs Jr CJ, Rodgers-Johnson P, et al. Human spongiform encephalopathy: the National Institutes of Health series of 300 cases of experimentally transmitted disease. *Ann Neurol*. 1994;35(5):513–529.
50. Irwin DJ, Abrams JY, Schonberger LB, et al. Evaluation of potential infectivity of Alzheimer and Parkinson disease proteins in recipients of cadaver-derived human growth hormone. *JAMA Neurol*. 2013;70(4):462–468.
51. Jaunmuktane Z, Mead S, Ellis M, et al. Evidence for human transmission of amyloid-beta pathology and cerebral amyloid angiopathy. *Nature*. 2015;525(7568):247–250.
52. Purro SA, Farrow MA, Linehan J, et al. Transmission of amyloid-β protein pathology from cadaveric pituitary growth hormone. *Nature*. 2018;564:415–419.
53. Bartz JC. Prion strain diversity. *Cold Spring Harb Perspect Med*. 2016;6(12).
54. Chien P, Weissman JS. Conformational diversity in a yeast prion dictates its seeding specificity. *Nature*. 2001;410(6825):223–227.

55. Verges KJ, Smith MH, Toyama BH, Weissman JS. Strain conformation, primary structure and the propagation of the yeast prion [PSI+]. *Nat Struct Mol Biol.* 2011;18(4):493–499.
56. Fraser H, Dickinson AG. Distribution of experimentally induced scrapie lesions in the brain. *Nature.* 1967;216(5122):1310–1311.
57. Brachmann A, Baxa U, Wickner RB. Prion generation in vitro: amyloid of Ure2p is infectious. *EMBO J.* 2005;24(17):3082–3092.
58. Derkatch IL, Chernoff YO, Kushnirov VV, Inge-Vechtomov SG, Liebman SW. Genesis and variability of [PSI] prion factors in Saccharomyces cerevisiae. *Genetics.* 1996;144(4):1375–1386.
59. Schlumpberger M, Prusiner SB, Herskowitz I. Induction of distinct [URE3] yeast prion strains. *Mol Cell Biol.* 2001;21(20):7035–7046.
60. Dickinson AG, Meikle VM, Fraser H. Identification of a gene which controls the incubation period of some strains of scrapie agent in mice. *J Comp Pathol.* 1968;78(3):293–299.
61. Fraser H, Dickinson AG. The sequential development of the brain lesions of scrapie in three strains of mice. *J Comp Pathol.* 1968;78:301–311.
62. Bousset L, Pieri L, Ruiz-Arlandis G, et al. Structural and functional characterization of two alpha-synuclein strains. *Nat Commun.* 2013;4:2575.
63. Guo JL, Covell DJ, Daniels JP, et al. Distinct α-synuclein strains differentially promote tau inclusions in neurons. *Cell.* 2013;154:103–117.
64. Narasimhan S, Guo JL, Changolkar L, et al. Pathological tau strains from human brains recapitulate the diversity of tauopathies in nontransgenic mouse brain. *J Neurosci.* 2017;37(47):11406–11423.
65. Tanaka M, Chien P, Naber N, Cooke R, Weissman JS. Conformational variations in an infectious protein determine prion strain differences. *Nature.* 2004;428(6980):323–328.
66. McGlinchey RP, Kryndushkin D, Wickner RB. Suicidal [PSI+] is a lethal yeast prion. *Proc Natl Acad Sci U S A.* 2011;108(13):5337–5341.
67. Edskes HK, McCann LM, Hebert AM, Wickner RB. Prion variants and species barriers among Saccharomyces Ure2 proteins. *Genetics.* 2009;181(3):1159–1167.
68. Ohhashi Y, Yamaguchi Y, Kurahashi H, et al. Molecular basis for diversification of yeast prion strain conformation. *Proc Natl Acad Sci U S A.* 2018;115(10):2389–2394.
69. Tanaka M, Collins SR, Toyama BH, Weissman JS. The physical basis of how prion conformations determine strain phenotypes. *Nature.* 2006;442(7102):585–589.
70. Bessen RA, Marsh RF. Distinct PrP properties suggest the molecular basis of strain variation in transmissible mink encephalopathy. *J Virol.* 1994;68(12):7859–7868.
71. Safar J, Wille H, Itri V, et al. Eight prion strains have PrP(Sc) molecules with different conformations. *Nat Med.* 1998;4(10):1157–1165.
72. Telling GC, Parchi P, DeArmond SJ, et al. Evidence for the conformation of the pathologic isoform of the prion protein enciphering and propagating prion diversity. *Science.* 1996;274(5295):2079–2082.
73. Kascsak RJ, Rubenstein R, Merz PA, Carp RI, Wisniewski HM, Diringer H. Biochemical differences among scrapie-associated fibrils support the biological diversity of scrapie agents. *J Gen Virol.* 1985;66(pt 8):1715–1722.
74. Bessen RA, Marsh RF. Identification of two biologically distinct strains of transmissible mink encephalopathy in hamsters. *J Gen Virol.* 1992;73(pt 2):329–334.
75. Bessen RA, Marsh RF. Biochemical and physical properties of the prion protein from two strains of the transmissible mink encephalopathy agent. *J Virol.* 1992;66(4):2096–2101.
76. Marsh RF, Bessen RA. Physicochemical and biological characterizations of distinct strains of the transmissible mink encephalopathy agent. *Philos Trans R Soc Lond B Biol Sci.* 1994;343(1306):413–414.

77. Deleault NR, Harris BT, Rees JR, Supattapone S. Formation of native prions from minimal components in vitro. *Proc Natl Acad Sci U S A*. 2007;104(23):9741–9746.
78. Wang F, Wang X, Yuan CG, Ma J. Generating a prion with bacterially expressed recombinant prion protein. *Science*. 2010;327(5969):1132–1135.
79. Deleault NR, Walsh DJ, Piro JR, et al. Cofactor molecules maintain infectious conformation and restrict strain properties in purified prions. *Proc Natl Acad Sci U S A*. 2012;109(28):E1938–E1946.
80. Burke CM, Walsh DJ, Steele AD, et al. Full restoration of specific infectivity and strain properties from pure mammalian prion protein. *PLoS Pathog*. 2019;15(3):e1007662.
81. Mahal SP, Baker CA, Demczyk CA, Smith EW, Julius C, Weissmann C. Prion strain discrimination in cell culture: the cell panel assay. *Proc Natl Acad Sci U S A*. 2007;104(52):20908–20913.
82. Oelschlegel AM, Fallahi M, Ortiz-Umpierre S, Weissmann C. The extended cell panel assay characterizes the relationship of prion strains RML, 79A, and 139A and reveals conversion of 139A to 79A-like prions in cell culture. *J Virol*. 2012;86(9):5297–5303.
83. Karapetyan YE, Saa P, Mahal SP, et al. Prion strain discrimination based on rapid in vivo amplification and analysis by the cell panel assay. *PLoS One*. 2009;4(5):e5730.
84. Peretz D, Scott MR, Groth D, et al. Strain-specified relative conformational stability of the scrapie prion protein. *Protein Sci*. 2001;10(4):854–863.
85. Cescatti M, Saverioni D, Capellari S, et al. Analysis of conformational stability of abnormal prion protein aggregates across the spectrum of Creutzfeldt-Jakob disease prions. *J Virol*. 2016;90(14):6244–6254.
86. Ayers JI, Schutt CR, Shikiya RA, Aguzzi A, Kincaid AE, Bartz JC. The strain-encoded relationship between PrP replication, stability and processing in neurons is predictive of the incubation period of disease. *PLoS Pathog*. 2011;7(3):e1001317.
87. Legname G, Nguyen HO, Peretz D, Cohen FE, DeArmond SJ, Prusiner SB. Continuum of prion protein structures enciphers a multitude of prion isolate-specified phenotypes. *Proc Natl Acad Sci U S A*. 2006;103(50):19105–19110.
88. Condello C, Lemmin T, Stöhr J, et al. Structural heterogeneity and intersubject variability of Aβ in familial and sporadic Alzheimer's disease. *Proc Natl Acad Sci U S A*. 2018;115:E782–E791.
89. Nilsberth C, Westlind-Danielsson A, Eckman CB, et al. The 'Arctic' APP mutation (E693G) causes Alzheimer's disease by enhanced Aβ protofibril formation. *Nat Neurosci*. 2001;4:887–893.
90. Lannfelt L, Bogdanovic N, Appelgren H, et al. Amyloid precursor protein mutation causes Alzheimer's disease in a Swedish family. *Neurosci Lett*. 1994;168(1–2):254–256.
91. Mullan M, Crawford F, Axelman K, et al. A pathogenic mutation for probable Alzheimer's disease in the APP gene at the N-terminus of beta-amyloid. *Nat Genet*. 1992;1(5):345–347.
92. Philipson O, Lord A, Lalowski M, et al. The Arctic *amyloid-β precursor protein* (*AβPP*) mutation results in distinct plaques and accumulation of N- and C-truncated Aβ. *Neurobiol Aging*. 2012;33:1010.e1011–1010.e1013.
93. Rasmussen J, Mahler J, Beschorner N, et al. Amyloid polymorphisms constitute distinct clouds of conformational variants in different etiological subtypes of Alzheimer's disease. *Proc Natl Acad Sci U S A*. 2017;114(49):13018–13023.
94. Qiang W, Yau WM, Lu JX, Collinge J, Tycko R. Structural variation in amyloid-β fibrils from Alzheimer's disease clinical subtypes. *Nature*. 2017;541:217–221.
95. Lee VM, Goedert M, Trojanowski JQ. Neurodegenerative tauopathies. *Annu Rev Neurosci*. 2001;24:1121–1159.
96. Clavaguera F, Akatsu H, Fraser G, et al. Brain homogenates from human Tauopathies induce tau inclusions in mouse brain. *Proc Natl Acad Sci U S A*. 2013;110(23):9535–9540.

97. Boluda S, Iba M, Zhang B, Raible KM, Lee VM, Trojanowski JQ. Differential induction and spread of tau pathology in young PS19 tau transgenic mice following intracerebral injections of pathological tau from Alzheimer's disease or corticobasal degeneration brains. *Acta Neuropathol.* 2015;129(2):221–237.
98. Gibbons GS, Banks RA, Kim B, et al. GFP-mutant human tau transgenic mice develop Tauopathy following CNS injections of Alzheimer's brain-derived pathological tau or synthetic mutant human tau fibrils. *J Neurosci.* 2017;37(47):11485–11494.
99. Sanders DW, Kaufman SK, DeVos SL, et al. Distinct tau prion strains propagate in cells and mice and define different tauopathies. *Neuron.* 2014;82(6):1271–1288.
100. Kaufman SK, Sanders DW, Thomas TL, et al. Tau prion strains dictate patterns of cell pathology, progression rate, and regional vulnerability in vivo. *Neuron.* 2016;92(4): 796–812.
101. Peelaerts W, Bousset L, Van der Perren A, et al. α-Synuclein strains cause distinct synucleinopathies after local and systemic administration. *Nature.* 2015;522:340–344.
102. Peelaerts W, Baekelandt V. α-Synuclein strains and the variable pathologies of synucleinopathies. *J Neurochem.* 2016;139(suppl 1):256–274.
103. Parchi P, Castellani R, Capellari S, et al. Molecular basis of phenotypic variability in sporadic Creutzfeldt-Jakob disease. *Ann Neurol.* 1996;39:767–778.
104. Benestad SL, Sarradin P, Thu B, Schonheit J, Tranulis MA, Bratberg B. Cases of scrapie with unusual features in Norway and designation of a new type, Nor98. *Vet Rec.* 2003;153(7):202–208.
105. Biacabe AG, Laplanche JL, Ryder S, Baron T. Distinct molecular phenotypes in bovine prion diseases. *EMBO Rep.* 2004;5(1):110–115.
106. Woerman AL, Stohr J, Aoyagi A, et al. Propagation of prions causing synucleinopathies in cultured cells. *Proc Natl Acad Sci U S A.* 2015;112(35):E4949–E4958.
107. Yamasaki TR, Holmes BB, Furman JL, et al. Parkinson's disease and multiple system atrophy have distinct alpha-synuclein seed characteristics. *J Biol Chem.* 2019;294(3): 1045–1058.
108. Cali I, Castellani R, Alshekhlee A, et al. Co-existence of scrapie prion protein types 1 and 2 in sporadic Creutzfeldt-Jakob disease: Its effect on the phenotype and prion-type characteristics. *Brain.* 2009;132(pt 10):2643–2658.
109. Uro-Coste E, Cassard H, Simon S, et al. Beyond PrP^{res} type 1/type 2 dichotomy in Creutzfeldt-Jakob disease. *PLoS Pathog.* 2008;4:e1000029.
110. Parchi P, Giese A, Capellari S, et al. Classification of sporadic Creutzfeldt-Jakob disease based on molecular and phenotypic analysis of 300 subjects. *Ann Neurol.* 1999;46(2): 224–233.
111. Puoti G, Giaccone G, Rossi G, Canciani B, Bugiani O, Tagliavini F. Sporadic Creutzfeldt-Jakob disease: co-occurrence of different types of PrP(Sc) in the same brain. *Neurology.* 1999;53(9):2173–2176.
112. Kovács GG, Head MW, Hegyi I, et al. Immunohistochemistry for the prion protein: comparison of different monoclonal antibodies in human prion disease subtypes. *Brain Pathol.* 2006;12(1):1–11.
113. Haik S, Peoc'h K, Brandel JP, et al. Striking PrPsc heterogeneity in inherited prion diseases with the D178N mutation. *Ann Neurol.* 2004;56(6):909–910. [author reply 910-901].
114. Head MW, Bunn TJ, Bishop MT, et al. Prion protein heterogeneity in sporadic but not variant Creutzfeldt-Jakob disease: UK cases 1991-2002. *Ann Neurol.* 2004;55(6): 851–859.
115. Lewis V, Hill AF, Klug GM, Boyd A, Masters CL, Collins SJ. Australian sporadic CJD analysis supports endogenous determinants of molecular-clinical profiles. *Neurology.* 2005;65(1):113–118.

116. Schoch G, Seeger H, Bogousslavsky J, et al. Analysis of prion strains by PrPSc profiling in sporadic Creutzfeldt-Jakob disease. *PLoS Med.* 2006;3(2):e14.
117. Milenkovic I, Petrov T, Kovacs GG. Patterns of hippocampal tau pathology differentiate neurodegenerative dementias. *Dement Geriatr Cogn Disord.* 2014;38(5–6):375–388.
118. Gibbons GS, Banks RA, Kim B, et al. Detection of Alzheimer disease (AD)-specific tau pathology in AD and NonAD tauopathies by immunohistochemistry with novel conformation-selective tau antibodies. *J Neuropathol Exp Neurol.* 2018;77(3):216–228.
119. Bartz JC, Bessen RA, McKenzie D, Marsh RF, Aiken JM. Adaptation and selection of prion protein strain conformations following interspecies transmission of transmissible mink encephalopathy. *J Virol.* 2000;74(12):5542–5547.
120. Dickinson AG, Outram GW. The scrapie replication-site hypothesis and its implications for pathogenesis. In: Hadlow WJ, Prusiner SB, eds. *Slow transmissible diseases of the nervous system.* New York: Academic Press, Inc; 1979:13–31. vol 2.
121. Kimberlin RH, Walker CA. Competition between strains of scrapie depends on the blocking agent being infectious. *Intervirology.* 1985;23(2):74–81.
122. Taylor DM, Dickinson AG, Fraser H, Marsh RF. Evidence that transmissible mink encephalopathy agent is biologically inactive in mice. *Neuropathol Appl Neurobiol.* 1986;12(2):207–215.
123. Bartz JC, Kramer ML, Sheehan MH, et al. Prion interference is due to a reduction in strain-specific PrPSc levels. *J Virol.* 2007;81(2):689–697.
124. Shikiya RA, Ayers JI, Schutt CR, Kincaid AE, Bartz JC. Coinfecting prion strains compete for a limiting cellular resource. *J Virol.* 2010;84(11):5706–5714.
125. Eckland TE, Shikiya RA, Bartz JC. Independent amplification of co-infected long incubation period low conversion efficiency prion strains. *PLoS Pathog.* 2018;14(10):e1007323.
126. Bradley ME, Edskes HK, Hong JY, Wickner RB, Liebman SW. Interactions among prions and prion "strains" in yeast. *Proc Natl Acad Sci U S A.* 2002;99(suppl 4):16392–16399.
127. Caughey B, Ernst D, Race RE. Congo red inhibition of scrapie agent replication. *J Virol.* 1993;67(10):6270–6272.
128. Caspi S, Halimi M, Yanai A, Sasson SB, Taraboulos A, Gabizon R. The anti-prion activity of Congo red. Putative mechanism. *J Biol Chem.* 1998;273(6):3484–3489.
129. Cronier S, Beringue V, Bellon A, Peyrin JM, Laude H. Prion strain- and species-dependent effects of antiprion molecules in primary neuronal cultures. *J Virol.* 2007;81(24):13794–13800.
130. Ingrosso L, Ladogana A, Pocchiari M. Congo red prolongs the incubation period in scrapie-infected hamsters. *J Virol.* 1995;69(1):506–508.
131. Milhavet O, Mange A, Casanova D, Lehmann S. Effect of Congo red on wild-type and mutated prion proteins in cultured cells. *J Neurochem.* 2000;74(1):222–230.
132. Herrmann US, Schutz AK, Shirani H, et al. Structure-based drug design identifies polythiophenes as antiprion compounds. *Sci Transl Med.* 2015;7(299):299ra123.
133. Margalith I, Suter C, Ballmer B, et al. Polythiophenes inhibit prion propagation by stabilizing prion protein (PrP) aggregates. *J Biol Chem.* 2012;287(23):18872–18887.
134. Kawasaki Y, Kawagoe K, Chen CJ, Teruya K, Sakasegawa Y, Doh-ura K. Orally administered amyloidophilic compound is effective in prolonging the incubation periods of animals cerebrally infected with prion diseases in a prion strain-dependent manner. *J Virol.* 2007;81(23):12889–12898.
135. Lu D, Giles K, Li Z, et al. Biaryl amides and hydrazones as therapeutics for prion disease in transgenic mice. *J Pharmacol Exp Ther.* 2013;347(2):325–338.
136. Matthes D, Gapsys V, Griesinger C, de Groot BL. Resolving the atomistic modes of Anle138b inhibitory action on peptide oligomer formation. *ACS Chem Nerosci.* 2017;8(12):2791–2808.

137. Wagner J, Ryazanov S, Leonov A, et al. Anle138b: a novel oligomer modulator for disease-modifying therapy of neurodegenerative diseases such as prion and Parkinson's disease. *Acta Neuropathol.* 2013;125(6):795–813.
138. Giles K, Berry DB, Condello C, et al. Different 2-aminothiazole therapeutics produce distinct patterns of Scrapie prion neuropathology in mouse brains. *J Pharmacol Exp Ther.* 2015;355(1):2–12.
139. Yamasaki T, Suzuki A, Hasebe R, Horiuchi M. Comparison of the anti-prion mechanism of four different anti-prion compounds, anti-PrP monoclonal antibody 44B1, pentosan polysulfate, chlorpromazine, and U18666A, in prion-infected mouse neuroblastoma cells. *PLoS One.* 2014;9(9):e106516.
140. Doh-ura K, Ishikawa K, Murakami-Kubo I, et al. Treatment of transmissible spongiform encephalopathy by intraventricular drug infusion in animal models. *J Virol.* 2004;78(10):4999–5006.
141. Bone I, Belton L, Walker AS, Darbyshire J. Intraventricular pentosan polysulphate in human prion diseases: an observational study in the UK. *Eur J Neurol.* 2008;15(5): 458–464.
142. Newman PK, Todd NV, Scoones D, et al. Postmortem findings in a case of variant Creutzfeldt-Jakob disease treated with intraventricular pentosan polysulfate. *J Neurol Neurosurg Psychiatry.* 2014;85(8):921–924.
143. Parry A, Baker I, Stacey R, Wimalaratna S. Long term survival in a patient with variant Creutzfeldt-Jakob disease treated with intraventricular pentosan polysulphate. *J Neurol Neurosurg Psychiatry.* 2007;78(7):733–734.
144. Honda H, Sasaki K, Minaki H, et al. Protease-resistant PrP and PrP oligomers in the brain in human prion diseases after intraventricular pentosan polysulfate infusion. *Neuropathology.* 2012;32(2):124–132.
145. Terada T, Tsuboi Y, Obi T, et al. Less protease-resistant PrP in a patient with sporadic CJD treated with intraventricular pentosan polysulphate. *Acta Neurol Scand.* 2010;121(2):127–130.
146. Tsuboi Y, Doh-Ura K, Yamada T. Continuous intraventricular infusion of pentosan polysulfate: clinical trial against prion diseases. *Neuropathology.* 2009;29(5):632–636.
147. Larramendy-Gozalo C, Barret A, Daudigeos E, et al. Comparison of CR36, a new heparan mimetic, and pentosan polysulfate in the treatment of prion diseases. *J Gen Virol.* 2007;88(pt 3):1062–1067.
148. Horiuchi M, Priola SA, Chabry J, Caughey B. Interactions between heterologous forms of prion protein: binding, inhibition of conversion, and species barriers. *Proc Natl Acad Sci U S A.* 2000;97(11):5836–5841.
149. Priola SA, Caughey B, Race RE, Chesebro B. Heterologous PrP molecules interfere with accumulation of protease-resistant PrP in scrapie-infected murine neuroblastoma cells. *J Virol.* 1994;68(8):4873–4878.
150. Skinner PJ, Kim HO, Bryant D, et al. Treatment of prion disease with heterologous prion proteins. *PLoS One.* 2015;10(7):e0131993.
151. Vorberg I, Groschup MH, Pfaff E, Priola SA. Multiple amino acid residues within the rabbit prion protein inhibit formation of its abnormal isoform. *J Virol.* 2003;77(3):2003–2009.
152. Berry DB, Lu D, Geva M, et al. Drug resistance confounding prion therapeutics. *Proc Natl Acad Sci U S A.* 2013;110(44):E4160–E4169.
153. Berry D, Giles K, Oehler A, Bhardwaj S, DeArmond SJ, Prusiner SB. Use of a 2-aminothiazole to treat chronic wasting disease in transgenic mice. *J Infect Dis.* 2015;212(suppl 1):S17–S25.
154. Gallardo-Godoy A, Gever J, Fife KL, Silber BM, Prusiner SB, Renslo AR. 2-Aminothiazoles as therapeutic leads for prion diseases. *J Med Chem.* 2011;54(4): 1010–1021.

155. Ghaemmaghami S, May BC, Renslo AR, Prusiner SB. Discovery of 2-aminothiazoles as potent antiprion compounds. *J Virol*. 2010;84(7):3408–3412.
156. Silber BM, Rao S, Fife KL, et al. Pharmacokinetics and metabolism of 2-aminothiazoles with antiprion activity in mice. *Pharm Res*. 2013;30(4):932–950.
157. McKenzie D, Kaczkowski J, Marsh R, Aiken J. Amphotericin B delays both scrapie agent replication and PrP-res accumulation early in infection. *J Virol*. 1994;68(11): 7534–7536.
158. Hartsel SC, Weiland TR. Amphotericin B binds to amyloid fibrils and delays their formation: a therapeutic mechanism? *Biochemistry*. 2003;42(20):6228–6233.
159. Mange A, Milhavet O, McMahon HE, Casanova D, Lehmann S. Effect of amphotericin B on wild-type and mutated prion proteins in cultured cells: putative mechanism of action in transmissible spongiform encephalopathies. *J Neurochem*. 2000;74(2):754–762.
160. Masullo C, Macchi G, Xi YG, Pocchiari M. Failure to ameliorate Creutzfeldt-Jakob disease with amphotericin B therapy. *J Infect Dis*. 1992;165(4):784–785.
161. Motoyoshi-Yamashiro A, Takano K, Kawabe K, et al. Amphotericin B induces glial cell line-derived neurotrophic factor in the rat brain. *J Vet Med Sci*. 2014;76(10): 1353–1358.
162. Motoyoshi A, Nakajima H, Takano K, Moriyama M, Kannan Y, Nakamura Y. Effects of Amphotericin B on the expression of neurotoxic and neurotrophic factors in cultured microglia. *Neurochem Int*. 2008;52(6):1290–1296.
163. Pocchiari M, Schmittinger S, Masullo C. Amphotericin B delays the incubation period of scrapie in intracerebrally inoculated hamsters. *J Gen Virol*. 1987;68(pt 1):219–223.
164. Xi YG, Ingrosso L, Ladogana A, Masullo C, Pocchiari M. Amphotericin B treatment dissociates in vivo replication of the scrapie agent from PrP accumulation. *Nature*. 1992;356(6370):598–601.
165. Nazor Friberg K, Hung G, Wancewicz E, et al. Intracerebral infusion of antisense oligonucleotides into prion-infected mice. *Mol Ther Nucleic Acids*. 2012;1:e9.
166. Raymond GJ, Zhao HT, Race B, et al. Antisense oligonucleotides extend survival of prion-infected mice. *JCI Insight*. 2019;5.
167. Ferreira NC, Ascari LM, Hughson AG, et al. A promising antiprion trimethoxychalcone binds to the globular domain of the cellular prion protein and changes its cellular location. *Antimicrob Agents Chemother*. 2018;62(2):e01441-17.
168. Ferreira NC, Marques IA, Conceicao WA, et al. Anti-prion activity of a panel of aromatic chemical compounds: in vitro and in silico approaches. *PLoS One*. 2014; 9(1):e84531.
169. Massignan T, Cimini S, Stincardini C, et al. A cationic tetrapyrrole inhibits toxic activities of the cellular prion protein. *Sci Rep*. 2016;6:23180.
170. Nicoll AJ, Trevitt CR, Tattum MH, et al. Pharmacological chaperone for the structured domain of human prion protein. *Proc Natl Acad Sci U S A*. 2010;107(41): 17610–17615.
171. Peretz D, Williamson RA, Kaneko K, et al. Antibodies inhibit prion propagation and clear cell cultures of prion infectivity. *Nature*. 2001;412(6848):739–743.
172. Enari M, Flechsig E, Weissmann C. Scrapie prion protein accumulation by scrapie-infected neuroblastoma cells abrogated by exposure to a prion protein antibody. *Proc Natl Acad Sci U S A*. 2001;98(16):9295–9299.
173. Heppner FL, Musahl C, Arrighi I, et al. Prevention of scrapie pathogenesis by transgenic expression of anti-prion protein antibodies. *Science*. 2001;294(5540):178–182.
174. Antonyuk SV, Trevitt CR, Strange RW, et al. Crystal structure of human prion protein bound to a therapeutic antibody. *Proc Natl Acad Sci U S A*. 2009;106(8):2554–2558.
175. White AR, Enever P, Tayebi M, et al. Monoclonal antibodies inhibit prion replication and delay the development of prion disease. *Nature*. 2003;422(6927):80–83.

176. Kim CL, Karino A, Ishiguro N, Shinagawa M, Sato M, Horiuchi M. Cell-surface retention of PrPC by anti-PrP antibody prevents protease-resistant PrP formation. *J Gen Virol.* 2004;85(pt 11):3473–3482.
177. Pankiewicz J, Prelli F, Sy MS, et al. Clearance and prevention of prion infection in cell culture by anti-PrP antibodies. *Eur J Neurosci.* 2006;23(10):2635–2647.
178. Perrier V, Solassol J, Crozet C, et al. Anti-PrP antibodies block PrPSc replication in prion-infected cell cultures by accelerating PrPC degradation. *J Neurochem.* 2004;89(2):454–463.
179. David MA, Jones DR, Tayebi M. Potential candidate camelid antibodies for the treatment of protein-misfolding diseases. *J Neuroimmunol.* 2014;272(1–2):76–85.
180. Jones DR, Taylor WA, Bate C, David M, Tayebi M. A camelid anti-PrP antibody abrogates PrP replication in prion-permissive neuroblastoma cell lines. *PLoS One.* 2010;5(3):e9804.
181. Herrmann US, Sonati T, Falsig J, et al. Prion infections and anti-PrP antibodies trigger converging neurotoxic pathways. *PLoS Pathog.* 2015;11(2):e1004662.
182. Imberdis T, Heeres JT, Yueh H, et al. Identification of anti-prion compounds using a novel cellular assay. *J Biol Chem.* 2016;291(50):26164–26176.
183. Miyazaki Y, Ishikawa T, Kamatari YO, et al. Identification of alprenolol hydrochloride as an anti-prion compound using surface plasmon resonance imaging. *Mol Neurobiol.* 2019;56(1):367–377.
184. Mashima T, Lee JH, Kamatari YO, et al. Development and structural determination of an anti-PrP(C) aptamer that blocks pathological conformational conversion of prion protein. *Sci Rep.* 2020;10(1):4934.
185. Mashima T, Nishikawa F, Kamatari YO, et al. Anti-prion activity of an RNA aptamer and its structural basis. *Nucleic Acids Res.* 2013;41(2):1355–1362.
186. Didonna A, Venturini AC, Hartman K, Vranac T, Curin Serbec V, Legname G. Characterization of four new monoclonal antibodies against the distal N-terminal region of PrP(c). *PeerJ.* 2015;3:e811.
187. Stincardini C, Massignan T, Biggi S, et al. An antipsychotic drug exerts anti-prion effects by altering the localization of the cellular prion protein. *PLoS One.* 2017;12(8): e0182589.
188. Barret A, Tagliavini F, Forloni G, et al. Evaluation of quinacrine treatment for prion diseases. *J Virol.* 2003;77(15):8462–8469.
189. Collinge J, Gorham M, Hudson F, et al. Safety and efficacy of quinacrine in human prion disease (PRION-1 study): a patient-preference trial. *Lancet Neurol.* 2009;8(4):334–344.
190. Doh-Ura K, Iwaki T, Caughey B. Lysosomotropic agents and cysteine protease inhibitors inhibit scrapie-associated prion protein accumulation. *J Virol.* 2000;74(10): 4894–4897.
191. Georgieva D, Schwark D, von Bergen M, Redecke L, Genov N, Betzel C. Interactions of recombinant prions with compounds of therapeutical significance. *Biochem Biophys Res Commun.* 2006;344(2):463–470.
192. Geschwind MD, Kuo AL, Wong KS, et al. Quinacrine treatment trial for sporadic Creutzfeldt-Jakob disease. *Neurology.* 2013;81:2015–2023.
193. Ghaemmaghami S, Ahn M, Lessard P, et al. Continuous quinacrine treatment results in the formation of drug-resistant prions. *PLoS Pathog.* 2009;5(11):e1000673.
194. Adjou KT, Demaimay R, Deslys JP, et al. MS-8209, a water-soluble amphotericin B derivative, affects both scrapie agent replication and PrPres accumulation in Syrian hamster scrapie. *J Gen Virol.* 1999;80(pt 4):1079–1085.
195. Adjou KT, Privat N, Demart S, et al. MS-8209, an amphotericin B analogue, delays the appearance of spongiosis, astrogliosis and PrPres accumulation in the brain of scrapie-infected hamsters. *J Comp Pathol.* 2000;122(1):3–8.

196. Adjou KT, Demaimay R, Lasmezas C, Deslys JP, Seman M, Dormont D. MS-8209, a new amphotericin B derivative, provides enhanced efficacy in delaying hamster scrapie. *Antimicrob Agents Chemother.* 1995;39(12):2810–2812.
197. Adjou KT, Demaimay R, Lasmezas CI, Seman M, Deslys JP, Dormont D. Differential effects of a new amphotericin B derivative, MS-8209, on mouse BSE and scrapie: implications for the mechanism of action of polyene antibiotics. *Res Virol.* 1996;147(4): 213–218.
198. Beringue V, Adjou KT, Lamoury F, et al. Opposite effects of dextran sulfate 500, the polyene antibiotic MS-8209, and Congo red on accumulation of the protease-resistant isoform of PrP in the spleens of mice inoculated intraperitoneally with the scrapie agent. *J Virol.* 2000;74(12):5432–5440.
199. Demaimay R, Adjou K, Lasmezas C, et al. Pharmacological studies of a new derivative of amphotericin B, MS-8209, in mouse and hamster scrapie. *J Gen Virol.* 1994; 75(pt 9):2499–2503.
200. Grigoriev VB, Adjou KT, Sales N, et al. Effects of the polyene antibiotic derivative MS-8209 on the astrocyte lysosomal system of scrapie-infected hamsters. *J Mol Neurosci.* 2002;18(3):271–281.
201. Ohsawa N, Song CH, Suzuki A, Furuoka H, Hasebe R, Horiuchi M. Therapeutic effect of peripheral administration of an anti-prion protein antibody on mice infected with prions. *Microbiol Immunol.* 2013;57(4):288–297.
202. Forloni G, Iussich S, Awan T, et al. Tetracyclines affect prion infectivity. *Proc Natl Acad Sci U S A.* 2002;99(16):10849–10854.
203. Gu Y, Singh N. Doxycycline and protein folding agents rescue the abnormal phenotype of familial CJD H187R in a cell model. *Brain Res Mol Brain Res.* 2004;123(1–2): 37–44.
204. Haïk S, Marcon G, Mallet A, et al. Doxycycline in Creutzfeldt-Jakob disease: A phase 2, randomised, double-blind, placebo-controlled trial. *Lancet Neurol.* 2014;13(2): 150–158.
205. Hannaoui S, Gougerot A, Privat N, et al. Cycline efficacy on the propagation of human prions in primary cultured neurons is strain-specific. *J Infect Dis.* 2014;209(7): 1144–1148.
206. Pocchiari M, Ladogana A. Rethinking of doxycycline therapy in Creutzfeldt-Jakob disease. *J Neurol Neurosurg Psychiatry.* 2015;86(7):705.
207. Moreno JA, Halliday M, Molloy C, et al. Oral treatment targeting the unfolded protein response prevents neurodegeneration and clinical disease in prion-infected mice. *Sci Transl Med.* 2013;5(206):206ra138.
208. Ertmer A, Gilch S, Yun SW, et al. The tyrosine kinase inhibitor STI571 induces cellular clearance of PrPSc in prion-infected cells. *J Biol Chem.* 2004;279(40):41918–41927.
209. Abdulrahman BA, Abdelaziz D, Thapa S, et al. The celecoxib derivatives AR-12 and AR-14 induce autophagy and clear prion-infected cells from prions. *Sci Rep.* 2017; 7(1):17565.
210. Heiseke A, Aguib Y, Riemer C, Baier M, Schatzl HM. Lithium induces clearance of protease resistant prion protein in prion-infected cells by induction of autophagy. *J Neurochem.* 2009;109(1):25–34.
211. Mead S, Tagliavini F. Clinical trials. *Handb Clin Neurol.* 2018;153:431–444.
212. Forloni G, Tettamanti M, Lucca U, et al. Preventive study in subjects at risk of fatal familial insomnia: Innovative approach to rare diseases. *Prion.* 2015;9(2):75–79.
213. Elbein AD, Solf R, Dorling PR, Vosbeck K. Swainsonine: An inhibitor of glycoprotein processing. *Proc Natl Acad Sci U S A.* 1981;78(12):7393–7397.
214. Tulsiani DR, Harris TM, Touster O. Swainsonine inhibits the biosynthesis of complex glycoproteins by inhibition of Golgi mannosidase II. *J Biol Chem.* 1982;257(14): 7936–7939.

215. Tulsiani DRP, Touster O. Swainsonine, a potent mannosidase inhibitor, elevates rat liver and brain lysosomal α-d-mannosidase, decreases Golgi α-d-mannosidase II, and increases the plasma levels of several acid hydrolases. *Arch Biochem Biophys.* 1983; 224(2):594–600.
216. Mahal SP, Browning S, Li J, Suponitsky-Kroyter I, Weissmann C. Transfer of a prion strain to different hosts leads to emergence of strain variants. *Proc Natl Acad Sci U S A.* 2010;107(52):22653–22658.
217. Li J, Mahal SP, Demczyk CA, Weissmann C. Mutability of prions. *EMBO Rep.* 2011;12(12):1243–1250.
218. Li J, Browning S, Mahal SP, Oelschlegel AM, Weissmann C. Darwinian evolution of prions in cell culture. *Science.* 2010;327(5967):869–872.
219. Oelschlegel AM, Weissmann C. Acquisition of drug resistance and dependence by prions. *PLoS Pathog.* 2013;9(2):e1003158.
220. Collinge J, Clarke AR. A general model of prion strains and their pathogenicity. *Science.* 2007;318(5852):930–936.
221. Langenfeld KA, Shikiya RA, Kincaid AE, Bartz JC. Incongruity between prion conversion and incubation period following coinfection. *J Virol.* 2016;90(12):5715–5723.
222. Bartz JC, Aiken JM, Bessen RA. Delay in onset of prion disease for the HY strain of transmissible mink encephalopathy as a result of prior peripheral inoculation with the replication-deficient DY strain. *J Gen Virol.* 2004;85(pt 1):265–273.
223. Dickinson AG, Fraser H, Meikle VM, Outram GW. Competition between different scrapie agents in mice. *Nat New Biol.* 1972;237(77):244–245.
224. Holec SAM, Yuan Q, Bartz JC. Alteration of prion strain emergence by nonhost factors. *mSphere.* 2019;4(5).
225. Bruce ME, Dickinson AG. Biological stability of different classes of scrapie agent. In: *Slow Transmissible Disease of the Nervous System.* Academic Press, Inc; 1979:71–86. vol. 2.
226. Büeler H, Fisher M, Lang Y, et al. Normal development and behaviour of mice lacking the neuronal cell-surface PrP protein. *Nature.* 1992;356:577–582.
227. Büeler H, Aguzzi A, Sailer A, et al. Mice devoid of PrP are resistant to scrapie. *Cell.* 1993;73:1339–1347.
228. Richt JA, Kasinathan P, Hamir AN, et al. Production of cattle lacking prion protein. *Nat Biotechnol.* 2007;25:132–138.
229. Mallucci G, Dickinson A, Linehan J, Klohn PC, Brandner S, Collinge J. Depleting neuronal PrP in prion infection prevents disease and reverses spongiosis. *Science.* 2003;302(5646):871–874.
230. Lewis V. Proteolytic processing of the prion protein. In: Collins S, Lawson V, eds. *The Cellular and Molecular Biology of Prion Disease.* Kerala: Research Signpost; 2011:53–71.
231. Chen SG, Teplow DB, Parchi P, Teller JK, Gambetti P, Autilio-Gambetti L. Truncated forms of the human prion protein in normal brain and in prion diseases. *J Biol Chem.* 1995;270(32):19173–19180.
232. Jiménez-Huete A, Lievens PMJ, Vidal R, et al. Endogenous proteolytic cleavage of normal and disease-associated isoforms of the human prion protein in neural and non-neural tissues. *Am J Pathol.* 1998;153(5):1561–1572.
233. Zanusso G, Righetti PG, Ferrari S, et al. Two-dimensional mapping of three phenotype-associated isoforms of the prion protein in sporadic Creutzfeldt-Jakob disease. *Electrophoresis.* 2002;23(2):347–355.
234. Owen JP, Rees HC, Maddison BC, et al. Molecular profiling of ovine prion diseases by using thermolysin-resistant PrPSc and endogenous C2 PrP fragments. *J Virol.* 2007;81(19):10532–10539.
235. Hope J, Reekie LJD, Hunter N, et al. Fibrils from brains of cows with new cattle disease contain scrapie-associated protein. *Nature.* 1988;336:390–392.

236. Hope J, Morton LJ, Farquhar CF, Multhaup G, Beyreuther K, Kimberlin RH. The major polypeptide of scrapie-associated fibrils (SAF) has the same size, charge distribution and N-terminal protein sequence as predicted for the normal brain protein (PrP). *EMBO J.* 1986;5(10):2591–2597.
237. Pan T, Wong P, Chang B, et al. Biochemical fingerprints of prion infection: accumulations of aberrant full-length and N-terminally truncated PrP species are common features in mouse prion disease. *J Virol.* 2005;79(2):934–943.
238. O'Brien RJ, Wong PC. Amyloid precursor protein processing and Alzheimer's disease. *Annu Rev Neurosci.* 2011;34:185–204.
239. Kim J, Onstead L, Randle S, et al. Abeta40 inhibits amyloid deposition in vivo. *J Neurosci.* 2007;27(3):627–633.
240. Deng Y, Tarassishin L, Kallhoff V, et al. Deletion of presenilin 1 hydrophilic loop sequence leads to impaired gamma-secretase activity and exacerbated amyloid pathology. *J Neurosci.* 2006;26(14):3845–3854.
241. Iijima K, Liu HP, Chiang AS, Hearn SA, Konsolaki M, Zhong Y. Dissecting the pathological effects of human Abeta40 and Abeta42 in drosophila: A potential model for Alzheimer's disease. *Proc Natl Acad Sci U S A.* 2004;101(17):6623–6628.
242. McGowan E, Pickford F, Kim J, et al. Abeta42 is essential for parenchymal and vascular amyloid deposition in mice. *Neuron.* 2005;47(2):191–199.
243. Zhang Z, Kang SS, Liu X, et al. Asparagine endopeptidase cleaves alpha-synuclein and mediates pathologic activities in Parkinson's disease. *Nat Struct Mol Biol.* 2017;24(8): 632–642.
244. Bremer J, Baumann F, Tiberi C, et al. Axonal prion protein is required for peripheral myelin maintenance. *Nat Neurosci.* 2010;13(3):310–318.
245. Vincent B, Paitel E, Saftig P, et al. The disintegrins ADAM10 and TACE contribute to the constitutive and phorbol ester-regulated normal cleavage of the cellular prion protein. *J Biol Chem.* 2001;276(41):37743–37746.
246. Cisse MA, Sunyach C, Lefranc-Jullien S, Postina R, Vincent B, Checler F. The disintegrin ADAM9 indirectly contributes to the physiological processing of cellular prion by modulating ADAM10 activity. *J Biol Chem.* 2005;280(49):40624–40631.
247. Linsenmeier L, Mohammadi B, Wetzel S, et al. Structural and mechanistic aspects influencing the ADAM10-mediated shedding of the prion protein. *Mol Neurodegener.* 2018;13(1):18.
248. Taylor DR, Parkin ET, Cocklin SL, et al. Role of ADAMs in the ectodomain shedding and conformational conversion of the prion protein. *J Biol Chem.* 2009;284(34): 22590–22600.
249. Altmeppen HC, Prox J, Puig B, et al. Lack of a-disintegrin-and-metalloproteinase ADAM10 leads to intracellular accumulation and loss of shedding of the cellular prion protein in vivo. *Mol Neurodegener.* 2011;6:36.
250. Altmeppen HC, Prox J, Krasemann S, et al. The sheddase ADAM10 is a potent modulator of prion disease. *Elife.* 2015;4.
251. Feraudet C, Morel N, Simon S, et al. Screening of 145 anti-PrP monoclonal antibodies for their capacity to inhibit PrPSc replication in infected cells. *J Biol Chem.* 2005;280(12):11247–11258.
252. Lefebvre-Roque M, Kremmer E, Gilch S, et al. Toxic effects of intracerebral PrP antibody administration during the course of BSE infection in mice. *Prion.* 2007;1(3): 198–206.
253. Solforosi L, Criado JR, McGavern DB, et al. Cross-linking cellular prion protein triggers neuronal apoptosis in vivo. *Science.* 2004;303(5663):1514–1516.
254. Sonati T, Reimann RR, Falsig J, et al. The toxicity of antiprion antibodies is mediated by the flexible tail of the prion protein. *Nature.* 2013;501(7465):102–106.

255. Tayebi M, David M, Bate C, et al. Epitope-specific anti-prion antibodies upregulate apolipoprotein E and disrupt membrane cholesterol homeostasis. *J Gen Virol.* 2010;91(pt 12):3105–3115.
256. Goni F, Knudsen E, Schreiber F, et al. Mucosal vaccination delays or prevents prion infection via an oral route. *Neuroscience.* 2005;133(2):413–421.
257. Goni F, Mathiason CK, Yim L, et al. Mucosal immunization with an attenuated Salmonella vaccine partially protects white-tailed deer from chronic wasting disease. *Vaccine.* 2015;33(5):726–733.
258. Pilon J, Loiacono C, Okeson D, et al. Anti-prion activity generated by a novel vaccine formulation. *Neurosci Lett.* 2007;429(2–3):161–164.
259. Hedlin PD, Cashman NR, Li L, et al. Design and delivery of a cryptic PrP(C) epitope for induction of PrP(Sc)-specific antibody responses. *Vaccine.* 2010;28(4):981–988.
260. Marciniuk K, Maattanen P, Taschuk R, et al. Development of a multivalent, PrP(Sc)-specific prion vaccine through rational optimization of three disease-specific epitopes. *Vaccine.* 2014;32(17):1988–1997.
261. Taschuk R, Van der Merwe J, Marciniuk K, et al. In vitro neutralization of prions with PrP(Sc)-specific antibodies. *Prion.* 2015;9(4):292–303.
262. Taschuk R, Scruten E, Woodbury M, et al. Induction of PrP(Sc)-specific systemic and mucosal immune responses in white-tailed deer with an oral vaccine for chronic wasting disease. *Prion.* 2017;11(5):368–380.
263. Wood ME, Griebel P, Huizenga ML, et al. Accelerated onset of chronic wasting disease in elk (Cervus canadensis) vaccinated with a PrP(Sc)-specific vaccine and housed in a prion contaminated environment. *Vaccine.* 2018;36(50):7737–7743.
264. Arsenault RJ, Li Y, Potter A, Griebel PJ, Kusalik A, Napper S. Induction of ligand-specific PrP (C) signaling in human neuronal cells. *Prion.* 2012;6(5):477–488.
265. Saborio GP, Permanne B, Soto C. Sensitive detection of pathological prion protein by cyclic amplification of protein misfolding. *Nature.* 2001;411(6839):810–813.
266. Wilham JM, Orru CD, Bessen RA, et al. Rapid end-point quantitation of prion seeding activity with sensitivity comparable to bioassays. *PLoS Pathog.* 2010;6(12): e1001217.
267. Lacroux C, Comoy E, Moudjou M, et al. Preclinical detection of variant CJD and BSE prions in blood. *PLoS Pathog.* 2014;10(6):e1004202.
268. Mathiason CK, Powers JG, Dahmes SJ, et al. Infectious prions in the saliva and blood of deer with chronic wasting disease. *Science.* 2006;314(5796):133–136.
269. McGuire LI, Peden AH, Orru CD, et al. Real time quaking-induced conversion analysis of cerebrospinal fluid in sporadic Creutzfeldt-Jakob disease. *Ann Neurol.* 2012;72(2):278–285.
270. Moda F, Gambetti P, Notari S, et al. Prions in the urine of patients with variant Creutzfeldt-Jakob disease. *N Engl J Med.* 2014;371(6):530–539.
271. Orru CD, Groveman BR, Hughson AG, Zanusso G, Coulthart MB, Caughey B. Rapid and sensitive RT-QuIC detection of human Creutzfeldt-Jakob disease using cerebrospinal fluid. *MBio.* 2015;6(1).
272. Safar JG, Lessard P, Tamguney G, et al. Transmission and detection of prions in feces. *J Infect Dis.* 2008;198(1):81–89.
273. Orru CD, Bongianni M, Tonoli G, et al. A test for Creutzfeldt-Jakob disease using nasal brushings. *N Engl J Med.* 2014;371(6):519–529.
274. Saa P, Castilla J, Soto C. Presymptomatic detection of prions in blood. *Science.* 2006;313(5783):92–94.
275. Sawyer EB, Edgeworth JA, Thomas C, Collinge J, Jackson GS. Preclinical detection of infectivity and disease-specific PrP in blood throughout the incubation period of prion disease. *Sci Rep.* 2015;5:17742.

276. Wang Z, Manca M, Foutz A, et al. Early preclinical detection of prions in the skin of prion-infected animals. *Nat Commun.* 2019;10(1):247.
277. Elder AM, Henderson DM, Nalls AV, et al. In vitro detection of prionemia in TSE-infected cervids and hamsters. *PLoS One.* 2013;8(11):e80203.
278. Elder AM, Henderson DM, Nalls AV, et al. Immediate and ongoing detection of prions in the blood of hamsters and deer following oral, nasal, or blood inoculations. *J Virol.* 2015;89(14):7421–7424.
279. Fairfoul G, McGuire LI, Pal S, et al. Alpha-synuclein RT-QuIC in the CSF of patients with alpha-synucleinopathies. *Ann Clin Transl Neurol.* 2016;3(10):812–818.
280. Saijo E, Ghetti B, Zanusso G, et al. Ultrasensitive and selective detection of 3-repeat tau seeding activity in Pick disease brain and cerebrospinal fluid. *Acta Neuropathol.* 2017;133(5):751–765.
281. Salvadores N, Shahnawaz M, Scarpini E, Tagliavini F, Soto C. Detection of misfolded Abeta oligomers for sensitive biochemical diagnosis of Alzheimer's disease. *Cell Rep.* 2014;7(1):261–268.
282. Shahnawaz M, Tokuda T, Waragai M, et al. Development of a biochemical diagnosis of Parkinson disease by detection of alpha-synuclein misfolded aggregates in cerebrospinal fluid. *JAMA Neurol.* 2017;74(2):163–172.
283. Saijo E, Groveman BR, Kraus A, et al. Ultrasensitive RT-QuIC seed amplification assays for disease-associated tau, alpha-synuclein, and prion aggregates. *Methods Mol Biol.* 2019;1873:19–37.

CHAPTER SIX

Strain variation in treatment and prevention of human prion diseases

Laura J. Ellett, Zoe T. Revill, Yong Qian Koo, and Victoria A. Lawson*
Department of Microbiology and Immunology, The University of Melbourne, Parkville, VIC, Australia
*Corresponding author: e-mail address: v.lawson@unimelb.edu.au

Contents

Abstract

Transmissible spongiform encephalopathies or prion diseases describe a number of different human disorders that differ in their clinical phenotypes, which are nonetheless united by their transmissible nature and common pathology. Clinical variation in the absence of a conventional infectious agent is believed to be encoded by different conformations of the misfolded prion protein. This misfolded protein is the target of methods designed to prevent disease transmission in a surgical setting and reduction of the misfolded seed or preventing its continued propagation have been the focus of therapeutic strategies. It is therefore possible that strain variation may influence the efficacy of prevention and treatment approaches.

Historically, an understanding of prion disease transmission and pathogenesis has been focused on research tools developed using agriculturally relevant strains of prion disease. However, an increased understanding of the molecular biology of human prion disorders has highlighted differences not only between different forms of the disease affecting humans and animals but also within diseases such as Creutzfeldt-Jakob

Progress in Molecular Biology and Translational Science, Volume 175
ISSN 1877-1173
https://doi.org/10.1016/bs.pmbts.2020.08.006

Disease (CJD), which is represented by several sporadic CJD specific conformations and an additional conformation associated with variant CJD. In this chapter we will discuss whether prion strain variation can affect the efficacy of methods used to decontaminate prions and whether strain variation in pre-clinical models of prion disease can be used to identify therapeutic strategies that have the best possible chance of success in the clinic.

1. Introduction

The transmissible spongiform encephalopathies, or prion diseases encapsulate a variety of disease phenotypes affecting both humans and animals. All forms of the disease share a common pathology that includes vacuolation, a reactive gliosis and the accumulation of a misfolded form of the normal host encoded prion protein. In animals the disease is fundamentally considered transmissible, although sporadic forms are likely to exist. In contrast most forms of human disease occur due to spontaneous or sporadic misfolding of the prion protein or arise from mutations in the prion protein gene. However; regardless of origin, the disease can be acquired in humans by consumption of contaminated tissue or through medical procedures.

Diagnosis of human prion disease is based on clinical signs with Creutzfeldt-Jakob Disease (CJD) presenting as a rapidly progressive dementia, Gerstmann–Sträussler–Scheinker Syndrome (GSS) as a cerebellar ataxia and Fatal Familial Insomnia (FFI) as intractable insomnia. The animal forms of prion disease also manifest with different clinical signs in naturally affected species and these differences are reflected in the terms used to describe them; a neuropathic itch in sheep and goats with scrapie, abnormal behavior in cattle with Bovine Spongiform Encephalopathy (BSE, referred to in the media as mad cows disease) and weight loss in cervids with Chronic Wasting Disease (CWD). The different manifestations of each disease are likely to reflect the predominant region of the brain affected. The molecular basis for this difference, in the absence of a nucleic acid genome, is likely to represent the region of the brain where protein misfolding and degeneration occurs. In addition to these disease specific differences within each disease there is variation that has been described as strains.

2. Prion strains

The first recognition that strains of prion diseases existed, and that these strains gave rise to different clinical phenotypes was made in 1961 when Pattison et al.[1] isolated prion disease from goats that presented with either a drowsy or scratching phenotype. When this inoculum was administered to uninfected goats, the phenotype of the inoculum was preserved. Since this finding, many independent researchers have further described various methods to differentiate between and identify scrapie strains. Historically, animal derived prion strains, often passaged in mice, have been differentiated by their incubation time, clinical presentation and pattern of vacuolation, or lesion profile[2–4]. Molecular approaches have correlated this variation with different properties of PrP^{Sc}, the typically protease resistant, misfolded form of the prion protein which conveys transmissibility and is closely associated with pathogenesis.

The isolation and further characterization of two prion strains of transmissible mink encephalopathy (TME) in Syrian hamsters has provided a molecular understanding of prion strains. In hamsters the Hyper (HY) and Drowsy (DY) of TME present with different incubation periods, have different clinical signs and vacuolation is observed in different regions of the brain[5]. Following protease digestion, the PrP^{Sc} from these strains has a distinct electrophoretic mobility on polyacrylamide gel. FTIR analysis revealed that this apparent difference in susceptibility to protease digestion could be attributed to a difference in the conformation of the misfolded protein with the HY strain possessing a higher β-sheet content than the DY strain[6].

The molecular differences in PrP^{Sc} strains are characterized by the difference in electrophoretic mobility following PK digestion coupled with differences in glycosylation; PrP has two potential N-linked glycosylation sites giving rise to three potential glycosylation patterns or glycoforms representing glycosylation at one, two or no sites, was seminal in the confirmation that variant CJD was a unique prion strain derived from BSE[7] and further led to the identification of 2–3 distinct sporadic forms of CJD, each with a unique electrophoretic mobility following PK-digestion[8,9].

Parchi et al.[8] described two main types of PrP^{Sc}, "type 1" and "type 2," that differed by electrophoretic mobility following PK digestion and were further differentiated into six molecular sub types by genotype at codon 129 (either MM, MV or VV). These molecular sub types were found to correspond with the clinical and pathological features of disease in patients[10]. In contrast to Parchi and collaborators' classification, Collinge et al.[9] and Zanusso et al.[11] identified three sCJD types based on electrophoretic mobility following PK digestion, with copper binding[12] and pH[11] identified as important determinants of the additional conformation. Like Parchi, Collinge's work further classified molecular strains based on genotype and correlated this with clinical presentation and pathology[13]. Regardless of classification system, these observations correlate clinical phenotype with the molecular characteristics of PrP^{Sc} and support the idea that human prion strains are enciphered by this.

The consequences of prion strain variation, beyond an impact on clinical phenotype and pathology has yet to be fully appreciated. Transmission of human prions to mice has been reported for familial and sporadic forms of CJD[14–16] and shown to maintain the electrophoretic mobility of the original inoculum. We have recently adapted two sporadic CJD strains to wild-type mice and demonstrated that the electrophoretic mobility of these sporadic CJD types is also maintained (ref. 15 and Lawson unpublished). Table 1 describes the origin of commonly used rodent adapted laboratory prion strains.

The hypothesis[22], now accepted, that a disease can be transmitted through the propagation of protein misfolding, is even more intriguing in the face of prion strains that suggest that not only does the misfolded protein faithfully propagate unique conformations but that those conformations also instil unique clinical characteristics. These features also have the potential to affect treatment and prevention of disease.

The treatment of prion diseases can involve preventing protein misfolding, attenuating the toxic consequences of protein misfolding or restoring neural function. There are a number of targets that might prevent or inhibit protein misfolding. The normal cellular form of the prion protein provides a substrate for misfolding and it is clear that targeting this molecule through pharmacological (stabilizing or redistributing) or molecular (knockdown) approaches has potential to slow disease progression[23,24]. An alternative target is the PrP^{Sc} that acts as the template for protein misfolding; by clearing PrP^{Sc} or stabilizing it into a conformation that cannot propagate misfolding. Although the mechanism by which misfolding occurs is poorly understood it could also be attenuated by preventing binding of PrP^{C} to PrP^{Sc} or preventing the conformational change associated with misfolding either directly or through an interaction with cofactors (reviewed in ref. 25).

The accumulation of protein misfolding is also likely to represent a disequilibrium between production and clearance of misfolded proteins and in this respect proteostasis can be restored by enhancing the function of molecular chaperones or directing misfolded proteins for degradation through the ubiquitin (Ub)-proteasome system or *via* autophagy[26]. Both the rate of prion propagation and rate of degradation are likely to contribute strain specific incubation times and it has been suggested that these properties also reflect the inherent stability of PrP^{Sc}, with Safar et al.[27] concluding that the different

Table 1 Origins of laboratory prion strains.

Strain	Origin	Species
SSBP/1	Pool of sheep scrapie serially passaged in Cheviot sheep	Sheep[a]
Drowsy goat	Sub passage of SSBP/1 in goats	Goats[a]
22A	SSBP/1 transmitted to wild-type mice	Mice[a]
22L		
22C		
ME7	Suffolk sheep derived natural scrapie transmitted to wild-type mice	Mice[a]
RML/Chandler	"drowsy goat" transmitted to wild-type mice	Mice[a]
79A	"drowsy goat" transmitted to wild-type mice	Mice[a]
79V		
139A		
263K	"drowsy goat" strain after transmission to mice, serially passaged in rats and then in hamsters	Hamsters[a]
237	Chandler strain of mouse-adapted scrapie passaged in rats and then hamsters	Hamsters[17]
HY	Stetsonville, Wisconsin, U.S.A. source of transmissible mink encephalopathy (TME) transmitted to outbred Syrian golden hamsters	Hamsters[18]
DY		
S.Co	CJD transmitted to guinea pigs	Guinea pigs[19]
FU-01/K.Fu	GSS transmitted to wild-type mice	Mice[16]
M1000	Sub passage of FU-01 in wild-type mice	Mice[20]
MU-02	T2MM[13] sporadic CJD transmitted to wild-type mice	Mice[15]

[a]Described in 21.

incubation times observed in prion diseases of different strains arise due to different rates of PrP^{Sc} clearance, while Gonzalez-Montalban et al.[28] suggested that a short incubation period of disease is associated with a prion strain of low stability.

Treatment may also include protecting cells from the damaging effects of prion induced toxicity, for example, by attenuating oxidative stress[20,29]. Even if protein misfolding can be slowed or halted it still remains to be determined whether lost neural function can be restored. In prion diseases, like so many of the protein misfolding disorders, diagnosis often occurs with the appearance and progression of clinical signs, at a time when significant neuronal loss has occurred. In all neurodegenerative disorders, early detection will be key to meaningful therapeutic intervention. However; the rapid clinical progression of prion diseases offers only a very small window for timely intervention. In this respect, familial forms of prion disease are likely to benefit most from early prophylactic intervention, whereas prevention, through universal infection control strategies, may offer the best approach to avoiding iatrogenic transmission.

3. Effect of prion strain variation on prevention of disease transmission

Although the susceptibility of the prion to denaturation and protease digestion defines its proteinaceous nature[22], PrP^{Sc} is remarkably resistant to both processes in environmental[30] and surgical[31] settings. Prevention of prion disease transmission, specifically in a surgical setting, involves elimination of the transmissible seed of PrP^{Sc}. Prions from different strains exhibit different susceptibilities to both denaturation and digestion which could affect the efficacy of decontamination and treatment methods.

The susceptibility of prions to unfolding in the presence of a chaotropic salt, such as guanidine, has been exploited in the conformation dependant immunoassay[32] and conformation stability assay[33] which measure the concentration of denaturant required to alter antibody epitope exposure or reduce the protease resistance of PrP^{Sc}, respectively. These assays have identified conformational variation in prion strains with similar electrophoretic mobilities following PK digestion, such as 237 and HY[32,33], and variation within human sporadic CJD molecular subtypes[34].

PrP^{Sc} is remarkably resistant to digestion with proteases. Differences in the long term resistance of various rodent adapted prion strains have been shown; with purified PrP^{Sc} from strains such as M1000[35]; 79A, ME7 and 87V[36] exhibiting robust resistance to prolonged (6 h) exposure to proteases whereas PrP^{Sc} from strains such as RML/Chandler and 22A are relatively more sensitive to digestion[36]. Similar trends in the sensitivity of whole brain homogenates have been observed with the PrP^{Sc} associated with Syrian hamster passaged ME7 appearing to have slightly greater resistance to protease digestion than that of Syrian hamster passaged RML[37].

3.1 Evidence for strain specific effects on chemical inactivation of prions

Prions pose a particular challenge for decontamination in a medical setting. Formaldehyde and alcohols which were commonly used in hospitals for sterilization purposes do not inactivate prions. In fact, formaldehyde and 68% of alcohol are ineffective in inactivating prions and may have contributed to iatrogenic disease transmission[19,38–41]. Consequently, a number of chemicals have been tested for their ability to reduce infectivity in prion brain homogenates of rodent adapted sheep scrapie (263K, 139A, 22A) and human (Fu-01, S.Co) prion strains. Chemicals that initially failed to reduce infectivity include phenolic disinfectants, chlorine dioxide, permanganate, iodine, Trixton X-100, sodium dodecyl sulfate (SDS), sodium deoxycholate and hydrogen peroxide[19,39,42].

Not unexpectedly the ability of chemicals to inactivate prions will reflect the final effective concentration of the active ingredient. In initial studies compounds containing less than 1% phenol failed to effectively inactivate rodent adapted scrapie strain 139A and 22A[42] and 263K[39], whereas higher concentrations of phenol were able to inactivate FU-01 (50% and 80% phenol)[43] and 263K (9–80% phenol)[44]. However, in the absence of direct comparisons it is difficult to determine whether some effects may reflect differences in the stability of the prion strain being tested.

The efficacy of tested compounds also depends on the conditions of treatment. For example, SDS has been shown to be ineffective against multiple prion strains (FU-01, 139A, 22A) at room temperature but if the temperature of treatment is sufficiently increased, SDS can inactivate both scrapie and human derived prions[19,42,43,45]. Although some strain specificity may relate to the temperature required to inactivate specific prion strains as FU-01 requires SDS treatment at 100 °C to remove detectable infectivity whereas 139A and 22A can be completely inactivated by SDS treatment at 70 °C[42].

The apparent resistance of the mouse adapted human prion strain FU-01 to chemical inactivation has also been reported for NaOH. Treatment for 1 h in 1 N NaOH was reported to remove all detectable infectivity from 263K scrapie[39] but not FU-01[45,46] derived brain homogenates. This does not seem to be the case for all human derived prion strains as Brown et al.[47] reported complete inactivation of the S.Co strain of guinea pig adapted CJD as well as 263K hamster scrapie prions by treatment with 1 M NaOH for 1 h.

The most universally effective chemical decontaminant is sodium hypochlorite which when used at more than 10,000 ppm (1% v/v) for 1 h has been reported to inactivate the K.Fu human prion isolate at 12.5%[48] and 1.3% sodium hypochlorite[19]; S.Co at 1.3%[49]; 139A and 22A at 1%[42] and 263K at 2.5%[49]. Although again concentration is important with incomplete inactivation of 263K reported 0.5% and 1%[49]; and 139A and 22A at 0.1%[42]. Recognizing the corrosive nature of sodium hypochlorite and NaOH WHO recommended either treatment with 1 N NaOH or 20,000 ppm (2%) sodium hypochlorite, although coupling with physical inactivation methods is also recommended.

3.2 Evidence for strain specific effects on heat inactivation of prions

The earliest decontamination studies tested the effectiveness of standard autoclaving procedure in inactivating prions. The results of these experiments suggested that autoclaving FU-01 and S.Co prions at 121 °C for 1 h would completely inactivate detectable infectivity of these human derived strains but not the scrapie derived strain 263K[39,48,49]. Autoclaving for 1-h at higher temperatures; 126 °C for 139A, 132 °C for 263K and 136 °C for 22A, was found to be effective against scrapie derived strains[39,48,49]. Although these results suggest strain variation it should be noted that these experiments were performed using different formats including dry heat, porous load and gravity displacement autoclaves. Nonetheless where strains were compared it was concluded that the thermal stability of 22A was greater than 139A[42] and that the relative resistance of 263K to thermal inactivation reflected the high infectious titre present in 263K infected hamster brains.

3.3 Evidence for strain specific effects on inactivation of surface bound prions

Typically approaches used to assess the ability to degrade prions have been based on brain homogenates or macerates which do not address the process of cleaning prior to sterilization that would be associated with iatrogenic transmission in a surgical setting. Contamination and bioassay of prions bound to surgical steel wires[50] highlighted the potential for surgical transmission of prion from stainless steel surfaces. In their original study Flechsig et al.[50] demonstrated complete inactivation of RML infectivity following treatment with 1 M NaOH for 1 h. Jackson et al.[51] similarly reported complete inactivation of RML prions with 2 N NaOH followed by autoclaving. Lemmer et al.[52] reported inactivation of 263K with 1 M NaOH for 1 h (0/6 Syrian hamsters), whereas Yan et al.[53] failed to completely inactivate 263K prions from contaminated wire after treatment with 1 M NaOH for 24 h followed by autoclaving for 134 °C for 18 min (2/6 in hamsters). It was previously reported that NaOH was unable to completely eliminate infectivity present in an FU-01 prion infected brain homogenate[45,46]. Using M1000 a prion strain derived from FU-01, we similarly showed that 1 M NaOH was not able to remove infectivity from steel wires[35]. The effect of heat treatment has been similarly variable with steel bound M1000 infectivity reduced but not eliminated by autoclaving at 121 °C for 20 min or 134 °C for 3 or 18 min[35] and Yan et al.[53] reporting incomplete inactivation of 263K following autoclaving at 134 °C for 18 min. These modest results are likely to reflect the importance of cleaning prior to sterilization.

To this end a number of detergent and enzymatic protocols have been assessed against both homogenate and steel wire bound prions[31]. We showed that for the M1000 strain, an enzymatic detergent was more effective than NaOH at reducing infectivity and when coupled with autoclaving could prevent disease in mice[35].

Further evidence of the resistance of steel bound prions, and in particular human derived prions, to decontamination was observed by Peretz et al.[54] who compared the effect of acidified SDS on steel wires contaminated with 237 hamster scrapie and sCJD. While this treatment of 263K infected brain homogenates was effective after only 30 min, the same treatment required 18 h to decontaminate the same strain from steel wires. Similarly, sCJD that was eliminated from homogenates after 18 h of treatment was not completely eliminated from wires contaminated with the same strain.

Perhaps the most comprehensive investigation of strain variation has been made possible by the RT-QuIC assay which assesses cell free prion propagation as a marker of infectivity[55]. Using this assay Hughson et al.[56] were able to screen the ability of hypochlorous acid to prevent propagation from PrP^{Sc} derived from 263K, sCJD, vCJD, CWD, BSE and sheep scrapie prions. The recent adaptation of the RT-QuIC and PMCA to detect PrP^{Sc} bound to steel wires will enable further assessment of strain variation on decontamination in a medical setting[56–58].

4. Effect of prion strain variation on treatment of disease

In contrast to most neurodegenerative disorders there are many pre-clinical models of prion disease that genuinely reflect the transmission and pathogenesis of this disease. Yet, despite identifying an extensive list of possible therapeutics only four have been tested in a clinical setting and none have provided meaningful therapeutic benefit[59]. This failure may reflect the relatively short therapeutic window available for treatment of prion diseases, which by their very nature are rapidly progressive and the likelihood that significant neuronal loss has occurred before patients develop symptoms that lead to a diagnosis of prion disease. In these respects, prion diseases are like many other neurodegenerative diseases which will require early diagnosis in order for treatment to preserve neurological function and provide meaningful clinical outcome. However, prion diseases also describe a group of disorders that affect animals and humans, include human diseases which present with different phenotypes and include different strains of prions within these disease states. This variation, enciphered by the misfolded prion, which is itself often a target of these therapeutics, may have contributed the failure to translate the promising results of pre-clinical tests to the clinic. The following section will consider whether the clinical performance of quinacrine, pentosan polysulfate, flupirtine and doxycycline could have been predicted by strain specific effects in pre-clinical studies.

4.1 Quinacrine

Quinacrine is arguably the most extensively studied therapeutic for prion diseases with its ability to clear PrP^{Sc} from prion infected cell lines repeatedly demonstrated yet failing to significantly improve outcome in animal models or clinical trials. Quinacrine is known to cross the blood brain barrier but appears to be actively transported through efflux pumps which effectively reduces the therapeutic dose in the central nervous system (CNS). Although this may explain the failure of quinacrine to delay disease progression *in vivo*, it is important to also consider the potential for strain specific effects.

Following reports of the successful clearance of PrP^{Sc} from scrapie infected cell lines[60], and the known safety of quinacrine as an antimalarial drug, *in vivo* experiments were undertaken to determine whether *in vitro* results would translate into prevention or delayed disease progression.

Collins et al.[61] was the first *in vivo* study to report the failure of quinacrine to delay disease progression in wild-type mice infected with the mouse adapted human M1000 prion strain. Subsequently quinacrine treatment failed to delay disease in naturally infected sheep[62], 263K infected hamster PrP transgenic mice[63] or RML infected mice[64] and clinical trials have failed sporadic, iatrogenic, familial and variant CJD patients[65–68].

The most extensively used *in vitro* model of prion disease is the scrapie cell model. Since the first description of the efficacy of quinacrine in this model based on the infection of mouse neuroblastoma cells (N2a) with the RML (or Chandler) strain of mouse adapted prions[60,69] quinacrine has been tested in both a chronic and acutely infected cells and has consistently cleared detectable PrP^{Sc} with a EC_{50} between 0.2 and 0.5 μM[70–75]. Quinacrine was also reported to clear PrP^{Sc} from RML infected rabbit kidney epithelial cells (RK13) that express mouse PrP^{C} [76].

Another extensively used *in vitro* model is the prion-infected GT1 hypothalamic mouse cell line. When treated with quinacrine PrP^{Sc} is reduced from RML infected GT1-7 cells, however the treatment time is typically longer than that required to clear RML from N2a cells. Barret et al.[77] reported reduction of RML from N2a cells within 3 days whereas a similar reduction in GT1-7 cells required a 6-day treatment and complete clearance was only observed after 3 weeks treatment. Sandberg et al.[72,78], reported similarly prolonged treatment times required to clear RML from GT1-7 cells, while other studies also typically report longer treatment times[79,80]. The difference in clearance of RML PrP^{Sc} *in vitro* would appear therefore to depend on the nature of the cell line infected and may indicate the mechanism of action. GT1-7 are slower growing than N2a cells and while treatment may prevent formation of *de novo* PrP^{Sc}, reduction in pre-existing PrP^{Sc} through dilution at passage may take longer in the slower growing GT1 cells. This is further supported by Ghaemmaghami et al.[64] who reported only a transient reduction of PrP^{Sc} in differentiated and therefore non-dividing N2a cells and the longer treatment time required to reduce PrP^{Sc} from RML infected mouse brain aggregates[81].

Although Barret et al.[77] reported that quinacrine is able to reduce PrP^{Sc} in N2a cells infected with 22 L strain of rodent adapted scrapie prions and conventional ScN2a (RML) without toxicity, Nguyen et al.[71] reported that full clearance of 22 L (99% reduction in detectable protease resistant PrP) was not achieved within a non-toxic dose range. In an effort to address the contribution of strain in developing therapeutics for prion disease Nguyen et al. tested a panel of quinacrine derivatives against N2a cells infected with different prion strains. In addition to N2a cells infected with RML (ScN2a) and 22 L (N167) they also tested N2a cells that had been genetically modified to overexpress PrP^{C} and infected with RML (Ch2) or FU-01 (F3) prions. Although quinacrine treatment was reported to reduce PrP^{Sc} in all strains, the EC_{50} for each strain differed. RML infection of wild-type and overexpressing cells required an EC_{50} of 0.23–0.46 μM which is consistent with the reported EC_{50} in other studies and 99% of detectable PrP^{Sc} was cleared within the non-toxic range of the drug. In contrast 22 L and FU-01 had higher EC_{50} concentrations of 0.59 and 1.88 μM, respectively and 99% of detectable PrP^{Sc} was not cleared within the drugs non-toxic range. This study suggests that there may be some effect of strain on therapeutic outcome. These strain specific therapeutic effects may further contribute to the failure of *in vivo* studies. Treatment of mice infected with the M1000 strain of prions with quinacrine at both early and late time points failed to slow disease development at a dose of 10 mg/kg/day which may not have been sufficient to inhibit replication of this strain which is derived from FU-01, which Nguyen et al. showed to be four times more resistant to quinacrine treatment than RML prions[61].

Perhaps of more concern was the observation that quinacrine treatment of cell lines infected with CWD prions, rather than reducing the amount of protease resistant PrP, rather increased it[76]. This effect was observed in two different cell lines infected with CWD and in both cases the increase in protease resistant PrP was associated with an increased conformational stability of PrP^{Sc} in the presence of Gdn-HCl and increased *in vitro* titre. Paradoxically, inoculation of mice with these quinacrine treated cells was associated with an increased incubation period *in vivo* and altered pattern of PrP^{Sc} deposition suggesting that the quinacrine treatment had changed the properties of the CWD prions. A similar change in the characteristics of PrP^{Sc} derived from RML infected neuroblastoma cells following quinacrine treatment was thought to reflect the development of a quinacrine resistant prion strain[64]. Whether this is the case or whether these observations reflect the selective advantage of an inherently quinacrine resistant prion quasi species within a strain still highlights the challenge of strain variation in the treatment of prion disease.

The ability of quinacrine to reduce PrP^{Sc} from cell lines propagating prions but failure to reduce the protease resistance of pre-existing PrP^{Sc} from a range of animal species and prion strains suggests that quinacrine prevents the formation of new PrP^{Sc}. This observation is further supported by observations in cell free assays where it has been observed that quinacrine reduces the large oligomer population and delays fibril formation in a shaking induced conversion of full-length recombinant hamster PrP[82] and that thermally induced aggregation of mouse PrP can be inhibited with quinacrine[83]. Conflicting results have been obtained from PrP^{Sc} seeded cell free conversion assays. Despite being based on 263K model of hamster scrapie prions, formation of protease resistant PrP in the cell free conversion assay based on purified hamster PrP^{C} and PrP^{Sc} was not inhibited by quinacrine[69] whereas amplification was inhibited in the PMCA assay which uses whole hamster brain homogenates as the source of PrP^{C} and PrP^{Sc} [77]. Such observations perhaps suggest that quinacrine interacts with cofactors present in the PMCA assay that are important to the conversion process.

Although it is clear that to date quinacrine treatment has had no lasting effect on disease progression, analysis of pathology in quinacrine treated animal models suggests that modification of PrP^{Sc} accumulation does occur in the asymptomatic phase of disease development. Ghaemmaghami et al.[64] reported that $MDR^{0/0}$ mice that lack expression of P-glycoprotein transporter reported to be involved in the efflux of quinacrine across the blood brain barrier (BBB) have 100 times higher levels of quinacrine in the brain and less PrP^{Sc} during disease development relative to wild-type CD1 mice. Spilman et al.[84] observed that there was less PrP^{Sc} accumulated in the neocortex and thalamus of quinacrine treatment of CD-1 mice infected with RML prions 50 days post inoculation. In this study the incubation period was not assessed. Changes in PrP^{Sc} without a significant change in incubation period poses the possibility that while a reduction in PrP^{Sc} reduces transmission it does not affect pathogenesis. In this regard a number of studies have investigated the ability of quinacrine to reduce the toxicity associated with misfolded forms of the prion protein. When folded into a ß-sheet conformation in the presence of quinacrine the toxicity of human PrP 90–231 was reduced[85]. Quinacrine treatment was also shown to attenuate the channel forming effects of the toxic PrP peptide 106–126[86] and reduce toxicity associated with the formation of free radicals by 106–126 in the presence of copper[87].

Collectively a number of observations can be made from the plethora of studies that have been performed to assess the efficacy of quinacrine in prion disease models. Meaningful bioavailability in the CNS is clearly important in the treatment of neurodegenerative disorders. Small strain specific differences in sensitivity to treatment such as those reported for EC_{50} differences between RML and FU-01/M1000, for example, could explain failure of *in vivo* studies if a strain specific therapeutic threshold is not reached.

Cell-culture models that measure outcome as a clearance of PrP^{Sc} may not accurately reflect the clearance of toxic species required to delay disease progression. Indeed, we have shown that acute toxicity is mediated by a modestly protease resistant PrP pool of PrP^{Sc} [88]. Alternatively, subtle changes in PrP^{Sc} accumulation during treatment as reported by Ghaemmaghami et al.[64] and Spilman et al.[84] may reflect an anti-prion effect, which is not sufficient to translate to clinical outcome.

4.2 Pentosan polysulfate

Pentosan polysulfate (PPS) is a synthetic sulfated glycan that was investigated for its anti-prion properties following reports of the efficacy of polyanions such as dextran sulfate to delay disease onset in rodent models of scrapie[89–91]. Despite its inability to cross the blood brain barrier intraperitoneal and oral treatment with PPS has been effective in extending the incubation period following both intraperitoneal and, in some cases, intracerebral inoculation of prions when administered prior to or soon after infection[92,93]. Overcoming the barrier to the brain by intraventricular infusion of PPS was successfully shown to extend the incubation period in hamster PrP transgenic mice infected intracerebrally with 263K prions even when administered late in the infection period; 35 days post infection at a time when PrP^{Sc} deposition has begun and 2 weeks before animals typically succumb to disease[63]. The low toxicity and relative efficacy of PPS led to the use of PPS in the clinic in patients with sporadic, variant, iatrogenic and familial forms of prion disease[94–100].

Clinical studies describe the treatment of patients with sporadic, variant and familial V180I-CJD, P102L-GSS and iatrogenic CJD following dura mater grafts or treatment with human growth hormone (hGH). While a number of patients succumbed to sepsis and pneumonia, all studies described an increase in survival relative to the historical mean and median incubation time for each disease group[94,96–99]. In cases where post-mortem was described, although pathology was present in the form of brain atrophy, there were examples where it was reported that there was less overall PrP^{Sc} in the brains of PPS treated patients[95,96]. The caveats to these promising observations is the potential for intensive care associated with the clinical trials to extend survival times, the potential that earlier diagnosis of disease and therefore longer apparent disease duration is associated with increased awareness of these disorders in recent times relative to historical data on disease duration and in the cases where less PrP^{Sc} is reported in patients who succumbed to intercurrent disease the possibility they had not acquired levels of PrP^{Sc} associated with terminal prion disease.

Animal studies provide a more controlled environment for clinical assessment of drug efficacy, although it again must be noted that the majority of prion studies are rooted in models based on scrapie prion strains. Nonetheless animal studies support the capacity of PPS treatment to extend disease duration across a range of prion strains (263K, RML, FU-01, 139A, ME7, 22A) when infected either intraperitoneally or intracerebrally[63,89–93,101,102]. Although prophylactic treatment prior to or immediately after treatment is most effective, both intraperitoneal and intracerebral treatment following intracerebral infection of hamster transgenic mice with 263K has been reported to extend incubation period[63,101]. Of note the only study to compare the efficacy of multiple prion strains was the seminal work of Doh-ura et al.[63] who compared FU-01 and RML in PrP overexpressing mice (Tga20) treated with PPS from 14 to 49 days post inoculation. Although the incubation period was extended in both strains the proportional increase was not as great for FU-01. It is interesting to speculate that FU-01 infected cells were also harder to cure than RML infected cells with quinacrine, which may reflect a generally more robust infection with this human derived strain.

A single intraperitoneal dose of PPS has been reported to delay disease induced by intraperitoneal infection with ME7 or 22 L prions when given within 24 h of infection[93]. In contrast, an oral doses of PPS were not effective against ME7 infection despite its efficacy when delivered intraperitoneally[93]. This may explain the failure of oral prophylaxis with PPS from 2 weeks before to 2 weeks following oral infection of mule deer with CWD prions[103]. Although a strain specific effect cannot be discounted, PPS was able to clear PrP^{Sc} from Mule Deer Brain (MDB) cells infected with CWD, suggested that CWD prions are inherently susceptible to PPS[104].

In vitro studies not only provide the opportunity to investigate mechanism of action but also strain specific affects. N2a cells infected with RML return an EC_{50} of between 10 and 100 ng/mL[105,106] and a similar EC_{50} for SMB (Chandler) of 150 ng/mL[107]. Kocisko et al. reported an EC_{50} for 22 L of 500 ng/mL, which is 50 times more than the EC_{50} reported by them for RML infection of the same cell line[106]. However, Yamasaki et al. reported an EC_{65} of 100 ng/mL for 22 L, suggesting that overall both strains a likely susceptible to PPS[108]. Larramendy-Gozalo et al.[102] reported that although treatment with PPS could reduce PrP^{Sc} in Chandler infected GT1-7, complete clearance was not achieved within the 4-day treatment even when the dose was escalated to 100 μg/mL. This observation is consistent with the likely action of PPS which is to prevent the generation of *de novo* PrP^{Sc} but not clear existing PrP^{Sc}. As with cell-based results using quinacrine the slower growing GT1-7 turnover prevents the dilution of existing PrP^{Sc} through cell growth. These observations may translate to a failure to clear existing PrP^{Sc} from terminally differentiated neurons.

A number of studies have pursued the mechanism of PPS activity *in vitro*. Although it does not deplete the overall pool of PrP^C that provides the substrate for PrP^{Sc} propagation PPS treatment has been shown to deplete PrP^C from the cell surface by endocytosis and redistribute it within the endosomal/lysosomal pathway[109]. These observations would suggest that *in vitro* and *in vivo* PPS depletes the available pool of PrP^C substrate rather than destabilizing the PrP^{Sc} template. Such observations may support a strain specific affect as some prion strains are thought to propagate at the cell surface while for other strains it may occur along the endosomal pathway. An alternative mechanism of action may involve binding of PrP^{Sc} to a receptor required for uptake of PrP^{Sc}. It has been suggested that PPS treatment inhibits binding of PrP^{Sc} to a putative laminin receptor[110]. Strain specific receptor mediated uptake is possible and supported by the observation that heparan sulfate can also act as a cellular receptor for prions[111]. In a cell free assay of prion propagation, the PMCA, the inclusion of PPS seemed to directedly interfere with propagation, which also suggest that PPS may interfere with the binding of PrP^C to PrP^{Sc}, the first step required for the template directed misfolding of PrP[112]. Finally, PPS treatment of cells has been shown to inhibit apoptosis induced by the toxic 106–126 peptide[113], although it is possible that this is due to internalization of PrP^C as knock-out cells are not susceptible to the toxic effects of 106–126 PrP peptide.

It is intriguing that in some case studies it was observed that PrP^{Sc} was reduced in the brains of PPS treated patients[95,96]. Such an observation is consistent with the clearance of PrP^{Sc} and infectivity from prion infected cell lines. It remains to be determined whether the continued atrophy reported in the brains of patients reflects the stage of disease when treatment began or whether clearance of PrP^{Sc} is a marker of infectivity and that a toxic species survives and continues to progress neurodegeneration, albeit at a reduced rate[98].

4.3 Flupirtine

Following *in vitro* studies reporting the ability of flupirtine maleate to attenuate toxicity associated with 106–126 peptide of PrP^C [114–116]; and in light of its known pharmacokinetics and safety profile, a case-controlled treatment study was run over a 4-year period recruiting 28 patients diagnosed with probable sporadic CJD[117]. Patients either received placebo ($N=13$) or flupirtine ($N=12$) with escalating dose reaching 300–400 mg/day over 3 days. Although the median survival time for patients receiving flupirtine was not significantly different to patients receiving the placebo (107 *vs* 106 days), cognitive testing indicated that patients receiving flupirtine did not decline as rapidly as those patients receiving the placebo control. The therapeutic effect of flupirtine is unlikely to be strain specific as the drug is proposed to up-regulate bcl-2 and therefore act to prevent the consequences of protein misfolding rather than prevent protein misfolding. However, the authors did note that this study had an unusually small proportion of T1MM patients enrolled across both arms. Patients with T1MM are associated with a faster clinical progression and in this regard strain variation could be expected to influence the window of opportunity for treatment and thus affect efficacy. Interestingly, unlike many of the cases treated with quinacrine, no patients enrolled in this study entered the study with akinetic mutism. This in itself may have contributed to the relative success of flupirtine in slowing disease progression by nature of enrolling patients earlier in disease. The duration of this treatment window is likely to be strain specific. The ability to diagnose and initiate treatment early remains elusive for many neurodegenerative disorders.

4.4 Doxycycline

Based on earlier work that demonstrated that the amyloid binding dye Congo Red (CR) could interfere with the conversion of PrP^C to PrP^{Sc} in cell based[118] and animal models[119] of prion disease structural analogues of CR, including tetracyclines and more specifically doxycycline were investigated for their therapeutic effect in prion disease, based on their lower toxicity and ability to cross the blood brain barrier. Initially *in vitro* studies using tetracycline demonstrated an ability to bind amyloid made from 106–126 and 82–146 fragments of the prion protein. In addition to binding to these fragments, tetracycline was also found to inhibit phosphate buffer induced fibrillization, increased sensitivity to protease digestion and reduced the *in vitro* toxicity of the 106–126 PrP peptide[120]. The clinical relevance of these observations was further tested when it was shown that tetracycline and doxycycline could reduce the protease resistance of PrP^{Sc} derived from the tissue of patients with type 1 or type 2 CJD, albeit with higher concentrations of tetracycline required to reduce the protease resistance of type 1 relative to type 2. Later studies also demonstrated that doxycycline and tetracycline treatment reduced the protease resistance of vCJD and BSE derived PrP^{Sc} [121].

The effect of reducing the protease resistance of PrP^{Sc} on infectivity was later confirmed with *in vivo* studies using 263K derived scrapie prions that had been pre-treated with tetracycline or doxycycline. In both cases pre-treatment delayed disease in hamsters after intracerebral infection[121,122]. Magnetic Resonance Imaging (MRI), biochemical and immunohistochemical analysis further demonstrated reduced pathology (T2 weighted signal, PrP^{Sc}) in hamsters during the incubation period, although there was no reduction in pathology in animals at the terminal stages of disease. Recognition that doxycycline may prove a useful prophylaxis following iatrogenic exposure to prions was demonstrated when intraperitoneal doxycycline treatment delivered shortly after intramuscular or subcutaneous inoculation of hamsters delayed disease progression[123]. Encouragingly, intraventricular administration of doxycycline delayed disease when treatment was commenced at the onset of symptoms. It was disappointing therefore when a randomized, double-blind study of doxycycline *vs* placebo in patients with CJD was stopped for futility when it showed no benefit[124]. Although the potential for a strain specific effect had been suggested by earlier *in vitro* studies; which had shown a difference in the dose of doxycycline required to reduce protease resistance in different PrP^{Sc} subtypes[120] and a strain specific effect reported from cell based studies using doxycycline to clear PrP^{Sc} from cerebellar granule cells infected with different human CJD strains[125], a post-hoc analysis of subgroups in the clinical trial did not identify a significant effect that could have been attributed to strain variation based on disease duration, codon 129 genotype or cause. It was concluded that the earlier diagnosis and treatment or higher doses of doxycycline may be more effective. In 2015 a 10-year preventative study was commenced for patients at genetic risk of developing FFI was commenced[126].

5. Concluding remarks

The existence of sub-clinical and pre-clinical prion disease infection highlights the need for asymptomatic screening for prion disease to prevent potential transmission in a surgical setting[127]. In the absence of such a test, a safe and effective universal infection control approach which can reduce prion infectivity without damaging sensitive surgical equipment is required to alleviate concerns of iatrogenic transmission. This chapter has highlighted the limitations of using a single prion strain to identify a universal decontamination method, with 263K prions, for example, being susceptible to decontamination with NaOH but relatively resistant to steam sterilization while the FU-01 strain appears relatively resistant to chemical inactivation with NaOH but sensitive to steam sterilization. This variation only addresses the tip of the iceberg with human prion strain specific differences attributable to prion conformation likely to add further complexity to this problem. The use of prion propagation models, such as the RT-QuIC and PMCA in conjunction with steel wire models of decontamination offers a solution to testing strain variation using patient derived tissue and mouse adapted human prion strains.

Effective treatment will also benefit from early diagnosis, with the results of trials in prion disease patients and other neurogenerative orders indicting that early intervention will have the most meaningful outcomes for patients. However, this review has also highlighted that differences in prion strains could contribute to resistance and persistence of existing PrP^{Sc} to degradation *in vivo*, subtle but important differences in EC_{50} that could affect the effective treatment thresholds and differences in propagation of prion strains that reflect either the site of propagation or cofactors that contribute to propagation.

Although challenging, these issues need to be addressed if we are to meaningfully change the outcome for people at risk or affected by prion diseases.

References

1. Pattison IH, Millson GC. Scrapie produced experimentally in goats with special reference to the clinical syndrome. *J. Comp Pathol.* 1961;71:101–109.
2. Bruce ME, McConnell I, Fraser H, Dickinson AG. The disease characteristics of different strains of scrapie in Sinc congenic mouse lines: implications for the nature of the agent and host control of pathogenesis. *J Gen Virol.* 1991;72(Pt. 3):595–603.
3. Fraser H, Dickinson AG. Scrapie in mice. Agent-strain differences in the distribution and intensity of grey matter vacuolation. *J Comp Pathol.* 1973;83(1):29–40.
4. Boyle A, Hogan K, Manson JC, Diack AB. Strain typing of prion diseases using in vivo mouse models. *Methods Mol Biol.* 1658;2017:263–283.
5. Bessen RA, Marsh RF. Biochemical and physical properties of the prion protein from two strains of the transmissible mink encephalopathy agent. *J Virol.* 1992;66(4): 2096–2101.
6. Caughey B, Raymond GJ, Bessen RA. Strain-dependent differences in beta-sheet conformations of abnormal prion protein. *J Biol Chem.* 1998;273(48):32230–32235.
7. Hill AF, Desbruslais M, Joiner S, et al. The same prion strain causes vCJD and BSE. *Nature.* 1997;389(6650):448–450 [526].

8. Parchi P, Castellani R, Capellari S, et al. Molecular basis of phenotypic variability in sporadic Creutzfeldt-Jakob disease. *Ann Neurol.* 1996;39(6):767–778.
9. Collinge J, Sidle KC, Meads J, Ironside J, Hill AF. Molecular analysis of prion strain variation and the aetiology of 'new variant' CJD. *Nature.* 1996;383(6602): 685–690.
10. Parchi P, Giese A, Capellari S, et al. Classification of sporadic Creutzfeldt-Jakob disease based on molecular and phenotypic analysis of 300 subjects. *Ann Neurol.* 1999;46(2): 224–233.
11. Zanusso G, Farinazzo A, Fiorini M, et al. pH-dependent prion protein conformation in classical Creutzfeldt-Jakob disease. *J Biol Chem.* 2001;276(44):40377–40380.
12. Wadsworth JD, Hill AF, Joiner S, Jackson GS, Clarke AR, Collinge J. Strain-specific prion-protein conformation determined by metal ions. *Nat Cell Biol.* 1999;1(1):55–59.
13. Hill AF, Joiner S, Wadsworth JD, et al. Molecular classification of sporadic Creutzfeldt-Jakob disease. *Brain.* 2003;126(Pt. 6):1333–1346.
14. Telling GC, Parchi P, DeArmond SJ, et al. Evidence for the conformation of the pathologic isoform of the prion protein enciphering and propagating prion diversity. *Science.* 1996;274(5295):2079–2082.
15. Lawson VA, Vella LJ, Stewart JD, et al. Mouse-adapted sporadic human Creutzfeldt-Jakob disease prions propagate in cell culture. *Int J Biochem Cell Biol.* 2008;40(12): 2793–2801.
16. Tateishi J, Ohta M, Koga M, Sato Y, Kuroiwa Y. Transmission of chronic spongiform encephalopathy with kuru plaques from humans to small rodents. *Ann Neurol.* 1979; 5(6):581–584.
17. Chandler RL, Turfrey BA. Inoculation of voles, Chinese hamsters, gerbils and guinea-pigs with scrapie brain material. *Res Vet Sci.* 1972;13(3):219–224.
18. Bessen RA, Marsh RF. Identification of two biologically distinct strains of transmissible mink encephalopathy in hamsters. *J Gen Virol.* 1992;73(Pt. 2):329–334.
19. Brown P, Gibbs Jr CJ, Amyx HL, et al. Chemical disinfection of Creutzfeldt-Jakob disease virus. *N Engl J Med.* 1982;306(21):1279–1282.
20. Brazier MW, Lewis V, Ciccotosto GD, et al. Correlative studies support lipid peroxidation is linked to PrP(res) propagation as an early primary pathogenic event in prion disease. *Brain Res Bull.* 2006;68(5):346–354.
21. Kimberlin RH, Walker CA, Fraser H. The genomic identity of different strains of mouse scrapie is expressed in hamsters and preserved on reisolation in mice. *J Gen Virol.* 1989;70(Pt. 8):2017–2025.
22. Prusiner SB. Novel proteinaceous infectious particles cause scrapie. *Science.* 1982; 216(4542):136–144.
23. Yamaguchi K, Kamatari YO, Ono F, et al. A designer molecular chaperone against transmissible spongiform encephalopathy slows disease progression in mice and macaques. *Nat Biomed Eng.* 2019;3(3):206–219.
24. Mallucci GR, White MD, Farmer M, et al. Targeting cellular prion protein reverses early cognitive deficits and neurophysiological dysfunction in prion-infected mice. *Neuron.* 2007;53(3):325–335.
25. Sim VL, Caughey B. Recent advances in prion chemotherapeutics. *Infect Disord Drug Targets.* 2009;9(1):81–91.
26. Milisav I, Suput D, Ribaric S. Unfolded protein response and macroautophagy in Alzheimer's, Parkinson's and prion diseases. *Molecules.* 2015;20(12):22718–22756.
27. Safar J, Cohen FE, Prusiner SB. Quantitative traits of prion strains are enciphered in the conformation of the prion protein. *Arch Virol Suppl.* 2000;16:227–235.
28. Gonzalez-Montalban N, Makarava N, Savtchenko R, Baskakov IV. Relationship between conformational stability and amplification efficiency of prions. *Biochemistry.* 2011;50(37):7933–7940.

29. Brazier MW, Doctrow SR, Masters CL, Collins SJ. A manganese-superoxide dismutase/catalase mimetic extends survival in a mouse model of human prion disease. *Free Radic Biol Med*. 2008;45(2):184–192.
30. Bartelt-Hunt SL, Bartz JC. Behavior of prions in the environment: implications for prion biology. *PLoS Pathog*. 2013;9(2), e1003113.
31. Sakudo A. Inactivation methods for prions. *Curr Issues Mol Biol*. 2020;36:23–32.
32. Safar J, Wille H, Itri V, et al. Eight prion strains have PrP(Sc) molecules with different conformations. *Nat Med*. 1998;4(10):1157–1165.
33. Peretz D, Scott MR, Groth D, et al. Strain-specified relative conformational stability of the scrapie prion protein. *Protein Sci*. 2001;10(4):854–863.
34. Choi YP, Peden AH, Groner A, Ironside JW, Head MW. Distinct stability states of disease-associated human prion protein identified by conformation-dependent immunoassay. *J Virol*. 2010;84(22):12030–12038.
35. Lawson VA, Stewart JD, Masters CL. Enzymatic detergent treatment protocol that reduces protease-resistant prion protein load and infectivity from surgical-steel monofilaments contaminated with a human-derived prion strain. *J Gen Virol*. 2007; 88(Pt. 10):2905–2914.
36. Kuczius T, Groschup MH. Differences in proteinase K resistance and neuronal deposition of abnormal prion proteins characterize bovine spongiform encephalopathy (BSE) and scrapie strains. *Mol Med*. 1999;5(6):406–418.
37. Scott MR, Groth D, Tatzelt J, et al. Propagation of prion strains through specific conformers of the prion protein. *J Virol*. 1997;71(12):9032–9044.
38. Pattison IH. Scrapie in the welsh mountain breed of sheep and its experimental transmission to goats. *Vet Rec*. 1965;77(47):1388–1390.
39. Brown P, Rohwer RG, Green EM, Gajdusek DC. Effect of chemicals, heat, and histopathologic processing on high-infectivity hamster-adapted scrapie virus. *J Infect Dis*. 1982;145(5):683–687.
40. Bernoulli C, Siegfried J, Baumgartner G, et al. Danger of accidental person-to-person transmission of Creutzfeldt-Jakob disease by surgery. *Lancet*. 1977;1(8009):478–479.
41. Gibbs Jr CJ, Asher DM, Kobrine A, Amyx HL, Sulima MP, Gajdusek DC. Transmission of Creutzfeldt-Jakob disease to a chimpanzee by electrodes contaminated during neurosurgery. *J Neurol Neurosurg Psychiatry*. 1994;57(6):757–758.
42. Kimberlin RH, Walker CA, Millson GC, et al. Disinfection studies with two strains of mouse-passaged scrapie agent. Guidelines for Creutzfeldt-Jakob and related agents. *J Neurol Sci*. 1983;59(3):355–369.
43. Tateishi J, Tashima T, Kitamoto T. Practical methods for chemical inactivation of Creutzfeldt-Jakob disease pathogen. *Microbiol Immunol*. 1991;35(2):163–166.
44. Ernst DR, Race RE. Comparative analysis of scrapie agent inactivation methods. *J Virol Methods*. 1993;41(2):193–201.
45. Tateishi J, Tashima T, Kitamoto T. Inactivation of the Creutzfeldt-Jakob disease agent. *Ann Neurol*. 1988;24(3):466.
46. Tamai Y, Taguchi F, Miura S. Inactivation of the Creutzfeldt-Jakob disease agent. *Ann Neurol*. 1988;24(3):466–467.
47. Brown P, Rohwer RG, Gajdusek DC. Sodium hydroxide decontamination of Creutzfeldt-Jakob disease virus. *N Engl J Med*. 1984;310(11):727.
48. Walker AS, Inderlied CB, Kingsbury DT. Conditions for the chemical and physical inactivation of the K. Fu. strain of the agent of Creutzfeldt-Jakob disease. *Am J Public Health*. 1983;73(6):661–665.
49. Brown P, Rohwer RG, Gajdusek DC. Newer data on the inactivation of scrapie virus or Creutzfeldt-Jakob disease virus in brain tissue. *J Infect Dis*. 1986;153(6):1145–1148.
50. Flechsig E, Hegyi I, Enari M, Schwarz P, Collinge J, Weissmann C. Transmission of scrapie by steel-surface-bound prions. *Mol Med*. 2001;7(10):679–684.

51. Jackson GS, McKintosh E, Flechsig E, et al. An enzyme-detergent method for effective prion decontamination of surgical steel. *J Gen Virol.* 2005;86(Pt. 3):869–878.
52. Lemmer K, Mielke M, Pauli G, Beekes M. Decontamination of surgical instruments from prion proteins: in vitro studies on the detachment, destabilization and degradation of PrPSc bound to steel surfaces. *J Gen Virol.* 2004;85(Pt. 12):3805–3816.
53. Yan ZX, Stitz L, Heeg P, Pfaff E, Roth K. Infectivity of prion protein bound to stainless steel wires: a model for testing decontamination procedures for transmissible spongiform encephalopathies. *Infect Control Hosp Epidemiol.* 2004;25(4):280–283.
54. Peretz D, Supattapone S, Giles K, et al. Inactivation of prions by acidic sodium dodecyl sulfate. *J Virol.* 2006;80(1):322–331.
55. Wilham JM, Orru CD, Bessen RA, et al. Rapid end-point quantitation of prion seeding activity with sensitivity comparable to bioassays. *PLoS Pathog.* 2010;6(12): e1001217.
56. Hughson AG, Race B, Kraus A, et al. Inactivation of prions and amyloid seeds with hypochlorous acid. *PLoS Pathog.* 2016;12(9):e1005914.
57. Mori T, Atarashi R, Furukawa K, et al. A direct assessment of human prion adhered to steel wire using real-time quaking-induced conversion. *Sci Rep.* 2016;6:24993.
58. Belondrade M, Nicot S, Beringue V, Coste J, Lehmann S, Bougard D. Rapid and highly sensitive detection of variant Creutzfeldt-Jakob disease abnormal prion protein on steel surfaces by protein misfolding cyclic amplification: application to prion decontamination studies. *PLoS One.* 2016;11(1):e0146833.
59. Teruya K, Doh-Ura K. Insights from therapeutic studies for PrP prion disease. *Cold Spring Harb Perspect Med.* 2017;7(3):a024430.
60. Korth C, May BC, Cohen FE, Prusiner SB. Acridine and phenothiazine derivatives as pharmacotherapeutics for prion disease. *Proc Natl Acad Sci U S A.* 2001;98(17): 9836–9841.
61. Collins SJ, Lewis V, Brazier M, Hill AF, Fletcher A, Masters CL. Quinacrine does not prolong survival in a murine Creutzfeldt-Jakob disease model. *Ann Neurol.* 2002; 52(4):503–506.
62. Gayrard V, Picard-Hagen N, Viguie C, Laroute V, Andreoletti O, Toutain PL. A possible pharmacological explanation for quinacrine failure to treat prion diseases: pharmacokinetic investigations in a ovine model of scrapie. *Br J Pharmacol.* 2005; 144:386–393.
63. Doh-ura K, Ishikawa K, Murakami-Kubo I, et al. Treatment of transmissible spongiform encephalopathy by intraventricular drug infusion in animal models. *J Virol.* 2004;78(10):4999–5006.
64. Ghaemmaghami S, Ahn M, Lessard P, et al. Continuous quinacrine treatment results in the formation of drug-resistant prions. *PLoS Pathog.* 2009;5(11):e1000673.
65. Nakajima M, Yamada T, Kusuhara T, et al. Results of quinacrine administration to patients with Creutzfeldt-Jakob disease. *Dement Geriatr Cogn Disord.* 2004;17(3):158–163.
66. Benito-Leon J. Combined quinacrine and chlorpromazine therapy in fatal familial insomnia. *Clin Neuropharmacol.* 2004;27(4):201–203.
67. Collinge J, Gorham M, Hudson F, et al. Safety and efficacy of quinacrine in human prion disease (PRION-1 study): a patient-preference trial. *Lancet Neurol.* 2009;8 (4):334–344.
68. Geschwind MD, Kuo AL, Wong KS, et al. Quinacrine treatment trial for sporadic Creutzfeldt-Jakob disease. *Neurology.* 2013;81(23):2015–2023.
69. Doh-Ura K, Iwaki T, Caughey B. Lysosomotropic agents and cysteine protease inhibitors inhibit scrapie-associated prion protein accumulation. *J Virol.* 2000;74(10): 4894–4897.
70. Nguyen T, Sakasegawa Y, Doh-Ura K, Go ML. Anti-prion activities and drug-like potential of functionalized quinacrine analogs with basic phenyl residues at the 9-amino position. *Eur J Med Chem.* 2011;46(7):2917–2929.

71. Nguyen TH, Lee CY, Teruya K, Ong WY, Doh-ura K, Go ML. Antiprion activity of functionalized 9-aminoacridines related to quinacrine. *Bioorg Med Chem.* 2008;16(14): 6737–6746.
72. Sandberg MK, Wallen P, Wikstrom MA, Kristensson K. Scrapie-infected GT1-1 cells show impaired function of voltage-gated N-type calcium channels (Ca(v) 2.2) which is ameliorated by quinacrine treatment. *Neurobiol Dis.* 2004;15(1):143–151.
73. Kishida H, Sakasegawa Y, Watanabe K, et al. Non-glycosylphosphatidylinositol (GPI)-anchored recombinant prion protein with dominant-negative mutation inhibits PrPSc replication in vitro. *Amyloid.* 2004;11(1):14–20.
74. Ryou C, Legname G, Peretz D, Craig JC, Baldwin MA, Prusiner SB. Differential inhibition of prion propagation by enantiomers of quinacrine. *Lab Investig.* 2003;83(6): 837–843.
75. Dollinger S, Lober S, Klingenstein R, Korth C, Gmeiner P. A chimeric ligand approach leading to potent antiprion active acridine derivatives: design, synthesis, and biological investigations. *J Med Chem.* 2006;49(22):6591–6595.
76. Bian J, Kang HE, Telling GC. Quinacrine promotes replication and conformational mutation of chronic wasting disease prions. *Proc Natl Acad Sci U S A.* 2014;111(16): 6028–6033.
77. Barret A, Tagliavini F, Forloni G, et al. Evaluation of quinacrine treatment for prion diseases. *J Virol.* 2003;77(15):8462–8469.
78. Sandberg MK, Low P. Altered interaction and expression of proteins involved in neurosecretion in scrapie-infected GT1-1 cells. *J Biol Chem.* 2005;280(2):1264–1271.
79. Didonna A, Legname G. Aberrant ERK 1/2 complex activation and localization in scrapie-infected GT1-1 cells. *Mol Neurodegener.* 2010;5:29.
80. Fasano C, Campana V, Griffiths B, Kelly G, Schiavo G, Zurzolo C. Gene expression profile of quinacrine-cured prion-infected mouse neuronal cells. *J Neurochem.* 2008; 105(1):239–250.
81. Bajsarowicz K, Ahn M, Ackerman L, Dearmond BN, Carlson G, DeArmond SJ. A brain aggregate model gives new insights into the pathobiology and treatment of prion diseases. *J Neuropathol Exp Neurol.* 2012;71(5):449–466.
82. Ladner-Keay CL, Ross L, Perez-Pineiro R, et al. A simple in vitro assay for assessing the efficacy, mechanisms and kinetics of anti-prion fibril compounds. *Prion.* 2018;12(5–6): 280–300.
83. de Moraes MC, Santos JB, Dos Anjos DM, et al. Prion protein-coated magnetic beads: synthesis, characterization and development of a new ligands screening method. *J Chromatogr A.* 2015;1379:1–8.
84. Spilman P, Lessard P, Sattavat M, et al. A gamma-secretase inhibitor and quinacrine reduce prions and prevent dendritic degeneration in murine brains. *Proc Natl Acad Sci U S A.* 2008;105(30):10595–10600.
85. Villa V, Tonelli M, Thellung S, et al. Efficacy of novel acridine derivatives in the inhibition of hPrP90-231 prion protein fragment toxicity. *Neurotox Res.* 2011;19(4): 556–574.
86. Farrelly PV, Kenna BL, Laohachai KL, et al. Quinacrine blocks PrP (106-126)-formed channels. *J Neurosci Res.* 2003;74(6):934–941.
87. Turnbull S, Tabner BJ, Brown DR, Allsop D. Quinacrine acts as an antioxidant and reduces the toxicity of the prion peptide PrP106-126. *Neuroreport.* 2003;14(13): 1743–1745.
88. Foliaki ST, Lewis V, Finkelstein DI, et al. Prion acute synaptotoxicity is largely driven by protease-resistant PrPSc species. *PLoS Pathog.* 2018;14(8):e1007214.
89. Ehlers B, Diringer H. Dextran sulphate 500 delays and prevents mouse scrapie by impairment of agent replication in spleen. *J Gen Virol.* 1984;65(Pt. 8):1325–1330.
90. Diringer H, Ehlers B. Chemoprophylaxis of scrapie in mice. *J Gen Virol.* 1991; 72(Pt. 2):457–460.

91. Ladogana A, Casaccia P, Ingrosso L, et al. Sulphate polyanions prolong the incubation period of scrapie-infected hamsters. *J Gen Virol.* 1992;73(Pt. 3):661–665.
92. Kocisko DA, Caughey WS, Race RE, Roper G, Caughey B, Morrey JD. A porphyrin increases survival time of mice after intracerebral prion infection. *Antimicrob Agents Chemother.* 2006;50(2):759–761.
93. Farquhar C, Dickinson A, Bruce M. Prophylactic potential of pentosan polysulphate in transmissible spongiform encephalopathies. *Lancet.* 1999;353(9147):117.
94. Newman PK, Todd NV, Scoones D, et al. Postmortem findings in a case of variant Creutzfeldt-Jakob disease treated with intraventricular pentosan polysulfate. *J Neurol Neurosurg Psychiatry.* 2014;85(8):921–924.
95. Honda H, Sasaki K, Minaki H, et al. Protease-resistant PrP and PrP oligomers in the brain in human prion diseases after intraventricular pentosan polysulfate infusion. *Neuropathology.* 2012;32(2):124–132.
96. Terada T, Tsuboi Y, Obi T, et al. Less protease-resistant PrP in a patient with sporadic CJD treated with intraventricular pentosan polysulphate. *Acta Neurol Scand.* 2010; 121(2):127–130.
97. Tsuboi Y, Doh-Ura K, Yamada T. Continuous intraventricular infusion of pentosan polysulfate: clinical trial against prion diseases. *Neuropathology.* 2009;29(5):632–636.
98. Bone I, Belton L, Walker AS, Darbyshire J. Intraventricular pentosan polysulphate in human prion diseases: an observational study in the UK. *Eur J Neurol.* 2008;15(5): 458–464.
99. Todd NV, Morrow J, Doh-ura K, et al. Cerebroventricular infusion of pentosan polysulphate in human variant Creutzfeldt-Jakob disease. *J Infect.* 2005;50(5):394–396.
100. Whittle IR, Knight RS, Will RG. Unsuccessful intraventricular pentosan polysulphate treatment of variant Creutzfeldt-Jakob disease. *Acta Neurochir.* 2006;148(6):677–679 [discussion 679].
101. Kocisko DA, Caughey B, Morrey JD, Race RE. Enhanced antiscrapie effect using combination drug treatment. *Antimicrob Agents Chemother.* 2006;50(10):3447–3449.
102. Larramendy-Gozalo C, Barret A, Daudigeos E, et al. Comparison of CR36, a new heparan mimetic, and pentosan polysulfate in the treatment of prion diseases. *J Gen Virol.* 2007;88(Pt. 3):1062–1067.
103. Wolfe LL, Kocisko DA, Caughey B, Miller MW. Assessment of prospective preventive therapies for chronic wasting disease in mule deer. *J Wildl Dis.* 2012;48(2):530–533.
104. Raymond GJ, Olsen EA, Lee KS, et al. Inhibition of protease-resistant prion protein formation in a transformed deer cell line infected with chronic wasting disease. *J Virol.* 2006;80(2):596–604.
105. Schonberger O, Horonchik L, Gabizon R, Papy-Garcia D, Barritault D, Taraboulos A. Novel heparan mimetics potently inhibit the scrapie prion protein and its endocytosis. *Biochem Biophys Res Commun.* 2003;312(2):473–479.
106. Kocisko DA, Engel AL, Harbuck K, et al. Comparison of protease-resistant prion protein inhibitors in cell cultures infected with two strains of mouse and sheep scrapie. *Neurosci Lett.* 2005;388(2):106–111.
107. Birkett CR, Hennion RM, Bembridge DA, et al. Scrapie strains maintain biological phenotypes on propagation in a cell line in culture. *EMBO J.* 2001;20(13):3351–3358.
108. Yamasaki T, Suzuki A, Hasebe R, Horiuchi M. Comparison of the anti-prion mechanism of four different anti-prion compounds, anti-PrP monoclonal antibody 44B1, pentosan polysulfate, chlorpromazine, and U18666A, in prion-infected mouse neuroblastoma cells. *PLoS One.* 2014;9(9), e106516.
109. Shyng SL, Lehmann S, Moulder KL, Harris DA. Sulfated glycans stimulate endocytosis of the cellular isoform of the prion protein, PrPC, in cultured cells. *J Biol Chem.* 1995;270(50):30221–30229.

110. Gauczynski S, Nikles D, El-Gogo S, et al. The 37-kDa/67-kDa laminin receptor acts as a receptor for infectious prions and is inhibited by polysulfated glycanes. *J Infect Dis*. 2006;194(5):702–709.
111. Horonchik L, Tzaban S, Ben-Zaken O, et al. Heparan sulfate is a cellular receptor for purified infectious prions. *J Biol Chem*. 2005;280(17):17062–17067.
112. Yokoyama T, Takeuchi A, Yamamoto M, Kitamoto T, Ironside JW, Morita M. Heparin enhances the cell-protein misfolding cyclic amplification efficiency of variant Creutzfeldt-Jakob disease. *Neurosci Lett*. 2011;498(2):119–123.
113. Deli MA, Sakaguchi S, Nakaoke R, et al. PrP fragment 106-126 is toxic to cerebral endothelial cells expressing PrP(C). *Neuroreport*. 2000;11(17):3931–3936.
114. Perovic S, Bohm M, Meesters E, Meinhardt A, Pergande G, Muller WE. Pharmacological intervention in age-associated brain disorders by Flupirtine: Alzheimer's and prion diseases. *Mech Ageing Dev*. 1998;101(1–2):1–19.
115. Perovic S, Pergande G, Ushijima H, Kelve M, Forrest J, Muller WE. Flupirtine partially prevents neuronal injury induced by prion protein fragment and lead acetate. *Neurodegeneration*. 1995;4(4):369–374.
116. Perovic S, Schroder HC, Pergande G, Ushijima H, Muller WE. Effect of flupirtine on Bcl-2 and glutathione level in neuronal cells treated in vitro with the prion protein fragment (PrP106-126). *Exp Neurol*. 1997;147(2):518–524.
117. Otto M, Cepek L, Ratzka P, et al. Efficacy of flupirtine on cognitive function in patients with CJD: a double-blind study. *Neurology*. 2004;62(5):714–718.
118. Caughey B, Race RE. Potent inhibition of scrapie-associated PrP accumulation by congo red. *J Neurochem*. 1992;59(2):768–771.
119. Ingrosso L, Ladogana A, Pocchiari M. Congo red prolongs the incubation period in scrapie-infected hamsters. *J Virol*. 1995;69(1):506–508.
120. Tagliavini F, Forloni G, Colombo L, et al. Tetracycline affects abnormal properties of synthetic PrP peptides and PrP(Sc) in vitro. *J Mol Biol*. 2000;300(5):1309–1322.
121. Forloni G, Iussich S, Awan T, et al. Tetracyclines affect prion infectivity. *Proc Natl Acad Sci U S A*. 2002;99(16):10849–10854.
122. Corato M, Ogliari P, Ceciliani F, et al. Doxorubicin and congo red effectiveness on prion infectivity in golden Syrian hamster. *Anticancer Res*. 2009;29(7):2507–2512.
123. De Luigi A, Colombo L, Diomede L, et al. The efficacy of tetracyclines in peripheral and intracerebral prion infection. *PLoS One*. 2008;3(3), e1888.
124. Haik S, Marcon G, Mallet A, et al. Doxycycline in Creutzfeldt-Jakob disease: a phase 2, randomised, double-blind, placebo-controlled trial. *Lancet Neurol*. 2014;13(2): 150–158.
125. Hannaoui S, Gougerot A, Privat N, et al. Cycline efficacy on the propagation of human prions in primary cultured neurons is strain-specific. *J Infect Dis*. 2014;209(7): 1144–1148.
126. Forloni G, Tettamanti M, Lucca U, et al. Preventive study in subjects at risk of fatal familial insomnia: innovative approach to rare diseases. *Prion*. 2015;9(2):75–79.
127. Collins S, Law MG, Fletcher A, Boyd A, Kaldor J, Masters CL. Surgical treatment and risk of sporadic Creutzfeldt-Jakob disease: a case-control study. *Lancet*. 1999;353(9154): 693–697.

CHAPTER SEVEN

Transgenic mouse models for the study of prion diseases

Alba Marín-Moreno, Juan Carlos Espinosa, and Juan María Torres*
Centro de Investigación en Sanidad Animal (CISA-INIA), Madrid, Spain
*Corresponding author: e-mail address: jmtorres@inia.es

Contents

Abstract

Prions are unique agents that challenge the molecular biology dogma by transmitting information on the protein level. They cause neurodegenerative diseases that lack of any cure or treatment called transmissible spongiform encephalopathies. The function of the normal form of the prion protein, the exact mechanism of prion propagation between species as well as at the cellular level and neuron degeneration remains elusive. However, great amount of information known for all these aspects has been achieved thanks to the use of animal models and more precisely to transgenic mouse models. In this chapter, the main contributions of these powerful research tools in the prion field are revised.

1. Introduction

Transmissible spongiform encephalopathies (TSEs) or prion diseases are fatal neurodegenerative diseases that affect several mammal species as humans and species included in the human food chain. Humans are affected by Creutzfeldt-Jakob disease (CJD), kuru, Gerstmann-Sträussler-Scheinker syndrome (GSS) and familial fatal insomnia (FFI); while scrapie, bovine spongiform encephalopathy (BSE) and chronic wasting disease (CWD)

Progress in Molecular Biology and Translational Science, Volume 175
ISSN 1877-1173
https://doi.org/10.1016/bs.pmbts.2020.08.007

primarily are described in small ruminants, cattle and cervid species, respectively. This group of diseases produce long incubation times and their main symptoms are neurological behavior abnormalities. In addition patients present motor dysfunction, cognitive impairment and cerebral ataxia.

All TSEs share the same molecular mechanism responsible for disease development. The normal form of the prion protein, also known as cellular form or PrP^C, is converted into a disease-associated form known as PrP^{Sc}.[1] PrP^{Sc} sustains its formation by recruiting PrP^C molecules and transforming them into new PrP^{Sc} particles in an exponential way. PrP^C transformation into PrP^{Sc} causes a change in the protein tridimensional structure characterized by an increase in the β-sheet content.[2] Such change is also reflected in the protein physicochemical features. PrP^C is monomeric, soluble in non-ionic detergents and sensitive to protease action whereas PrP^{Sc} tends to aggregate, is not soluble in non-ionic detergents and partially resistant to proteases.[3] Indeed, PrP^{Sc} treatment with proteinase K produces a protease-resistant core called PrP^{res} that is the most used prion surrogate marker.[4]

Given its long incubation times and their lack of treatment, modeling TSEs for laboratory studies is of utmost importance. Some of these diseases have animals as their primary host but species like small ruminants or cattle are not suitable for laboratory studies due to their size, generation time, difficult accessibility and manipulation and the high cost of their maintenance in animal facilities. For human prion diseases, primates might appear desirable models and some studies have been done.[5–18] However, some of the points mentioned above also apply for primates and there is also an ethical controversy with the use of primates for investigation. Mice are small, have both short lifespan and generation time, are easy to manipulate and well-implemented molecular biology techniques allow genetic manipulation to produce transgenic animals. Thus, mice and transgenic mice have been slowly implemented as the preferred animal model for the study of prion diseases in several aspects that will be separately discussed below.

2. Transgenic mouse models for the study of PrP^C function

The prion protein amino acid sequence is well conserved throughout evolution in mammal species.[19] This fact suggests that the prion protein in its cellular form has an important function. A common strategy in molecular biology for the study of protein function is the generation of knock-out transgenic mice. In knock-out transgenic mice, the expression of the target protein is abrogated disrupting the coding sequence of the respective gene by the use of the gene targeting method based on homologous recombination. In the absence of the target protein, the phenotype of the produced knock-out mice is studied and the function of the target protein then inferred.

The first knock-out mice for the prion protein (*Prnp*$^{-/-}$), the *Zurich I* line, was generated in 1992.[20] Surprisingly, the generated mice were fully developed till the end of their lifespan without any obvious detriment phenotypes.[20] This was soon validated by the production of other mice line, the *Npu* line.[21] Since then, many other *Prnp*$^{-/-}$ mice have been generated (Table 1) and different functions were inferred for the prion protein (Table 1). There is controversy on which of these inferred functions are real and the role that the different genetic backgrounds of the animal models may have played in the observed phenotypes.[28] Seven *Prnp*$^{-/-}$ mouse lines have been generated by the classical gene targeting approach using embryonic stem cells derived from the 129 strain of the laboratory mouse (Table 1). From them, only the *Npu* line has been kept in the same genetic background. The others have been maintained in other non-129 backgrounds thus 129 background alleles of genes flanking the *Prnp* locus will remain in the *Prnp*$^{-/-}$ mouse line despite extensive backcrossing to other non-129 genetic background because of the low probability of recombination between close genes.[29,30] To end this conundrum, in 2016 a new *Prnp*$^{-/-}$ mouse line, the *Zurich III*, was generated using the transcription activator-like effector nuclease (TALEN)-based gene editing.[27] This technique relies on gene edition thus allowing the disruption of the target coding sequence without altering the gene architecture.[31] Indeed, a phenotype attributed to the *Zurich I* and *Ngsk* [23] lines related to an increased macrophage phagocytosis of apoptotic cells was proved to be exerted by a *Prnp* close gene that encodes signal-regulatory protein α-1.[30] However, this was not the only *Prnp*$^{-/-}$ phenotype that was later found to be an artifact. *Prnp*$^{-/-}$ mouse lines *Rcm0* [22], *Ngsk* [23], *Rikn* [24] and *Zurich II* [25] presented late-onset ataxia due to massive death of Purkinje neurons of the cerebellum.[28] All these mouse lines were generated by gene targeting that deleted *Prnp* exon 3 while spanning a splice acceptor site that led to the ectopic expression of Dpl protein (encoded by the nearby *Prnd* gene) under the control of the *Prnp* regulatory sequences. The same problem appeared on *Ki-Prnp-GFP* mouse line (in which *Prnp* gene was disrupted by inserting the green fluorescent protein coding sequence) but at lower levels.[26] The death of the neurons was linked to Dpl-mediated neurotoxicity and not caused by *Prnp* absence.[28]

Table 1 *Prnp*$^{-/-}$ mouse models generated and inferred prion protein functions.

Knock-out mice line	References	Generation technique	Genetic background	Ectopic Dpl expression	Inferred PrP^C functions
Zurich I	20	Gene targeting	Mixed	No	Resistance to prion infection, synaptic and electrophysiological deficits, circadian rhythm abnormalities, processing of sensory information in the olfactory system, CPD
Npu	21	Gene targeting	129/Ola	No	Resistance to prion infection, circadian rhythm abnormalities, processing of sensory information in the olfactory system, hematopoietic stem cell renewal, neural stem cell proliferation in adult neurogenesis, CPD
Rcm0	22	Gene targeting	129/Ola	Yes	Dpl mediated-neurotoxicity
Ngsk	23	Gene targeting	Mixed	Yes	CPD, Dpl mediated-neurotoxicity
Rikn	24	Gene targeting	Mixed	Yes	Dpl mediated-neurotoxicity
Zurich II	25	Gene targeting	Mixed	Yes	Dpl mediated-neurotoxicity
Ki-Prnp-GFP	26	Gene targeting	Mixed	Yes (low levels)	CPD
Zurich III	27	TALEN-based gene editing	C57BL/6J	No	Resistance to prion infection, CPD

Other phenotypes attributed to *Prnp* depletion are subtle and lack of a proper molecular explanation on how PrP^C may exert such effect like synaptic and electrophysiological deficits, hematopoietic stem cell renewal, circadian rhythm regulation, processing of sensory information in the olfactory system or neural stem cell proliferation in adult neurogenesis (Table 1). Apart from resistance to prion infection,[32–35] the only phenotype caused by PrP abrogation that was maintained in several mouse lines was chronic demyelinating polyneuropathy (CDP) in peripheral nerves[27,36–38] (Table 1). CDP phenotype can be rescued by neuronspecific re-introduction of PrP^C expression in knock-out mice.[36] All these evidences suggest that neuronal PrP^C plays a critical role in the conservation of peripheral nerve myelination. More studies are needed to fully decipher the whole picture of PrP^C function and its understanding may improve the knowledge of the molecular basis of TSEs.

3. Transgenic mouse models for the study of the prion transmission barrier

Certain prion diseases can be transmitted from one species to a different one. This term is known as interspecies prion transmission and is mainly affected by the homology between sequences of inoculum PrP^{Sc} and host PrP^C.[39] In this line, mismatches between both sequences on different species and intraspecies PrP polymorphisms also influence TSE progression.[39] Apart from the species barrier, prion strains also play a major role in prion transmission. Prions strains are well-defined agents that differ on their incubation times in animals, the lesions they produce in the CNS, the neuroanatomic distribution of PrP^{Sc} deposits and the biochemical features of PrP^{res}.[40] Variety on these characteristics that reflect the existence of different strains is thought to be based on different PrP^{Sc} tertiary structures.[41] Thus, the study of prion transmission barrier (including both aspects of species barrier and strains) is of crucial importance from the point of view of human health and food safety. For that purpose, transgenic mouse lines expressing the prion protein from different species of interest would provide useful models for the study on how certain TSEs may jump from one species to another. Transgenic mouse lines were generated using two different approaches: transgenic mice and knock-in mice. Transgenic mice were generated by microinjection of fertilized mouse embryos with DNA encoding a transgene cassette that drives expression of mutant PrP in the brain. The transgene includes Syrian hamster or mouse PrP promoter elements to achieve neuronal expression of the mutant PrP. Usually, several copies of the transgene are inserted into the mouse genome, resulting in overexpression of the transgenic PrP. As a result, disease progression can be accelerated and phenotypes appear within the mouse lifespan but integration artifacts can occur due to the random insertion of transgenes into the genome resulting sometimes in lethality. In addition, high levels of PrP overexpression can trigger non-prion-disease-specific pathology.[42] In knock-in mouse models, the endogenous mouse PrP locus is replaced with the PrP protein of interest via gene targeting based on homologous recombination in embryonic stem cells. Knock-in mice express the transgenic PrP at physiological levels and always under the control of their endogenous promoter and regulatory elements thus achieving the same spatiotemporal expression pattern of endogenous mouse PrP. However, prion diseases have a long and slow progression and sometimes the incubation times may exceed the mouse lifespan for this type of transgenic lines.[43]

Nevertheless, before studying how a TSE may infect other hosts, transgenic mouse lines were developed for the study of prion diseases in their principal hosts (Tables 2 and 3). Thus, sheep/goat PrP transgenic lines were generated and found susceptible to classical and atypical scrapie [42,59–65,89] bovine PrP transgenic lines were found susceptible to classical and atypical BSE,[44–49,90–92] mink PrP transgenic lines were developed and found susceptible to transmissible mink encephalopathy (TME)[78] and cervid PrP transgenic lines were generated and found susceptible to CWD.[55,66–71] In the same line, different human PrP transgenic lines have also been produced and found susceptible to kuru and sCJD.[49,52–57,93] These generic models were of striking importance for prion strain characterization, especially when new atypical strains like atypical scrapie or atypical L- and H-type BSE arised.[94–96] Apart from these models, other species that do not have proper TSEs but may be affected by them or are somehow interest for studying prion transmission have been produced. Thus there also exists transgenic mouse models harboring hamster[75–77] and mouse[79–82] PrP on which general aspects of prion transmission have been deciphered. In order to study their possible susceptibility/resistance to prion disease affecting other mammals found also on the human food chain, also models for pig PrP have been developed.[87,88,97] Clues on how some species are naturally resistant to prions have been studied using models expressing rabbit,[84] horse[85] and dog[86] PrP. Finally, some transgenic mouse models have been produced to compare or validate the usefulness of certain species on the study of prion disease as is the case of bank vole[72–74] and macaques[83] thus comparing the original species and the transgenic mouse lines produced.

Table 2 Mouse models expressing bovine, human, sheep/goat or cervid PrP^C.

Species	PrP sequence	Generation technique	Mouse line	PrP expression level[a]	Spontaneous disease (onset)	Neuropathology	PrP^{res}	Transmissibility	References
Bovine	Bovine	Transgenic	boTg110	$8\times^{(+/+)}$	No	No	No	No	44
		Transgenic	boTg035	$6\times^{(+/+)}$	No	No	No	No	44
		Transgenic	boTg078	$2\times^{(+/+)}$	No	No	No	No	44
		Transgenic	boTg022	$0.5\times^{(+/+)}$	No	No	No	No	44
		Transgenic	BoPrP4125	$8–16\times^{(+/-)}$	No	No	No	No	45
		Transgenic	BoPrP4092	$4–8\times^{(+/-)}$	No	No	No	No	46
		Transgenic	Tg540	$8\times^{(+/-)}$	No	No	No	No	47
		Transgenic	TgbovXV	$8\times^{(+/-)}$	No	No	No	No	48
		Knock-in	BovTg	$1\times^{(+/+)}$	No	No	No	No	49
	Bovine 7 octa-repeats	Transgenic	bo7OR-Tg007	$6\times^{(+/+)}$	No	No	No	No	50
	Bovine 5 octa-repeats	Transgenic	bo5OR-Tg008	$1\times^{(+/+)}$	No	No	No	No	51

Continued

Table 2 Mouse models expressing bovine, human, sheep/goat or cervid PrP^C.—cont'd

Species	PrP sequence	Generation technique	Mouse line	PrP expression level[a]	Spontaneous disease (onset)	Neuropathology	PrP^{res}	Transmissibility	References
Human	Human Val_{129}	Transgenic	Tg110	$0.5\times^{(+/-)}$	No	No	No	No	52
		Transgenic	Tg152	$4–8\times^{(+/-)}$	No	No	No	No	52
		Transgenic	Tg361	$4\times^{(+/+)}$	No	No	No	No	53
	Human Met_{129}	Transgenic	Tg440	$2\times^{(+/-)}$	No	No	No	No	52
		Knock-in	HuMTg	$1\times^{(+/+)}$	No	No	No	No	49
		Knock-in	HuVTg	$1\times^{(+/+)}$	No	No	No	No	49
		Transgenic	Tg35	$2\times^{(+/+)}$	No	No	No	No	54
		Transgenic	Tg45	$4\times^{(+/+)}$	No	No	No	No	54
		Transgenic	Tg1	$2\times^{(N/A)}$	No	No	No	No	55
		Transgenic	Tg40	$1\times^{(N/A)}$	No	No	No	No	55
		Transgenic	Tg340	$4\times^{(+/+)}$	No	No	No	No	56
		Transgenic	Tg650	$6\times^{(+/+)}$	No	No	No	No	57
	Human Val_{127}–Met_{129}	Transgenic	Tg183	$2\times^{(+/+)}$	No	No	No	No	58
		Transgenic	Tg190	$1\times^{(+/+)}$	No	No	No	No	58

Sheep/ Goat	Sheep-goat	Transgenic	TgshpXI	2–4 $\times^{(+/-)}$	No	No	No	No	59
		Transgenic	TgshpIX	4–8 $\times^{(+/-)}$	No	No	No	No	59
		Transgenic	TgShePrPA1855	High$^{(+/-)}$	Yes	Yes	Yes	N/D	42
		Transgenic	TgShePrP1849	1 $\times^{(+/-)}$	No	No	No	N/D	42
		Transgenic	TgSShpPrP	4 $\times^{(N/A)}$	No	No	No	No	60
		Transgenic	TgOvA	1 $\times^{(+/-)}$	No	No	No	No	61
		Transgenic	TgovPrP4	3 $\times^{(+/+)}$	No	No	No	No	62
		Transgenic	TgovPrP59	3 $\times^{(+/+)}$	No	No	No	No	63
		Transgenic	Tg501	1 $\times^{(+/-)}$	No	No	No	No	64
	Sheep V_{136}	Transgenic	Tg301	8 $\times^{(+/-)}$	No	No	No	No	65
		Transgenic	Tg328	3.5 $\times^{(+/-)}$	No	No	No	No	65
		Transgenic	Tg338	8–10 $\times^{(+/+)}$	No	No	No	No	65
		Transgenic	TgOvV	1 $\times^{(+/-)}$	No	No	No	No	61
	Goat K_{222}	Transgenic	Tg516	1 $\times^{(+/-)}$	No	No	No	No	64
Cervids	Deer	Transgenic	TgCerPrP1536	5 $\times^{(+/-)}$	No	No	No	No	66
		Transgenic	TgCerPrP1534	3 $\times^{(+/-)}$	No	No	No	No	66
		Transgenic	TgDePrP 10,945	1 $\times^{(+/-)}$	No	No	No	No	67

Continued

Table 2 Mouse models expressing bovine, human, sheep/goat or cervid PrP^C.—cont'd

Species	PrP sequence	Generation technique	Mouse line	PrP expression level[a]	Spontaneous disease (onset)	Neuropathology	PrP^{res}	Transmissibility	References
		Transgenic	TgDePrP 10,969	$1\times^{(+/-)}$	No	No	No	No	67
		Knock-in	GtQ226	$1\times^{(+/+)}$	No	No	No	No	68
		Transgenic	Tg Line 33	$1\times^{(+/-)}$	No	No	No	No	69
		Transgenic	Tg Line 15	$0.5\times^{(+/-)}$	No	No	No	No	69
	Deer S_{96}	Transgenic	Tg Line 60	$0.7\times^{(+/-)}$	No	No	No	No	69
	Elk	Transgenic	TgCerPrP5037	$5\times^{(+/-)}$	No	No	No	No	70
		Transgenic	TgCerPrP5029	$1\times^{(+/-)}$	No	No	No	No	70
		Transgenic	TgElk	$2.5\times^{(+/+)}$	No	No	No	No	71
		Transgenic	TgElkPrP 12,577	$2\times^{(+/-)}$	No	No	No	No	67
		Transgenic	TgElkPrP 12,580	$2\times^{(+/-)}$	No	No	No	No	67
		Transgenic	TgElkPrP 3934	$1.5\times^{(+/-)}$	No	No	No	No	67
		Transgenic	TgElkPrP 12,584	$3\times^{(+/-)}$	No	No	No	No	67
		Knock-in	GtE226	$1\times^{(+/+)}$	No	No	No	No	68
	Elk L_{132}	Transgenic	Tg12	$2\times^{(N/A)}$	No	No	No	No	55

[a]Relative to brain PrP^C expression in the original PrP^C sequence species. (+/+) homozygous animals and (+/−) hemizygous animals.
N/D, not determined; N/A, not available.

Table 3 Mouse models expressing bank vole, hamster, mouse, macaque, rabbit, horse, dog or pig PrP^C.

Species	PrP sequence	Generation technique	Mouse line	PrP expression level[a]	Spontaneous disease (onset)	Neuropathology	PrP^{res}	Transmissibility	References
Bank vole	Bank vole I_{109}	Transgenic	Tg3574	$3.6\times^{(+/-)}$	Yes (≈340)	Yes	Yes	Yes	72
		Transgenic	Tg3581	$5.3\times^{(+/-)}$	Yes (≈218)	Yes	Yes	Yes	72
	Bank vole M_{109}	Transgenic	Tg407	$1\times^{(+/+)}$	No	No	No	No	73
		Transgenic	Tg3118	$2.5\times^{(+/-)}$	No	No	No	No	74
		Transgenic	Tg22019	$5\times^{(+/-)}$	No	No	No	No	74
Hamster	Hamster	Transgenic	Tg7	$4–8\times^{(+/-)}$	No	No	No	No	75
		Transgenic	Tg10	ND	No	No	No	No	76
		Transgenic	Tg81	$2–4\times^{(+/-)}$	No	No	No	No	77
		Transgenic	Tg71	$1–2\times^{(+/-)}$	No	No	No	No	77
		Transgenic	Tg52NSE	ND	No	No	No	No	76
Mink	Mink	Transgenic	Mk7	$2\times^{(+/-)}$	No	No	No	No	78
		Transgenic	Mk45	$1\times^{(+/-)}$	No	No	No	No	78
Mouse	MousePrP-A	Transgenic	Tga20	$10\times^{(+/+)}$	No	No	No	No	79
	MousePrP-B	Transgenic	Tg 94	$4\times^{(+/-)}$	No	No	No	No	75
	MousePrP-B	Transgenic	Tg 117	$4\times^{(+/-)}$	No	No	No	No	75
	MousePrP-A	Transgenic	Tg4053/ FVB	$8\times^{(+/-)}$	No	No	No	No	80

Continued

Table 3 Mouse models expressing bank vole, hamster, mouse, macaque, rabbit, horse, dog or pig PrP^C.—cont'd

Species	PrP sequence	Generation technique	Mouse line	PrP expression level[a]	Spontaneous disease (onset)	Neuropathology	PrP^{res}	Transmissibility	References
	MousePrP-A	Non modified	129/Ola PrP A	$1\times^{(+/+)}$	No	No	No	No	81
	MousePrP-B	Knock-in	129/Ola PrP gtB	$1\times^{(+/+)}$	No	No	No	No	81
	MousePrP-A loxP	Transgenic	Tg46 MloxP	$1\times^{(+/-)}$	No	No	No	No	82
	MousePrP-A loxP	Transgenic	Tg37 MloxP	$3\times^{(+/-)}$	No	No	No	No	82
Macaque	Macaque	Transgenic	Tg700	$2\times^{(+/+)}$	No	No	No	No	83
Rabbit	Rabbit	Transgenic	TgRab	$5–6\times^{(+/-)}$	No	No	No	No	84
Horse	Horse	Transgenic	TgEq	$2\times^{(N/A)}$	No	No	No	No	85
Dog	Dog D_{163}	Transgenic	TgDogN163	$2\times^{(+/-)}$	No	No	No	No	86
	Dog Q_{163}	Transgenic	TgDogE163	$2\times^{(+/-)}$	No	No	No	No	86
Pig	Pig	Transgenic	Tg001	$4\times^{(+/+)}$	No	No	No	No	87
		Transgenic	Tg027	$8\times^{(+/-)}$	No	No	No	No	87
		Transgenic	Tg002	$1\times^{(N/A)}$	No	No	No	No	88

[a]Relative to brain PrP^C expression in the original PrP^C sequence species. (+/+) homozygous animals and (+/−) hemizygous animals.
N/A, not available.

Transgenic mouse lines also provide useful information on how certain PrP polymorphisms affect prion transmission within a certain species (Table 2). For instance, some sheep and goat genotypes proved to confer resistance to classical scrapie like $A_{136}R_{154}R_{171}$ of sheep PrP while classical scrapie susceptible genotypes are related to low susceptibility to atypical scrapie like $A_{136}R_{154}Q_{171}$ and $V_{136}R_{151}Q_{171}$ of sheep PrP.[98] Ovine transgenic mouse lines harboring those genotypes have been produced and seem to adequately reflect what happens in sheep [42,59–63,65,89] (Table 2). This information was useful for the generation of breeding selection programs in Europe implemented to decrease classical scrapie incidence. For goats, both epidemiological studies and the generation of transgenic mouse models have suggested several polymorphic variants as interesting for breeding selection programs like $K_{222.}$[64,99–106] This mutation seems to be protective against cattle BSE and small ruminant classical scrapie in transgenic mice.[64]

In the case of cattle, transgenic mice have been generated harboring different number of octapeptide repeats[50,51] (Table 2). Cattle usually harbor 5 or 6 octapeptide repeats but variations have been detected. Mice harboring 7 octapeptide repeats are more susceptible to cattle BSE showing shorter incubation times while mice harboring just 5 octapeptide repeats are less susceptible to this prion disease.

Cervid polymorphism like $G_{96}S$ of deer PrP, $M_{132}L$ polymorphic variant of elk PrP and $E_{226}Q$ that differs between deer and elk have been modeled.[55,68,69] The S_{96}[69] and L_{132}[55] variants seem to be protective against CWD. However, some discrepancies have been reported in the case of S_{96} variant since CWD has been reported in S_{96} homozygous cervids.[107] E_{226} elk mice were resistant to classical scrapie although once the prions were adapted to the deer Q_{226} sequence infection was successful.[68] In the case of humans, the polymorphic heterozygous variant $G_{127}V$ was found to provide resistance against kuru and CJD.[58]

In the case of interspecies prion transmission, the most important point is the study of human transmission barrier to animal prion strains. To date, the only recognized zoonotic prion disease is variant CJD (vCJD). This disease firstly appeared in the United Kingdom (UK) in 1996[108] and was soon proved to be caused by the consumption of meat products contaminated with BSE.[109] The $M_{129}V$ polymorphism of human PrP is of special value for the human PrP prion transmission barrier. All vCJD definitely diagnosed cases are homozygous for methionine allele except for one recently found heterozygous case.[110] Transgenic mouse lines overexpressing human PrP harboring the M_{129} allele (Table 2) are successfully infected with BSE prions but showing a high transmission barrier reflected on the long survival times and partial attack rates that animals present on the first passage.[54,56,57,111] These models mimic perfectly the real situation since virtually all UK population was exposed to BSE contaminated meat, although just 232 cases of vCJD have been reported.[112] Knock-in mouse models for human PrP (Table 2) were found resistant to cattle BSE also pointing to the existence of a great barrier for this prion agent.[49] Transgenic mice overexpressing human PrP harboring the heterozygous $M_{129}V$ and homozygous V_{129} alleles (Table 2) have not been infected with BSE but vCJD transmission is possible on these animals thus pointing the risk that human to human secondary vCJD infection would pose for the population.[113] Nevertheless, BSE prion has a wide host range. This agent can also infect sheep,[114] goat,[115,116] pig,[117] deer,[118] cat[119,120] and many other zoo species.[121] Transgenic mouse models expressing sheep/goat or pig PrP (Tables 2 and 3) have also been successfully infected with BSE prions.[63,64,92,122,123] Sheep-passaged BSE prions can easily infect M_{129} human PrP transgenic mouse lines when compared to cattle BSE and the same applies to M129 knock-in models.[56,124] In the case of atypical BSE prions, L-type BSE was found to infect transgenic mice over-expressing M_{129} human PrP with no transmission barrier.[125] In knock-in mouse models, no transmission was achieved but subclinical infection in M_{129} mice have been detected.[126] H-type BSE was found no transmissible in transgenic mice overexpressing M_{129} human PrP.[125]

There are no reported cases of classical scrapie transmission to humans and epidemiological studies discourage any possible link between classical scrapie and sCJD. However, transgenic mouse models overexpressing human PrP (Table 2) were found to be susceptible to certain classical scrapie cases that produced prions with features resembling those of sCJD.[53] In knock-in mice no transmission has been reported (Table 2).[124] The studies performed to date have not transmitted atypical scrapie prions in any transgenic mouse line expressing human PrP.[124,126,127]

CWD is an epidemic disease in North America but cases have also been reported in South Korea and Europe.[128–130] Indeed, some European cases seem to be a different strain to the ones reported in North America according to transmission data in bank voles.[131] Epidemiological studies conducted in the USA have failed to find any link between CWD and any human TSE.[132] However, given that human TSEs show long incubation times and that little is known about the European CWD strains, the zoonotic potential of these strains would be better studied by the use of transgenic mouse models expressing human PrP although some inoculations have also been done in primates. Primary passage of different CWD isolates in transgenic mice expressing both M_{129} and V_{129} polymorphic variants did not produce any prion disease.[55,67,127] However, second passages have not been published and would be necessary to finally discard the development of prion disease on these animals. One study done in cynomolgus macaques showed no transmission.[18] However, it must be taken into account that all this information refers to American CWD and that the European isolates may produce different results.

4. Transgenic mouse models for the study of genetic prion diseases

Prion diseases have different etiologies. They can be infectious, iatrogenic, sporadic o genetic. Infectious and iatrogenic prion diseases are caused by the entry of an external PrP^{Sc} source that starts transforming host PrP^{C} into new, ascent PrP^{Sc}. Thus, vCJD is an infectious TSE caused by the consumption of BSE contaminated meat products while iatrogenic CJD, for example, may be caused by cornea transplant from a CJD infected donor.[133] Sporadic human TSEs have no apparent explanation since any infectious/iatrogenic events nor genetic mutations in the human PrP coding gene (*PRNP*) can be found in diseased individuals. Finally, genetic TSEs are due to the existence of mutations in the *PRNP* gene of the affected individuals. In humans, approximately 10–15% of TSEs are genetic and therefore heritable.[134] The mutations that cause the genetic TSEs are thought to promote the misfolding of PrP^{C} into PrP^{Sc} spontaneously and/or stabilize the PrP^{Sc} molecules once they are formed.[134] There are three different human genetic prion diseases based on their clinical and pathological features: familial Creutzfeldt-Jakob disease (fCJD), Gerstmann-Sträussler-Scheinker syndromes (GSS), and fatal familial insomnia (FFI).[134]

fCJD is a rapidly progressive dementia which patients present cerebral spongiform degeneration, PrP^{Sc} deposits lacking PrP amyloid and classical brain PrP^{res}. By contrast, GSS is a slowly progressive disease in which patients show ataxia, few spongiform degeneration and abundant PrP^{Sc} amyloid plaques. Proteinase K-resistant PrP^{Sc} (PrP^{res}) found in the brains of these patients is similar to that of atypical scrapie,[96] with molecular weights ranging from 7 to 11 kDa. Finally, the hallmark of FFI symptoms is progressive insomnia that derives in hallucinations and dementia. Patients present high spongiosis in the thalamus and low levels of classical PrP^{res} characterized by the presence of three protein bands corresponding to di-, mono- and non-glycosylated PrP states from which the non-glycosylated band has a molecular weight ranging between 19 and 21 kDa.

Both for the study of the onset, progression and molecular mechanisms involved of these diseases as well as for the test of possible treatments, transgenic mouse models of genetic human TSEs would be of great usefulness. Like for the study of prion transmission, two different strategies have been employed to generate transgenic mouse models of the human genetic prion diseases, transgenic mice and knock-in mice.

Something important to note is that practically all attempts to produce transgenic mouse models for human genetic TSEs using the human PrP sequence had been unsuccessful with the exception of one recent model for GSS and the $A_{117}V$ mutation.[135] The rest of successful transgenic mouse models for these disorders have been made in the mouse or bank vole PrP sequences or in chimeric mouse/human PrP molecules. Both mouse and bank vole PrP are more prone to spontaneous misfolding while somehow human PrP is more resistant. An alternative explanation may be that interactions between mouse PrP and other than PrP mouse factors are important for the generation of prions.[134] Bank voles PrP is especially interesting because can be converted into an infectious form by different PrP^{Sc} donors [74,136–138] However, it should not be ruled out that models involving human mutations in rodent PrPs may be only generating novel experimental prion strains instead of reproducing the strains causing the human diseases.[135]

Transgenic mouse models generated up to now to model human genetic TSEs can be found in Table 4. For modeling fCJD, several mutations have been used but only two produced successful models: $E_{200}K$ and $D_{178}N$ combined with V_{129} polymorphism (the same mutation in conjunction with M129 polymorphism produces FFI). For $E_{200}K$ mutation, both knock-in and transgenic mice have been generated. Knock-in mouse model harboring $E_{199}K$ mouse PrP plus the 3F4 epitope tag only showed mild symptoms like worse rotarod test performance and decreased burrowing behavior as well as spongiosis, brain PrP deposits and brain PrP^{res} that was mildly PK-resistant.[139] These prions were efficiently transmitted into Tga20 mice and wild-type mice.[139] Transgenic models overexpressing $E_{199}K$ mouse PrP at 8 × levels,[141] as well as $E_{200}K$ human PrP at 3 × levels,[140] showed no signs of spontaneous prion disease. By contrast, transgenic mice expressing a mutant $E_{199}K$ chimeric human/mouse PrP 2 × levels showed clinical symptoms of neurological disease at around 200 days of age and later died at approximately 1 year of age.[142] Mice showed few PrP deposits and

Table 4 Transgenic mouse models for the study of human genetic TSEs.

Disease	Mutation	Mouse line	Generation technique	PrP sequence	PrP expression level[b]	Spontaneous disease (onset)	Neuropathology	PrP^{res}	Transmissibility	References
fCJD	E200K	Ki-3F4-CJD	Knock-in	Mouse 3F4	$1^{(+/+)}$	No	Yes	No	Yes	139
		Tg23	Transgenic	Human	$3^{(+/+)}$	No	No	No	N/A	140
		Tg5182	Transgenic	Mouse	$8^{(+/-)}$	No	No	No	N/A	141
		TgMHu2ME199K	Transgenic	Mouse/human	$2^{(+/-)}$	Yes (~200 d)	Yes	Yes	Yes	142
		Tg14210	Transgenic	Bank vole I_{109}	$1.2^{(+/-)}$	Yes (~460 d)	Yes	Yes	Yes	72
		Tg7253	Transgenic	Bank vole I_{109}	$1.7^{(+/-)}$	Yes (~270 d)	Yes	Yes	Yes	72
		Tg4253	Transgenic	Bank vole I_{109}	$2.4^{(+/-)}$	Yes (~160 d)	Yes	Yes	Yes	72
		Tg7271	Transgenic	Bank vole I_{109}	$2.7^{(+/-)}$	Yes (~120 d)	Yes	Yes	Yes	72
	$D_{178}N/V_{129}$	CJD-A21	Transgenic	Mouse	$2^{(+/+)}$	Yes (~150 d)	Yes	No	No	143

Continued

Table 4 Transgenic mouse models for the study of human genetic TSEs.—cont'd

Disease	Mutation	Mouse line	Generation technique	PrP sequence	PrP expression level[b]	Spontaneous disease (onset)	Neuropathology	PrP^{res}	Transmissibility	References
GSS	$P_{102}L$	101LL	Knock-in	Mouse	$1^{(+/+)}$	No	No	No	N/A	144
		Tg174[a]	Transgenic	Mouse	$8^{(+/-)}$	Yes (~200 d)	Yes	No	Yes	75
		Tg87[a]	Transgenic	Mouse	$8^{(+/-)}$	Yes (~150 d)	Yes	No	Yes	145
		Tg196[a]	Transgenic	Mouse	$2^{(+/-)}$	No	No	No	N/A	145
		Tg2866	Transgenic	Mouse	$8^{(+/-)}$	Yes (~150 d)	Yes	No	Yes	141
		Tg2247[a]	Transgenic	Mouse	$8^{(+/-)}$	Yes (~230 d)	Yes	No	Yes	141
		Tg2862[a]	Transgenic	Mouse	$32^{(+/-)}$	Yes (~320 d)	Yes	No	N/D	141
		Tg69	Transgenic	Mouse/human	$2^{(+/-)}$	Yes (~360 d)	Yes	No	N/D	52
		Tg(GSS)2	Transgenic	Mouse	$0.5–1^{(+/-)}$	No	No	No	N/A	146
		Tg(GSS)6	Transgenic	Mouse	$3^{(+/-)}$	Yes (~600 d)	Yes	No	N/D	146
		Tg(GSS)12	Transgenic	Mouse	$6^{(+/-)}$	Yes (~430 d)	Yes	No	N/D	146
		Tg(GSS)22	Transgenic	Mouse	$12^{(+/-)}$	Yes (~160 d)	Yes	No	Yes	146
		113LBoPrP-Tg009	Transgenic	Cow	$1^{(+/-)}$	No	No	No	N/A	147
		113LBoPrP-Tg037	Transgenic	Cow	$6^{(+/-)}$	Yes (~190 d)	Yes	No	Yes	147
		Tg27	Transgenic	Human M_{129}	$3^{(+/+)}$	No	No	No	N/A	140

	$A_{117}V$	E15727	Transgenic	Hamster	$4^{(N/A)}$	Yes (~570 d)	N/D	No	No	148
		Tg(A116V)	Transgenic	Mouse MV_{128}	$4–6^{(N/A)}$	Yes (~150 d)	Yes	No	N/D	149
		Tg31	Transgenic	Human V_{129}	$3^{(+/+)}$	No	No	No	N/A	150
		Tg30	Transgenic	Human V_{129}	$2^{(+/+)}$	Yes (~475 d)	Yes	No	Yes	135,150
	9-OR Ins	PG14	Transgenic	Mouse 3F4	$1^{(N/A)}$	Yes (~240 d)	No	No	No	151
	10-OR In	bo10OR-Tg006	Transgenic	Cow	$4^{(+/+)}$	Yes (≈118)	Yes	No	No	152
	GPI-less	$Tg44^{+/+}$	Transgenic	Mouse	$0.13^{(+/+)}$	No	No	No	N/A	153
		Tg8015	Transgenic	Mouse Myc	$1.7^{(+/-)}$	Yes (~600 d)	Yes	Yes	Yes	154
		Tg24600	Transgenic	Bank vole I_{109}	$0.5^{(+/-)}$	Yes (~420 d)	Yes	Yes	Yes	137

Continued

Table 4 Transgenic mouse models for the study of human genetic TSEs.—cont'd

Disease	Mutation	Mouse line	Generation technique	PrP sequence	PrP expression level[b]	Spontaneous disease (onset)	Neuropathology	PrP^{res}	Transmissibility	References
FFI	$D_{178}N/M_{129}$	Ki-3F4-FFI	Knock-in	Mouse 3F4	$1^{(+/+)}$	Yes (N/D)	Yes	No	Yes	155
		FFI-K5	Transgenic	Mouse 3F4	$0.7^{(+/-)}$	No	Yes	N/D	N/D	156
		FFI-10	Transgenic	Mouse	$1^{(+/-)}$	Yes (~540 d)	N/D	N/D	N/D	156
		FFI-15	Transgenic	Mouse	$0.5^{(+/-)}$	No	No	N/A	N/D	156
		FFI-26	Transgenic	Mouse	$2^{(+/-)}$	Yes (~200 d)	Yes	No	No	156
		Tg15972	Transgenic	Bank vole I_{109}	$0.4^{(+/-)}$	Yes (~240 d)	Yes	Yes	Yes	72
		Tg15464	Transgenic	Bank vole I_{109}	$0.4^{(+/-)}$	Yes (~220 d)	Yes	Yes	N/D	72
		Tg15465	Transgenic	Bank vole I_{109}	$0.4^{(+/-)}$	Yes (~200 d)	Yes	Yes	N/D	72
		Tg15965	Transgenic	Bank vole I_{109}	$0.5^{(+/-)}$	Yes (~180 d)	Yes	Yes	Yes	72

[a]These mice express both mutant and mouse PrP. The other mouse lines express only mutant PrP.
[b]Relative to brain PrP^C expression in the original PrP^C sequence species, except for Tg14210, Tg7253, Tg4253, Tg7271, Tg24600, Tg15972, Tg15464, Tg15465 and Tg15965 on which PrP expression levels are relative to brain PrP^C expression in mice. (+/+) homozygous animals and (+/−) hemizygous animals.
N/A, not available; N/D, not determined.

spongiform degeneration and some atypical PK-resistant PrP species were found in their brains.[142,157] When inoculated in wild-type mice, these prions were sometimes transmitted but the efficiency of the process was variable.[157] In addition, the $E_{200}K$ mutation has been modeled in the bank vole PrP sequence harboring the homozygous I_{109} polymorphism that spontaneously induces bank vole PrP misfolding.[72] All generated lines developed a spontaneous prion disease including the ones overexpressing the lowest mutant PrP levels. Mice showed brain atypical PrP^{res} and neuropathological hallmarks of prion disease like PrP^{Sc} deposits and spongiform degeneration. For the $D_{178}N$ mutation, a transgenic mouse model overexpressing 2 × levels the mutation $D_{177}N$ in combination with $M_{128}V$ substitution in mouse PrP were generated.[143] Spontaneous disease characterized by ataxia, kyphosis and foot clasping appeared at 150 days of age, causing mice dead at 300 days. PrP^{res} were found in the brains of these animals as well as PrP^{Sc} deposition and reactive astrocytic gliosis but no spongiform degeneration. The generated prions were not transmissible to wild-type mice, mice overexpressing wild-type mouse PrP or other mice expressing lower levels of $D_{177}N$ mutation.[143]

In the case of GSS, successful models developing spontaneous prion disease were produced mainly using the mutations $P_{102}L$ and $A_{117}V$ although there are some for insertions in the octapeptide region (9-OR ins) and GPI-anchorless mutations like $Y_{226}X$ or $Q_{227}X$ (GPI-less) (Table 4). However, the generation of GSS-like atypical brain PrP^{res} was not always achieved. The $P_{102}L$ mutation is the most common cause of GSS.[158] Both knock-in mice and transgenic mice expressing low levels of $P_{101}L$ mouse PrP did not develop any spontaneous prion disease.[144–146] Once the mutation is greatly overexpressed on mouse PrP or introduced in chimeric mouse/human PrP and cattle PrP prion disease can be found (Table 4). These mice showed mean incubation periods ranging from 150 to 600 days and all showed neuropathological changes but any of them showed highly PK-resistant PrP^{res} apart from Tg174 mice in which brain disease-specific PrP conformers were found by digestion with PK at 4 °C plus precipitation with phosphotungstic acid.[146,159] Using cattle PrP sequence, transgenic mice expressing the equivalent $P_{113}L$ mutation at low levels (1 ×) did not develop a spontaneous prion illness but once overexpression was achieved (3 × or 6 ×) mice developed spontaneous disease in 272 or 190 days respectively.[147] When tried, transmissibility of all these GSS prions was achieved in Tg196 animals expressing low levels of the mouse mutation but not in wild-type mice or transgenic mice overexpressing wild-type mouse PrP.[142,145,159] In the case of the cattle P113L, it was transmissible on transgenic mice expressing wild-type bovine PrP producing a prion agent undistinguishable from epidemic BSE, thus a cattle genetic disease may have been the origin of the BSE epidemics.[147] In hamsters, just 9 out of 348 challenged animals developed the disease.[145] Finally, transgenic mice overexpressing 3 × levels the mutant $P_{102}L$ human PrP failed to develop any prion illness.[140] In the case of $A_{117}V$ mutation, transgenic mice overexpressing 4 × levels this mutation in the hamster PrP sequence developed a spontaneous neurodegenerative disease characterized by a long incubation period of around 570 days[148] (Table 4). When the mutation was introduced in the mouse sequence ($A_{116}V$) and overexpressed at 4–6 × levels, transgenic mice showed a prion disease much earlier, around 150 days.[149] These animals exhibited prion neuropathology markers like light vacuolation and PrP plaques in the cerebellar cortex but no highly resistant PrP^{res} were found. $A_{117}V$ GSS cases were successfully transmitted into $A_{117}V$ human PrP transgenic mice Tg30 and Tg31 (2 × and 3 × overexpression respectively),[150] from which Tg30 developed a spontaneous disease by its own characterized by a molecular and neuropathological phenotype congruent with that seen after transmission of brain isolates human $A_{117}V$ patients to the same mice.[135] Brain PrP^{res} was not detected although brain PrP assemblies were identified by means of sodium phosphotungstic acid (NaPTA) precipitation.[135] In addition, mice had immunoreactive PrP plaques that were localized to the anterior commissure of the brain.[135] This disease was transmissible to both Tg30 and Tg31 mice as well as to wild-type V129 human PrP transgenic mice.[135] For the other mutations (Table 4), PG14 transgenic mice harboring 9-OR were generated in 1998.[151] They develop a spontaneous disease, which its main symptom is ataxia in around 240 days.[160] Granule cell loss in the cerebellum, synaptic PrP^{Sc} deposits, reactive astrocytosis and mildly PK-resistant PrP^{res} are found on this mice[150,151] but the generated disease is not transmissible at all.[161] Transgenic mice overexpressing bovine PrP harboring 10-OR developed a spontaneous disease approximately at 118 days of age characterized by ataxia, vacuolization in different brain areas, gliosis, and loss of cerebellar granule cells.[152] This disease was non-transmissible to transgenic mice overexpressing wild-type bovine PrP.[161] First transgenic mice expressing GPI-less mouse PrP were produced for deciphering the role that GPI anchor play in prion propagation.[153] Their expression levels were low and they not exhibit any signs of neurological disease.[162] By contrast, 50% of the mice with 1.7 × higher expression levels[154] developed a late-onset spontaneous prion disease characterized by high amounts of PrP amyloid deposits and an atypical brain PrP^{res} that was transmissible in younger mice of the same line and transgenic mice overexpressing wild-type mouse PrP.[154] GPI-less mutation was also introduced in the bank vole I_{109} PrP sequence and the generated transgenic mice developed a prion disease around 400 days of age.[72] This prion disease generates abundant PrP deposits and an atypical brain PrP^{res} and is transmitted in both younger animals of the same line and transgenic mice expressing not mutant bank vole I_{109} PrP.[72]

Finally, for FFI only one causing mutation has been reported. This is the $D_{178}N$ mutation combined with homozygous M_{129} polymorphism. One knock-in model using the mouse PrP sequence harboring the equivalent mutation $D_{177}N$ plus the 3F4 epitope tag was generated (Table 4) and only some animals developed a late-onset prion disease.[155] Neither PrP^{Sc} deposits nor PK-resistant PrP^{res} were observed in the brain of these mice but neuronal loss and gliosis in the thalamus were reported and the disease was transmissible into Tga20 mice.[155] Transgenic mice expressing $D_{177}N$ mouse PrP have also been created. The expression level of the mutant protein was key to the development of prion disease. Mice expressing the mutant protein at 0.5 × did not develop any disease but animals with expression levels of 1 × and 2 × developed a disease characterized by ataxia and kyphosis.[156] No highly PK-resistant PrP^{res} species were present and this disease was not transmitted after inoculation in wild-type mice or Tga20 mice.[156] Finally, transgenic mice introducing the $D_{178}N$ mutation in the I_{109} bank vole PrP sequence have been generated.[72] Although presenting low expression levels of the mutant protein, mice showed a spontaneous neurological illness with incubation times around 200 days.[72] The brains of the animals displayed spongiform degeneration, reactive astrocytic gliosis and PrP deposits as well as an atypical brain PrP^{res}.[72] The disease could be passaged to mice expressing not mutant I_{109} bank vole PrP and to mice overexpressing wild-type mouse PrP.

5. Conclusion

Advances in the study of prion transmission and replication have paralleled the development of appropriate tools and technologies, like transgenic mouse lines that gave insights on the still undeciphered information of TSEs. Transgenic mouse models will help to clarify the function of PrP^C as well as other important aspects like the evolution of prions and the zoonotic potential of newly emerged prion strains.

References

1. Prusiner SB. Molecular biology of prion diseases. *Science*. 1991;252:1515–1522.
2. Vázquez-Fernández E, Vos MR, Afanasyev P, et al. The structural architecture of an infectious mammalian prion using electron cryomicroscopy. *PLoS Pathog*. 2016;8:12.
3. Prusiner SB. Prions. *Proc Natl Acad Sci U S A*. 1998;95:13363–13383.
4. Gielbert A, Davis LA, Sayers AR, Hope J, Gill AC, Sauer MJ. High-resolution differentiation of transmissible spongiform encephalopathy strains by quantitative n-terminal amino acid profiling (n-taap) of pk-digested abnormal prion protein. *J Mass Spectrom*. 2009;44:384–396.
5. Gibbs Jr CJ, Gajdusek DC. Transmission of scrapie to the cynomolgus monkey (Macaca fascicularis). *Nature*. 1972;236:73–74.
6. Gibbs Jr CJ, Gajdusek DC. Experimental subacute spongiform virus encephalopathies in primates and other laboratory animals. *Science*. 1973;182:67–68.
7. Lasmézas CI, Deslys JP, Demaimay R, et al. BSE transmission to macaques. *Nature*. 1996;381:743–744.
8. Lasmézas CI, Fournier JG, Nouvel V, et al. Adaptation of the bovine spongiform encephalopathy agent to primates and comparison with Creutzfeldt—Jakob disease: implications for human health. *Proc Natl Acad Sci U S A*. 2001;98:4142–4147.
9. Herzog C, Rivière J, Lescoutra-Etchegaray N, et al. PrP^{TSE} distribution in a primate model of variant, sporadic, and iatrogenic Creutzfeldt-Jakob disease. *J Virol*. 2005; 79:14339–14345.
10. Lasmézas CI, Comoy E, Hawkins S, et al. Risk of oral infection with bovine spongiform encephalopathy agent in primates. *Lancet*. 2005;365:781–783.
11. Comoy EE, Casalone C, Lescoutra-Etchegaray N, et al. Atypical BSE (BASE) transmitted from asymptomatic aging cattle to a primate. *PLoS One*. 2008;3:e3017.
12. Ono F, Tase N, Kurosawa A, et al. Atypical L-type bovine spongiform encephalopathy (L-BSE) transmission to cynomolgus macaques, a non-human primate. *Jpn J Infect Dis*. 2011;64:81–84.
13. Ono F, Terao K, Tase N, et al. Experimental transmission of bovine spongiform encephalopathy (BSE) to cynomolgus macaques, a non-human primate. *Jpn J Infect Dis*. 2011;64:50–54.
14. Comoy EE, Mikol J, Ruchoux MM, et al. Evaluation of the zoonotic potential of transmissible mink encephalopathy. *Pathogens*. 2013;2:520–532.
15. Race B, Meade-White KD, Phillips K, Striebel J, Race R, Chesebro B. Chronic wasting disease agents in nonhuman primates. *Emerg Infect Dis*. 2014;20:833–837.
16. Comoy EE, Mikol J, Luccantoni-Freire S, et al. Transmission of scrapie prions to primate after an extended silent incubation period. *Sci Rep*. 2015;5:11573.
17. Comoy EE, Mikol J, Jaffré N, et al. Experimental transfusion of variant CJD-infected blood reveals previously uncharacterised prion disorder in mice and macaque. *Nat Commun*. 2017;8:1268.
18. Race B, Williams K, Orrú CD, Hughson AG, Lubke L, Chesebro B. Lack of transmission of chronic wasting disease to cynomolgus macaques. *J Virol*. 2018;92(14):e00550-18.
19. Pastore A, Zagari A. A structural overview of the vertebrate prion proteins. *Prion*. 2007;1:185–197.
20. Büeler H, Fischer M, Lang Y, et al. Normal development and behavior of mice lacking the neuronal cell-surface PrP protein. *Nature*. 1992;356:577–582.
21. Manson JC, Clarke AR, Hooper ML, Aitchison L, McConnell I, Hope J. 129/Ola mice carrying a null mutation in PrP that abolishes messenger-RNA production are developmentally normal. *Mol Neurobiol*. 1994;8:121–127.
22. Moore RC, Redhead NJ, Selfridge J, Hope J, Manson JC, Melton DW. Double replacement gene targeting for the production of a series of mouse strains with different prion protein gene alterations. *Biotechnology (N Y)*. 1995;13:999–1004.

23. Sakaguchi S, Katamine S, Nishida N, et al. Loss of cerebellar Purkinje cells in aged mice homozygous for a disrupted Prp gene. *Nature.* 1996;380:528–531.
24. Yokoyama T, Kimura KM, Ushiki Y, et al. In vivo conversion of cellular prion protein to pathogenic isoforms, as monitored by conformation-specific antibodies. *J Biol Chem.* 2001;276:11265–11271.
25. Rossi D, Cozzio A, Flechsig E, et al. Onset of ataxia and Purkinje cell loss in PrP null mice inversely correlated with Dpl level in brain. *EMBO J.* 2001;20:694–702.
26. Heikenwalder M, Kurrer MO, Margalith I, et al. Lymphotoxin-dependent prion replication in inflammatory stromal cells of granulomas. *Immunity.* 2008;29:998–1008.
27. Nuvolone M, Hermann M, Sorce S, et al. Strictly co-isogenic C57BL/6J-Prnp(-/-) mice: a rigorous resource for prion science. *J Exp Med.* 2016;213:313–327.
28. Steele AD, Lindquist S, Aguzzi A. The prion protein knockout mouse: a phenotype under challenge. *Prion.* 2007;1:83–93.
29. Gerlai R. Gene-targeting studies of mammalian behavior: is it the mutation or the background genotype? *Trends Neurosci.* 1996;19:177–181.
30. Nuvolone M, Kana V, Hutter G, et al. SIRP α polymorphisms, but not the prion protein, control phagocytosis of apoptotic cells. *J Exp Med.* 2013;210:2539–2552.
31. Hermann M, Cermak T, Voytas DF, Pelczar P. Mouse genome engineering using designer nucleases. *J Vis Exp.* 2014;86:50930.
32. Büeler H, Aguzzi A, Sailer A, et al. Mice devoid of PrP are resistant to scrapie. *Cell.* 1993;73:1339–1347.
33. Manson JC, Clarke AR, Mcbride PA, Mcconnell I, Hope J. Prp gene dosage determines the timing but not the final intensity or distribution of lesions in scrapie pathology. *Neurodegeneration.* 1994;3:331–340.
34. Prusiner SB, Groth D, Serban A, et al. Ablation of the prion protein (PrP) gene in mice prevents scrapie and facilitates production of anti-PrP antibodies. *Proc Natl Acad Sci U S A.* 1993;90:10608–10612.
35. Nuvolone M, Sorce S, Paolucci M, Aguzzi A. Extended characterization of the novel co-isogenic C57BL/6J $Prnp^{-/-}$ mouse line. *Amyloid.* 2017;24(Suppl. 1):36–37.
36. Bremer J, Baumann F, Tiberi C, et al. Axonal prion protein is required for peripheral myelin maintenance. *Nat Neurosci.* 2010;13:310–318.
37. Kufer A, Lakkaraju AK, Mogha A, et al. The prion protein is an agonistic ligand of the G protein-coupled receptor Adgrg6. *Nature.* 2016;536:464–468.
38. Nishida N, Tremblay P, Sugimoto T, et al. A mouse prion protein transgene rescues mice deficient for the prion protein gene from Purkinje cell degeneration and demyelination. *Lab Invest.* 1999;79:689–697.
39. Béringue V, Vilotte JL, Laude H. Prion agent diversity and species barrier. *Vet Res.* 2008;39:47.
40. Bruce ME. TSE strain variation: an investigation into prion disease diversity. *Br Med Bull.* 2003;66:99–108.
41. Collinge J, Clarke AR. A general model of prion strains and their pathogenicity. *Science.* 2007;318:930–936.
42. Westaway D, DeArmond SJ, Cayetano-Canlas J, et al. Degeneration of skeletal muscle, peripheral nerves, and the central nervous system in transgenic mice overexpressing wild-type prion proteins. *Cell.* 1994;76:117–129.
43. Wadsworth JD, Asante EA, Collinge J. Review: contribution of transgenic models to understanding human prion disease. *Neuropathol Appl Neurobiol.* 2010;36:576–597.
44. Castilla J, Gutiérrez Adán A, Brun A, et al. Early detection of PrP^{res} in BSE-infected bovine PrP transgenic mice. *Arch Virol.* 2003;148:677–691.
45. Scott MR, Will R, Ironside J, et al. Compelling transgenetic evidence for transmission of bovine spongiform encephalopathy prions to humans. *Proc Natl Acad Sci U S A.* 1999;96(26):15137–15142.

46. Scott MR, Safar J, Telling G, et al. Identification of a prion protein epitope modulating transmission of bovine spongiform encephalopathy prions to transgenic mice. *Proc Natl Acad Sci U S A*. 1997;94:14279–14284.
47. Béringue V, Bencsik A, Le Dur A, et al. Isolation from cattle of a prion strain distinct from that causing bovine spongiform encephalopathy. *PLoS Pathog*. 2006;2 (10):e112.
48. Buschmann A, Pfaff E, Reifenberg K, Müller HM, Groschup MH. Detection of cattle-derived BSE prions using transgenic mice overexpressing bovine PrP(C). *Arch Virol Suppl*. 2000;16:75–86.
49. Bishop MT, Hart P, Aitchison L, et al. Predicting susceptibility and incubation time of human-to-human transmission of vCJD. *Lancet Neurol*. 2006;5:393–398.
50. Castilla J, Gutiérrez-Adán A, Brun A, et al. Different behavior toward bovine spongiform encephalopathy infection of bovine prion protein transgenic mice with one extra repeat octapeptide insert mutation. *J Neurosci*. 2004;24:2156–2164.
51. Brun A, Gutiérrez-Adán A, Castilla J, et al. Reduced susceptibility to bovine spongiform encephalopathy prions in transgenic mice expressing a bovine PrP with five octapeptide repeats. *J Gen Virol*. 2007;88:1842–1849.
52. Telling GC, Scott M, Mastrianni J, et al. Prion propagation in mice expressing human and chimeric PrP transgenes implicates the interaction of cellular PrP with another protein. *Cell*. 1995;83:79–90.
53. Cassard H, Torres JM, Lacroux C, et al. Evidence for zoonotic potential of ovine scrapie prions. *Nat Commun*. 2014;5:5821.
54. Asante EA, Linehan JM, Desbruslais M, et al. BSE prions propagate as either variant CJD-like or sporadic CJD-like prion strains in transgenic mice expressing human prion protein. *EMBO J*. 2002;21:6358–6366.
55. Kong Q, Huang S, Zou W, et al. Chronic wasting disease of elk: transmissibility to humans examined by transgenic mouse models. *J Neurosci*. 2005;25:7944–7949.
56. Padilla D, Béringue V, Espinosa JC, et al. Sheep and goat BSE propagate more efficiently than cattle BSE in human PrP transgenic mice. *PLoS Pathog*. 2011;7(3): e1001319.
57. Béringue V, Le Dur A, Tixador P, et al. Prominent and persistent extraneural infection in human PrP transgenic mice infected with variant CJD. *PLoS One*. 2008;3(1):e1419.
58. Asante EA, Smidak M, Grimshaw A, et al. A naturally occurring variant of the human prion protein completely prevents prion disease. *Nature*. 2015;522:478–481.
59. Kupfer L, Eiden M, Buschmann A, Groschup MH. Amino acid sequence and prion strain specific effects on the in vitro and in vivo convertibility of ovine/murine and bovine/murine prion protein chimeras. *Biochim Biophys Acta*. 1772;2007:704–713.
60. Rubenstein R, Bulgin MS, Chang B, Sorensen-Melson S, Petersen RB, LaFauci G. $PrP^{(Sc)}$ detection and infectivity in semen from scrapie-infected sheep. *J Gen Virol*. 2012;93:1375–1383.
61. Saijo E, Kang HE, Bian J, et al. Epigenetic dominance of prion conformers. *PLoS Pathog*. 2013;9(10):e1003692.
62. Crozet C, Flamant F, Bencsik A, Aubert D, Samarut J, Baron T. Efficient transmission of two different sheep scrapie isolates in transgenic mice expressing the ovine PrP gene. *J Virol*. 2001;75:5328–5334.
63. Cordier C, Bencsik A, Philippe S, et al. Transmission and characterization of bovine spongiform encephalopathy sources in two ovine transgenic mouse lines (TgOvPrP4 and TgOvPrP59). *J Gen Virol*. 2006;87:3763–3771.
64. Aguilar-Calvo P, Espinosa JC, Pintado B, et al. Role of the goat K222-$PrP^{(C)}$ polymorphic variant in prion infection resistance. *J Virol*. 2014;88:2670–2676.
65. Vilotte JL, Soulier S, Essalmani R, et al. Markedly increased susceptibility to natural sheep scrapie of transgenic mice expressing ovine prp. *J Virol*. 2001;75:5977–5984.

66. Browning SR, Mason GL, Seward T, et al. Transmission of prions from mule deer and elk with chronic wasting disease to transgenic mice expressing cervid PrP. *J Virol.* 2004;78:13345–13350.
67. Tamgüney G, Giles K, Bouzamondo-Bernstein E, et al. Transmission of elk and deer prions to transgenic mice. *J Virol.* 2006;80:9104–9114.
68. Bian J, Christiansen JR, Moreno JA, et al. Primary structural differences at residue 226 of deer and elk PrP dictate selection of distinct CWD prion strains in gene-targeted mice. *Proc Natl Acad Sci U S A.* 2019;116:12478–12487.
69. Meade-White K, Race B, Trifilo M, et al. Resistance to chronic wasting disease in transgenic mice expressing a naturally occurring allelic variant of deer prion protein. *J Virol.* 2007;81:4533–4539.
70. Angers RC, Seward TS, Napier D, et al. Chronic wasting disease prions in elk antler velvet. *Emerg Infect Dis.* 2009;15:696–703.
71. LaFauci G, Carp RI, Meeker HC, et al. Passage of chronic wasting disease prion into transgenic mice expressing Rocky Mountain elk (Cervus elaphus nelsoni) PrP^C. *J Gen Virol.* 2006;87:3773–3780.
72. Watts JC, Giles K, Bourkas ME, et al. Towards authentic transgenic mouse models of heritable PrP prion diseases. *Acta Neuropathol.* 2016;132:593–610.
73. Espinosa JC, Nonno R, Di Bari M, et al. PrP^C governs susceptibility to prion strains in bank vole, while other host factors modulate strain features. *J Virol.* 2016; 90:10660–10669.
74. Watts JC, Giles K, Patel S, Oehler A, DeArmond SJ, Prusiner SB. Evidence that bank vole PrP is a universal acceptor for prions. *PLoS Pathog.* 2014;10(4):e1003990.
75. Hsiao KK, Scott M, Foster D, Groth DF, DeArmond SJ, Prusiner SB. Spontaneous neurodegeneration in transgenic mice with mutant prion protein. *Science.* 1990; 250:1587–1590.
76. Race RE, Priola SA, Bessen RA, et al. Neuron-specific expression of a hamster prion protein minigene in transgenic mice induces susceptibility to hamster scrapie agent. *Neuron.* 1995;15:1183–1191.
77. Scott M, Foster D, Mirenda C, et al. Transgenic mice expressing hamster prion protein produce species-specific scrapie infectivity and amyloid plaques. *Cell.* 1989; 59:847–857.
78. Windl O, Buchholz M, Neubauer A, et al. Breaking an absolute species barrier: transgenic mice expressing the mink PrP gene are susceptible to transmissible mink encephalopathy. *J Virol.* 2005;79:14971–14975.
79. Fischer M, Rülicke T, Raeber A, et al. Prion protein (PrP) with amino-proximal deletions restoring susceptibility of PrP knockout mice to scrapie. *EMBO J.* 1996;15:1255–1264.
80. Carlson GA, Ebeling C, Yang SL, et al. Prion isolate specified allotypic interactions between the cellular and scrapie prion proteins in congenic and transgenic mice. *Proc Natl Acad Sci U S A.* 1994;91:5690–5694.
81. Moore RC, Hope J, McBride PA, et al. Mice with gene targetted prion protein alterations show that Prnp, Sinc and Prni are congruent. *Nat Genet.* 1998;18:118–125.
82. Mallucci GR, Ratté S, Asante EA, et al. Post-natal knockout of prion protein alters hippocampal CA1 properties, but does not result in neurodegeneration. *EMBO J.* 2002;21:202–210.
83. Espinosa JC, Comoy EE, Marin-Moreno A, et al. Transgenic mouse models expressing human and macaque prion protein exhibit similar prion susceptibility on a strain-dependent manner. *Sci Rep.* 2019;9(1):15699.
84. Vidal E, Fernández-Borges N, Pintado B, et al. Transgenic mouse bioassay: evidence that rabbits are susceptible to a variety of prion isolates. *PLoS Pathog.* 2015;11(8): e1004977.

85. Bian J, Khaychuk V, Angers RC, et al. Prion replication without host adaptation during interspecies transmissions. *Proc Natl Acad Sci U S A*. 2017;114:1141–1146.
86. Vidal E, Fernández-Borges N, Eraña H, et al. The amino acid residue in position 163 of canine PrPC is critical to the exceptional resistance of dogs to prion infections: evidence from transgenic mouse models. *FASEB J*. 2020;34(3):3969–3982. https://doi.org/10.1096/fj.201902646R.
87. Castilla J, Gutiérrez-Adán A, Brun A, et al. Subclinical bovine spongiform encephalopathy infection in transgenic mice expressing porcine prion protein. *J Neurosci*. 2004; 24:5063–5069.
88. Moore SJ, West Greenlee MH, Kondru N, et al. Experimental transmission of the chronic wasting disease agent to swine after oral or intracranial inoculation. *J Virol*. 2017;91(19):e00926-17.
89. Le Dur A, Béringue V, Andréoletti O, et al. A newly identified type of scrapie agent can naturally infect sheep with resistant PrP genotypes. *Proc Natl Acad Sci U S A*. 2005;102:16031–16036.
90. Buschmann A, Gretzschel A, Biacabe AG, et al. Atypical BSE in Germany—proof of transmissibility and biochemical characterization. *Vet Microbiol*. 2006;117:103–116.
91. Capobianco R, Casalone C, Suardi S, et al. Conversion of the BASE prion strain into the BSE strain: the origin of BSE? *PLoS Pathog*. 2007;3(3):e31.
92. Béringue V, Andréoletti O, Le Dur A, et al. A bovine prion acquires an epidemic bovine spongiform encephalopathy strain-like phenotype on interspecies transmission. *J Neurosci*. 2007;27:6965–6971.
93. Telling GC, Scott M, Hsiao KK, et al. Transmission of Creutzfeldt-Jakob disease from humans to transgenic mice expressing chimeric human-mouse prion protein. *Proc Natl Acad Sci U S A*. 1994;91:9936–9940.
94. Casalone C, Zanusso G, Acutis P, et al. Identification of a second bovine amyloidotic spongiform encephalopathy: molecular similarities with sporadic Creutzfeldt-Jakob disease. *Proc Natl Acad Sci U S A*. 2004;101:3065–3070.
95. Biacabe AG, Laplanche JL, Ryder S, Baron T. Distinct molecular phenotypes in bovine prion diseases. *EMBO Rep*. 2004;5:110–115.
96. Benestad SL, Sarradin P, Thu B, Schönheit J, Tranulis MA, Bratberg B. Cases of scrapie with unusual features in Norway and designation of a new type, Nor98. *Vet Rec*. 2003;153:202–208.
97. Espinosa JC, Marín-Moreno A, Aguilar-Calvo P, Benestad SL, Andreoletti O, Torres JM. Porcine PrP as a paradigm of limited susceptibility to prion strain propagation. *J Infect Dis*. 2020;jiz646. https://doi.org/10.1093/infdis/jiz646.
98. Saunders GC, Cawthraw S, Mountjoy SJ, Hope J, Windl O. PrP genotypes of atypical scrapie cases in Great Britain. *J Gen Microbiol*. 2006;87:3141–3149.
99. Vaccari G, Di Bari MA, Morelli L, et al. Identification of an allelic variant of the goat PrP gene associated with resistance to scrapie. *J Gen Virol*. 2006;87:1395–1402.
100. Acutis PL, Bossers A, Priem J, et al. Identification of prion protein gene polymorphisms in goats from Italian scrapie outbreaks. *J Gen Virol*. 2006;87:1029–1033.
101. Barillet F, Mariat D, Amigues Y, et al. Identification of seven haplotypes of the caprine PrP gene at codons 127, 142, 154, 211, 222 and 240 in French Alpine and Saanen breeds and their association with classical scrapie. *J Gen Virol*. 2009; 90:769–776.
102. Bouzalas IG, Dovas CI, Banos G, et al. Caprine PRNP polymorphisms at codons 171, 211, 222 and 240 in a Greek herd and their association with classical scrapie. *J Gen Virol*. 2010;91:1629–1634.
103. Acutis PL, Martucci F, D'Angelo A, et al. Resistance to classical scrapie in experimentally challenged goats carrying mutation K222 of the prion protein gene. *Vet Res*. 2012;43:8.

104. White SN, Reynolds JO, Waldron DF, Schneider DA, O'Rourke KI. Extended scrapie incubation time in goats singly heterozygous for PRNP S146 or K222. *Gene.* 2012;501:49–51.
105. Lacroux C, Perrin-Chauvineau C, Corbière F, et al. Genetic resistance to scrapie infection in experimentally challenged goats. *J Virol.* 2014;88:2406–2413.
106. Aguilar-Calvo P, Fast C, Tauscher K, et al. Effect of Q211 and K222 PRNP polymorphic variants in the susceptibility of goats to oral infection with goat bovine spongiform encephalopathy. *J Infect Dis.* 2015;212:664–672.
107. Keane DP, Barr DJ, Bochsler PN, et al. Chronic wasting disease in a Wisconsin white-tailed deer farm. *J Vet Diagn Invest.* 2008;20:698–703.
108. Collinge J, Sidle KC, Meads J, Ironside J, Hill AF. Molecular analysis of prion strain variation and the aetiology of 'new variant' CJD. *Nature.* 1996;383:685–690.
109. Hill AF, Desbruslais M, Joiner S, et al. The same prion strain causes vCJD and BSE. *Nature.* 1997;389:448–450. 526.
110. Mok T, Jaunmuktane Z, Joiner S, et al. Variant Creutzfeldt-Jakob disease in a patient with heterozygosity at PRNP codon 129. *N Engl J Med.* 2017;376:292–294.
111. Kong Q, Zheng M, Casalone C, et al. Evaluation of the human transmission risk of an atypical bovine spongiform encephalopathy prion strain. *J Virol.* 2008;82:3697–3701.
112. http://www.eurocjd.ed.ac.uk/surveillance%20data%201.html.
113. Fernández-Borges N, Espinosa JC, Marín-Moreno A, et al. Protective effect of Val129-PrP against bovine spongiform encephalopathy but not variant Creutzfeldt-Jakob disease. *Emerg Infect Dis.* 2017;23:1522–1530.
114. Houston F, Foster JD, Chong A, Hunter N, Bostock CJ. Transmission of BSE by blood transfusion in sheep. *Lancet.* 2000;356:999–1000.
115. Eloit M, Adjou K, Coulpier M, et al. BSE agent signatures in a goat. *Vet Rec.* 2005;156:523–524.
116. Spiropoulos J, Lockey R, Sallis RE, et al. Isolation of prion with BSE properties from farmed goat. *Emerg Infect Dis.* 2011;17:2253–2261.
117. Wells GA, Hawkins SA, Austin AR, et al. Studies of the transmissibility of the agent of bovine spongiform encephalopathy to pigs. *J Gen Virol.* 2003;84:1021–1031.
118. Dagleish MP, Martin S, Steele P, et al. Experimental transmission of bovine spongiform encephalopathy to European red deer (Cervus elaphus elaphus). *BMC Vet Res.* 2008;4:17.
119. Pearson GR, Gruffydd-Jones TJ, Wyatt JM, et al. Feline spongiform encephalopathy. *Vet Rec.* 1991;128:532.
120. Wyatt JM, Pearson GR, Smerdon TN, Gruffydd-Jones TJ, Wells GA, Wilesmith JW. Naturally occurring scrapie-like spongiform encephalopathy in five domestic cats. *Vet Rec.* 1991;129:233–236.
121. Fernández-Borges N, Marín-Moreno A, Konold T, Espinosa JC, Torres JM. Bovine spongiform encephalopathy (BSE). In: Stein J, ed. *Reference Module in Neuroscience and Biobehavioral Psychology.* Elsevier; 2017:1–10.
122. Baron T, Bencsik A, Biacabe AG, Morignat E, Bessen RA. Phenotypic similarity of transmissible mink encephalopathy in cattle and L-type bovine spongiform encephalopathy in a mouse model. *Emerg Infect Dis.* 2007;13:1887–1894.
123. Espinosa JC, Herva ME, Andréoletti O, et al. Transgenic mice expressing porcine prion protein resistant to classical scrapie but susceptible to sheep bovine spongiform encephalopathy and atypical scrapie. *Emerg Infect Dis.* 2009;15:1214–1221.
124. Plinston C, Hart P, Chong A, et al. Increased susceptibility of human-PrP transgenic mice to bovine spongiform encephalopathy infection following passage in sheep. *J Virol.* 2011;85:1174–1181.
125. Béringue V, Herzog L, Reine F, et al. Transmission of atypical bovine prions to mice transgenic for human prion protein. *Emerg Infect Dis.* 2008;14:1898–1901.

126. Wilson R, Dobie K, Hunter N, Casalone C, Baron T, Barron RM. Presence of subclinical infection in gene-targeted human prion protein transgenic mice exposed to atypical bovine spongiform encephalopathy. *J Gen Virol.* 2013;94:2819–2827.
127. Wilson R, Plinston C, Hunter N, et al. Chronic wasting disease and atypical forms of bovine spongiform encephalopathy and scrapie are not transmissible to mice expressing wild-type levels of human prion protein. *J Gen Virol.* 2012;93:1624–1629.
128. Williams ES. Chronic wasting disease. *Vet Pathol.* 2005;42:530–549.
129. Kim TY, Shon HJ, Joo YS, Mun UK, Kang KS, Lee YS. Additional cases of Chronic Wasting disease in imported deer in Korea. *J Vet Med Sci.* 2005;67:753–759.
130. Benestad SL, Mitchell G, Simmons M, Ytrehus B, Vikøren T. First case of chronic wasting disease in Europe in a Norwegian free-ranging reindeer. *Vet Res.* 2016;47(1):88.
131. Pirisinu L, Tran L, Chiappini B, et al. Novel type of Chronic Wasting disease detected in moose (Alces alces), Norway. *Emerg Infect Dis.* 2018;24:2210–2218.
132. Anderson CA, Bosque P, Filley CM, et al. Colorado surveillance program for chronic wasting disease transmission to humans: lessons from 2 highly suspicious but negative cases. *Arch Neurol.* 2007;64:439–441.
133. Hammersmith KM, Cohen EJ, Rapuano CJ, Laibson PR. Creutzfeldt-Jakob disease following corneal transplantation. *Cornea.* 2004;23:406–408.
134. Watts JC, Prusiner SB. Experimental models of inherited PrP prion diseases. *Cold Spring Harb Perspect Med.* 2017;7:a027151.
135. Asante EA, Linehan JM, Tomlinson A, et al. Spontaneous generation of prions and transmissible PrP amyloid in a humanised transgenic mouse model of A117V GSS. *PLoS Biol.* 2020;18(6):e3000725.
136. Nonno R, Di Bari MA, Cardone F, et al. Efficient transmission and characterization of Creutzfeldt-Jakob disease strains in bank voles. *PLoS Pathog.* 2006;2(2):e12.
137. Watts JC, Giles K, Stöhr J, et al. Spontaneous generation of rapidly transmissible prions in transgenic mice expressing wild-type bank vole prion protein. *Proc Natl Acad Sci U S A.* 2012;109:3498–3503.
138. Orrú CD, Groveman BR, Raymond LD, et al. Bank vole prion protein as an apparently universal substrate for RT-QuIC-based detection and discrimination of prion strains. *PLoS Pathog.* 2015;11(6):e1004983.
139. Jackson WS, Borkowski AW, Watson NE, et al. Profoundly different prion diseases in knock-in mice carrying single PrP codon substitutions associated with human diseases. *Proc Natl Acad Sci U S A.* 2013;110:14759–14764.
140. Asante EA, Gowland I, Grimshaw A, et al. Absence of spontaneous disease and comparative prion susceptibility of transgenic mice expressing mutant human prion proteins. *J Gen Virol.* 2009;90:546–558.
141. Telling GC, Haga T, Torchia M, Tremblay P, DeArmond SJ, Prusiner SB. Interactions between wild-type and mutant prion proteins modulate neurodegeneration in transgenic mice. *Genes Dev.* 1996;10:1736–1750.
142. Friedman-Levi Y, Meiner Z, Canello T, et al. Fatal prion disease in a mouse model of genetic E200K Creutzfeldt-Jakob disease. *PLoS Pathog.* 2011;7(11):e1002350.
143. Dossena S, Imeri L, Mangieri M, et al. Mutant prion protein expression causes motor and memory deficits and abnormal sleep patterns in a transgenic mouse model. *Neuron.* 2008;60:598–609.
144. Manson JC, Jamieson E, Baybutt H, et al. A single amino acid alteration (101L) introduced into murine PrP dramatically alters incubation time of transmissible spongiform encephalopathy. *EMBO J.* 1999;18:6855–6864.
145. Hsiao KK, Groth D, Scott M, et al. Serial transmission in rodents of neurodegeneration from transgenic mice expressing mutant prion protein. *Proc Natl Acad Sci U S A.* 1994;91:9126–9130.

146. Nazor KE, Kuhn F, Seward T, et al. Immunodetection of disease-associated mutant PrP, which accelerates disease in GSS transgenic mice. *EMBO J*. 2005;24:2472–2480.
147. Torres JM, Castilla J, Pintado B, et al. Spontaneous generation of infectious prion disease in transgenic mice. *Emerg Infect Dis*. 2013;19:1938–1947.
148. Hegde RS, Tremblay P, Groth D, DeArmond SJ, Prusiner SB, Lingappa VR. Transmissible and genetic prion diseases share a common pathway of neurodegeneration. *Nature*. 1999;402:822–826.
149. Yang W, Cook J, Rassbach B, Lemus A, DeArmond SJ, Mastrianni JA. A new transgenic mouse model of Gerstmann-Straussler-Scheinker syndrome caused by the A117V mutation of PRNP. *J Neurosci*. 2009;29:10072–10080.
150. Asante EA, Linehan JM, Smidak M, et al. Inherited prion disease A117V is not simply a proteinopathy but produces prions transmissible to transgenic mice expressing homologous prion protein. *PLoS Pathog*. 2013;9(9):e1003643.
151. Chiesa R, Piccardo P, Ghetti B, Harris DA. Neurological illness in transgenic mice expressing a prion protein with an insertional mutation. *Neuron*. 1998;21:1339–1351.
152. Castilla J, Gutiérrez-Adán A, Brun A, et al. Transgenic mice expressing bovine PrP with a four extra repeat octapeptide insert mutation show a spontaneous, non-transmissible, neurodegenerative disease and an expedited course of BSE infection. *FEBS Lett*. 2005;579:6237–6246.
153. Chesebro B, Trifilo M, Race R, et al. Anchorless prion protein results in infectious amyloid disease without clinical scrapie. *Science*. 2005;308:1435–1439.
154. Stöhr J, Watts JC, Legname G, et al. Spontaneous generation of anchorless prions in transgenic mice. *Proc Natl Acad Sci U S A*. 2011;108:21223–21228.
155. Jackson WS, Borkowski AW, Faas H, et al. Spontaneous generation of prion infectivity in fatal familial insomnia knockin mice. *Neuron*. 2009;63:438–450.
156. Bouybayoune I, Mantovani S, Del Gallo F, et al. Transgenic fatal familial insomnia mice indicate prion infectivity-independent mechanisms of pathogenesis and phenotypic expression of disease. *PLoS Pathog*. 2015;11(4):e1004796.
157. Friedman-Levi Y, Mizrahi M, Frid K, Binyamin O, Gabizon R. PrP(ST), a soluble, protease resistant and truncated PrP form features in the pathogenesis of a genetic prion disease. *PLoS One*. 2013;8(7):e69583.
158. Hsiao K, Baker HF, Crow TJ, et al. Linkage of a prion protein missense variant to Gerstmann-Sträussler syndrome. *Nature*. 1989;338:342–345.
159. Tremblay P, Ball HL, Kaneko K, et al. Mutant PrPSc conformers induced by a synthetic peptide and several prion strains. *J Virol*. 2004;78:2088–2099.
160. Chiesa R, Drisaldi B, Quaglio E, et al. Accumulation of protease-resistant prion protein (PrP) and apoptosis of cerebellar granule cells in transgenic mice expressing a PrP insertional mutation. *Proc Natl Acad Sci U S A*. 2000;97:5574–5579.
161. Chiesa R, Piccardo P, Quaglio E, et al. Molecular distinction between pathogenic and infectious properties of the prion protein. *J Virol*. 2003;77:7611–7622.
162. Chesebro B, Race B, Meade-White K, et al. Fatal transmissible amyloid encephalopathy: a new type of prion disease associated with lack of prion protein membrane anchoring. *PLoS Pathog*. 2010;6(3):e1000800.

CHAPTER EIGHT

The utility of bank voles for studying prion disease

Hamza Arshad[a,b], Matthew E.C. Bourkas[a,b], and Joel C. Watts[a,b,*]
[a]Tanz Centre for Research in Neurodegenerative Diseases, University of Toronto, Toronto, ON, Canada
[b]Department of Biochemistry, University of Toronto, Toronto, ON, Canada
*Corresponding author: e-mail address: joel.watts@utoronto.ca

Contents

Abstract

The transmission of prions between species is typically an inefficient process due to the species barrier, which represents incompatibility between prion seed and substrate molecules. Bank voles (*Myodes glareolus*) are an exception to this rule, as they are susceptible to a diverse range of prion strains from many different animal species. In particular, bank voles can be efficiently infected with most types of human prions and have played a critical role in validating variably protease-sensitive prionopathy (VPSPr) and certain forms of Gerstmann-Sträussler-Scheinker (GSS) disease as *bona fide* prion disorders rather than non-transmissible proteinopathies. The bank vole prion protein (BVPrP) confers a "universal prion acceptor" phenotype when expressed in mice and when used as a substrate for *in vitro* prion amplification assays, indicating that the

Progress in Molecular Biology and Translational Science, Volume 175
ISSN 1877-1173
https://doi.org/10.1016/bs.pmbts.2020.08.009

unique prion transmission properties of bank voles are mediated by BVPrP. Overexpression of BVPrP in mice can also promote the spontaneous development of prion disease, indicating that BVPrP is intrinsically prone to both spontaneous and template-directed misfolding. Here, we discuss the utility of bank voles and BVPrP for prion research and how they have provided new tools for establishing rapid animal bioassays, modeling spontaneous prion disease, standardizing prion diagnostics, and understanding the molecular basis of the species barrier.

1. An introduction to prions and prion transmission

Prion diseases are a group of fatal neurodegenerative illnesses that affect several species of animals in addition to humans.[1] In sheep, the disease is called scrapie, which was the first prion disease recognized in animals. In cervids such as deer, elk, and moose it is called Chronic Wasting Disease (CWD), and in cattle it is referred to as bovine spongiform encephalopathy (BSE) or "mad cow disease." In humans, prion disease can manifest as either Creutzfeldt-Jakob Disease (CJD), Fatal Familial Insomnia (FFI) or Gerstmann-Sträussler-Scheinker (GSS) disease. Prion diseases can be sporadic, inherited through a mutation in the *PRNP* gene, or acquired through exposure to a pre-existing source of infectious prions. In humans, inherited prion disease accounts for only 5–10% of all cases, and the specific *PRNP* mutation dictates whether the disease presents clinically as FFI, familial CJD (fCJD), or GSS. Approximately 90% of human prion disease cases are sporadic, meaning there is no known genetic or infectious cause. The most common sporadic human prion disease is sporadic CJD (sCJD). The acquired forms of human prion diseases are rare, accounting for less than 1% of all cases.[2] Examples include iatrogenic CJD due to exposure to prion-contaminated growth hormone, kuru resulting from ritualistic cannibalism in the Fore tribe of Papua New Guinea, and variant CJD (vCJD), which is linked to the consumption of material contaminated with BSE. Several animal prion diseases, such as CWD and scrapie, are naturally infectious disorders that can be transmitted from animal-to-animal. In contrast, sCJD does not appear to be readily transmissibly between individuals, although there have been a few cases of vCJD linked to blood transfusions.[3]

The underlying pathogenesis of prion disease is characterized by the spread of infectious proteins called prions in the brain, leading to profound degeneration of brain tissue, neuronal loss, and eventual death of the person or animal.[4–6] Prions are composed of aggregates of a single host-encoded protein called the prion protein (PrP), which is encoded by the *PRNP* gene.

In its normal state PrP is referred to as *cellular* PrP, or PrP^C. PrP^C is a predominantly alpha-helical neuronal glycoprotein that is GPI-anchored and thus is found on the outer leaflet of the plasma membrane. PrP^C has been implicated in many cellular processes, including the maintenance of myelin in the peripheral nervous system, but its precise function(s) in the central nervous remains a topic of debate.[7] In prion disease, PrP^C undergoes a profound conformational rearrangement to form an infectious and neurotoxic conformer called PrP *scrapie*, or PrP^{Sc}. The two versions of PrP differ only in their structural arrangement, as PrP^{Sc} is enriched in beta-sheet content and readily polymerizes to form aggregates. Conformational alterations between PrP^C and PrP^{Sc} allow these two proteins to be biochemically differentiated. For instance, due to its aggregated nature, PrP^{Sc} is insoluble in non-ionic detergents and is partially resistant to proteolytic digestion. The presence of proteinase K (PK)-resistant PrP (PrP^{res}) is the biochemical signature of prion disease. In the most common animal and human prion diseases, PK digestion produces a PrP^{res} banding pattern on an immunoblot with three dominant bands between ~19 and ~30 kDa in molecular weight, which is commonly referred to as PrP27–30. Each band corresponds to a different glycoform of PrP^{Sc}.

PrP^{Sc} is a self-propagating protein and can template the conversion of PrP^C into additional copies of PrP^{Sc}, thus leading to the accumulation of PrP^{Sc} in the brain. The templated conversion of PrP^C by PrP^{Sc} is the central event in the pathogenesis of prion disease. This property also underlies the infectious nature of prions, as pre-formed PrP^{Sc} seeds can initiate the conversion of host PrP^C when introduced into an animal. The protein-only hypothesis, which was originally put forth by Stanley Prusiner, argues that PrP^{Sc} is the sole component of infectious prions.[8] Not all PrP^{Sc} molecules are alike, and variability in the structural arrangement of PrP^{Sc} gives rise to the existence of prion strains.[9] Prion strains can lead to differences in clinical presentation, rates of disease progression, patterns of PrP^{Sc} deposition within the brain, and the electrophoretic mobility of PrP^{res} species. In humans, a methionine/valine polymorphism at codon 129 of PrP influences the generation of prions in patients with sCJD, leading to unique prion strains.[10] For example, the sCJD(MM1), sCJD(MM2), sCJD(VV2) disease subtypes as well as vCJD each constitute distinct strains of prions that exhibit unique biochemical, pathological, and transmission properties.

The study of prion disease has been greatly facilitated by the development of bioassays and molecular tools that exploit the template-directed misfolding phenomenon associated with prions. Animal bioassays are particularly useful as injection of laboratory rodents with prions produces all the clinical, biochemical, and neuropathological features of authentic prion diseases in humans or larger animals. Mice and hamsters are most commonly used for studying prion propagation in a laboratory animal as the disease incubation periods are comparatively short, ~4–5 months following inoculation in mice and as a little as ~75 days in hamsters.[11] However, the "species barrier" limits the types of prions that can be studied using normal hamsters or mice.[12] While all prion strains are potentially infectious, transmission of prions from one species to another is typically an inefficient process, and this is due to structural incompatibilities between PrP^C and PrP^{Sc} conferred by differences in amino acid sequence. This sequence mismatch prolongs incubation periods, decreases transmission efficiency (or "attack rate"), and often results in the formation of a species-adapted prion strain with properties distinct from the original prion isolate. For example, transmission of CJD prions to wild-type mice is very inefficient, and mouse-adapted CJD prions do not exhibit the same propagation properties as authentic human prions.[13] In many cases, prion transmission barriers can be overcome by using genetically modified mice that express PrP^C from a species of interest. For instance, transgenic mice expressing elk PrP^C are susceptible to CWD prions, and transgenic mice expressing human PrP are susceptible to CJD prions.[14,15]

While mice and hamsters are susceptible to prion strains that have been adapted to the mouse PrP or hamster PrP sequence, respectively, other animals are believed to be nearly or entirely resistant to prion transmission. These include rabbits, dogs, and horses, and it is thought that species-specific variations in the PrP amino acid sequence are responsible for producing an enhanced species barrier effect.[16] At the other end of the susceptibility spectrum, bank voles do not appear to impose a species barrier during prion transmission experiments and are thus susceptible to a wide range of prion strains from many different species. Over the past 15 years, bank voles have emerged as a powerful model organism for studying prion transmission and understanding the molecular basis of the species barrier.

2. Bank voles display an enhanced susceptibility to a diverse range of prion strains

The vole was first documented as a model organism for prion transmission as early as 1971 when it was tested as part of a battery of rodents inoculated with either mouse- or rat-passaged scrapie prions.[17,18]

The field voles (*Microtus agrestis*) developed prion disease within 3–4 months following inoculation, which was faster than any of the other species tested. Despite this interesting observation, the vole was not extensively used by the prion field over the next three decades, likely due to the logistical difficulties in acquiring and maintaining these rodents. Voles re-emerged as a tool for prion research in the mid-2000s when it was discovered that bank voles (*Myodes glareolus*) are susceptible to a diverse range of prion strains derived from multiple species (Table 1). Much of this pioneering work has been conducted by Romolo Nonno, Umberto Agrimi, and colleagues at the Istituto Superiore di Sanità in Rome, Italy. Other vole species, such as meadow voles (*Microtus pennsylvanicus*) and red-backed voles (*Myodes gapperi*) also appear to exhibit an enhanced susceptibility to prions.[26–28]

2.1 Susceptibility to non-human prions

Residue 109 in the bank vole prion protein (BVPrP) is polymorphic, where either a methionine (M109) or isoleucine (I109) residue can be present. Thus, the purpose of initial prion transmission studies in bank voles was to assess the potential influence of various codon 109 combinations. M109-homozygote and M/I109-heterozygote bank voles were challenged with two natural strains of sheep scrapie, SS3 and SS5, as well as one strain of goat scrapie, SG1.[29] These experiments revealed that the bank vole can be efficiently infected with scrapie prions, with 100% transmission rates and relatively short incubation periods (~200 days) observed for each strain. Not surprisingly, M/I109-heterozygotes exhibited incubation periods that were approximately 100 days longer than M109-homozygotes. Although it was hypothesized that BVPrP(I109) might convert at a slower rate, future transmission studies would refute this notion,[19,21,24] suggesting that the inhibitory effect was caused by the co-expression of different variants rather than either variant *per se*. Bank voles proved to be susceptible to a variety of other scrapie isolates as well, including those with varying PrP genotypes.[30] In a related study, it was shown that M109 bank voles can be efficiently infected with the ME7 strain of mouse-passaged scrapie prions as well as the 263 K strain of hamster-passaged scrapie prions.[25]

Animal bioassays represent the gold standard for quantifying prion infectivity, and the susceptibility of bank voles to sheep scrapie prions suggested that they might be useful for measuring prion levels. Groups of bank voles were challenged with serial dilutions of the scrapie isolates SS7B and SCR1 to determine infectivity titres.[30] The SS7B isolate was specifically chosen

Table 1 Susceptibility of bank voles and Tg(BVPrP,M109) mice to a variety of prion strains.

		M109 bank voles				Tg(BVPrP,M109) mice				
		Passage 1		Passage 2		Passage 1		Passage 2		
Prion strain	PrP^{Sc} sequence	Mean incubation period (d)	Clinical Symptoms (n/n_0)	Mean incubation period (d)	Clinical Symptoms (n/n_0)	Mean incubation period (d)	Clinical Symptoms (n/n_0)	Mean incubation period (d)	Clinical Symptoms (n/n_0)	Reference(s)
sCJD(MM1)	Human (MM129)	188	16/16	129	14/14	196	8/8	175	8/8	19
sCJD(MV1)	Human (MV129)	179	20/20	128	15/15	n.d.	n.d.	n.d.	n.d.	20
sCJD(MM2)	Human (MM129)	408	9/9	339	20/20	244	7/7	89	8/8	19,20
sCJD(MV2)	Human (MV129)	>500	0/20	n.d.	n.d.	n.d.	n.d.	n.d.	n.d.	20
sCJD(VV2)	Human (VV129)	>500	0/18	n.d.	n.d.	394	7/7	103	7/7	19,20
fCJD (E200K)	Human (MM129)	158	19/19	143	13/13	n.d.	n.d.	n.d.	n.d.	20
VPSPr(MM)	Human (MM129)	356	1/15	148	11/11	n.d.	n.d.	n.d.	n.d.	21
VPSPr(MV)	Human (MV129)	439	2/30	142	11/11	n.d.	n.d.	n.d.	n.d.	21

vCJD	Human (MM129)	760	7/12	n.d.	n.d.	330	7/7	40	7/7	19,22
Classical BSE	Cattle	>1044	0/6	483	6/6	368	7/7	64	8/8	19,23
Scrapie SS3	Sheep (ARQ/ARQ)	199	9/9	92	7/7	n.d.	n.d.	n.d.	n.d.	23
Scrapie SSBP/1	Sheep (VRQ/VRQ)	n.d.	n.d.	n.d.	n.d.	383	8/8	82	7/7	19
CWD	Elk	280	18/18	n.d.	n.d.	249	7/8	146	8/8	19,24
263 K/Sc237	Hamster	195	12/12	n.d.	n.d.	96	7/7	77	8/8	19,25
139A	Mouse	134	19/19	75	7/7	n.d.	n.d.	n.d.	n.d.	23

Different isolates were used to inoculate the bank voles and Tg(BVPrP,M109) mice. n, number of animals that developed disease; n_0, number of animals inoculated; n.d., not determined.

because it had proven difficult to transmit to the ovine PrP-overexpressing Tg338 line.[31] In Tg338 mice, the SS7B strain produced a mean incubation period of more than 500 days, significantly longer than the 188-day incubation period observed in bank voles.[30] Transmission efficiencies of 100% in bank voles were observed with both the SS7B and SCR1 isolates with dilutions up to 10^{-3}, whereas none of the wild-type mice inoculated with a 10^{-1} dilution of these isolates developed clinical prion disease. Thus, this study demonstrated that bank voles can be used as a sensitive bioassay for sheep scrapie prions.

In addition to sheep, hamster, and mouse scrapie prions, M109 bank voles can also be efficiently infected with CWD prions from elk and deer, demonstrating 100% attack rates and mean incubation periods between 250 and 300 days.[24] Challenge of I109 bank voles with CWD also resulted in efficient disease transmission and, for most isolates tested, substantially shorter incubation periods. Strikingly, serial passage of CWD prions in I109 voles led to incubation periods of around 35 days, which is by far the most rapid disease onset observed for prions in a non-transgenic animal. M109 bank voles challenged with classical BSE prions presented no clinical, neuropathological, or even biochemical evidence of prion transmission.[23] However, following serial passaging, 100% of the inoculated bank voles developed severe neurological illness, with a mean incubation period of just under 500 days. Third passage further reduced the mean incubation period to 92 days, suggestive of significant strain adaptation. In fact, many prion strains exhibit incubation periods of under 100 days when serially propagated in bank voles,[20,24] revealing that these animals can be used for rapid prion bioassays. While M109 bank voles were not particularly susceptible to classical BSE prions, isolates of atypical BSE (BSE-L and BSE-H) propagated more efficiently in these animals.[22]

2.2 Susceptibility to human prions

In a landmark study, it was demonstrated that M109 bank voles could be efficiently infected with the majority of human prion isolates tested, including several cases of sCJD and fCJD.[20] Conversely, none of the wild-type mice challenged with the same prion isolates developed clinical disease. Most intriguingly, there was very little transmission barrier for human prions in bank voles, as judged by the small differences in disease incubation period between initial and secondary passages. Further supporting this notion were the observations that strain-specific PrP^{res} properties were maintained upon propagation in bank voles, even after serial passaging, and that the disease incubation periods were comparable to those observed in mice expressing human PrP.[32–35] The sCJD(MV2) and sCJD(VV2) subtypes did not propagate well in M109 bank voles, indicating that bank voles cannot propagate all types of prions efficiently.[20] However, these isolates also proved refractory to propagation in humanized mice: the sCJD(MV2) and sCJD(VV2) subtypes exhibited low attack rates and long incubation periods when inoculated into human PrP knock-in mice that were homozygous for methionine 129,[34] a relevant comparison as methionine is present at codon 129 of the BVPrP sequence.[23] Collectively, these data suggest that BVPrP may be a better substrate for propagating certain human prion strains than even human PrP, despite the existence of a PrP^{Sc}-PrP^{C} sequence mismatch on initial transmission. Indeed, bank voles have also proven useful for characterizing atypical sCJD subtypes, such as the MV^{AG} variant with an altered glycosylation profile[36] and CJD cases with white matter amyloid plaques.[37]

Bank voles have been particularly useful for assessing the transmissibility of less common genetic forms of human prion diseases. The heritable prion disorder GSS, which is caused by PrP mutations such as P102L, A117V, and F198S, is biochemically characterized by atypical forms of PrP^{res}, including a characteristic 6–8 kDa PK-resistant PrP core that is thought to comprise GSS amyloids.[38] The P102L mutation can also manifest with classical PrP27–30, similar to the PrP^{res} observed in most types of sCJD.[39] Whether GSS cases with only 6–8 kDa PrP^{res} fragments should be classified as true prion disorders or non-transmissible proteinopathies has been controversial since these cases fail to transmit or only weakly transmit disease to wild-type and transgenic mouse models,[40–42] whereas P102L isolates with PrP27–30 transmit successfully.[40,43] I109 bank voles were challenged with isolates taken from F198S GSS patients with 8 kDa PrP^{res} fragments, A117V GSS patients with 6–7 kDa fragments, and a P102L GSS patient with the 8 kDa fragment.[44] Successful disease transmission was obtained with each of these isolates, albeit with variable transmission efficiencies, indicating that GSS is indeed an infectious prion disorder. Interestingly, despite being refractory to propagation in other animal models, the mean incubation periods for A117V and F198S GSS prions in I109 bank voles were between 95 and 153 days. Short ~7–8 kDa PrP^{res} fragments similar to those observed in the human isolates were found in the brains of the GSS-inoculated voles, indicative of at least partial maintenance of prion strain properties.

Variably protease-sensitive prionopathy (VPSPr) is an atypical form of human prion disease that accounts for 0.7–1.7% of all sporadic cases.[45,46] VPSPr is characterized by unglycosylated and monoglycosylated PrP fragments that produce a unique 5-band pattern after PK digestion, including short fragments similar to those found in GSS cases. Inoculation of VPSPr isolates into transgenic mice expressing human PrP did not result in clinical disease, and only half of the mice developed small, plaque-like PrP deposits.[47,48] Furthermore, no disease was observed upon second passage, calling into question the identity of VPSPr as a prion disorder. A variety of VPSPr isolates were next inoculated into either M109 or I109 bank voles.[21] Although disease transmission rates were low in both lines (5% in M109 and 35% in I109 voles), the disease could be serially transmitted with 100% efficiency, indicating that VPSPr is an infectious prion disorder. Three different classes of PrP^{res} were observed in the VPSPr-inoculated I109 voles: two of which resembled the phenotype in sCJD-inoculated bank voles and one which resembled the pattern observed in GSS-inoculated I109 voles. GSS-like, short PrP^{res} fragments were never observed in the VPSPr-inoculated M109 voles. It should be noted that none of the three PrP^{res} signatures precisely matched the profile observed in the original VPSPr isolates. With regards to strain typing, it is interesting that GSS and VPSPr display similar vole-specific transmissibility, with a preference for propagation in I109 bank voles, since VPSPR is hypothesized to be a sporadic form of GSS due to the presence of similar short PrP^{res} fragments.[45]

Collectively, these studies demonstrate that bank voles display an enhanced susceptibility to a diverse range of prion strains compared to other laboratory rodents such as mice and hamsters. This indicates that species barrier effects are significantly dampened in bank voles. Indeed, bank voles are susceptible to most strains of prions that have been tested, suggesting that they may constitute universal or near-universal acceptors for prions. While the role of polymorphic residue 109 of BVPrP in prion transmission remains to be fully elucidated, it appears as though I109 bank voles are superior at propagating certain types of prion strains, such as those with shorter PrP^{res} fragments as exemplified by GSS and VPSPr.

3. The bank vole prion protein functions as a universal acceptor for prions

3.1 Evidence from studies using genetically modified mice

The unique susceptibility of bank voles to a diverse range of prion strains prompted an investigation into the factors responsible for this phenomenon. Since the predominant determinant of the species barrier is the sequence of PrP,[49] the parsimonious explanation was that the enhanced prion susceptibility was enciphered within the amino acid sequence of BVPrP. To test this hypothesis, transgenic mice that over-express BVPrP(M109) in the absence of endogenous mouse PrP (Tg(BVPrP,M109) mice) were generated.[19,22] Tg(BVPrP,M109) mice were highly susceptible to prion isolates from voles, mice, hamsters, humans, elk, sheep, and cattle (Table 1), confirming that expression of BVPrP is sufficient to recreate the minimal species barrier effects observed in bank voles.[19,22] Knock-in mice expressing physiological levels of BVPrP(M109) have also been generated and exhibit susceptibility to several human prion isolates.[50] For the vast majority of prion isolates, 100% disease transmission was observed in Tg(BVPrP, M109) mice, indicating that BVPrP may function as a "universal acceptor" for prions. While Tg(BVPrP,I109) mice were also susceptible to a diverse series of prion isolates,[19] these mice unexpectedly developed spontaneous disease,[51] rendering them less useful for prion transmission studies.

Tg(BVPrP,M109) mice proved to be susceptible to many different prion strains, including some that were refractory to propagation in M109 bank voles, such as sCJD(VV2) and classical BSE[19] (Table 1). These mice exhibited 100% clinical attack rates upon inoculation with either strain, whereas no clinical symptoms were observed in M109 bank voles challenged with similar prion isolates. This discrepancy may reflect the fact that these Tg(BVPrP,M109) mice express PrP at levels approximately five times higher than non-transgenic animals. In another study that employed a distinct line of Tg(BVPrP,M109) mice, only 20% of animals inoculated with classical BSE prions developed prion disease, perhaps because BVPrP expression levels are lower in this line.[22] For other strains, such as sCJD(MM1) and CWD, only minor differences in incubation period were obtained when comparing PrP over-expressing Tg(BVPrP,M109) mice to M109 bank voles, suggesting that substrate availability is only rate-limiting for the transmission of certain strains.[19,20,23,24]

As impressive as the primary-passage incubation periods were for prion-inoculated Tg(BVPrP,M109) mice, the greatest benefit associated with this paradigm was arguably the rapid incubation periods displayed by many BVPrP-adapted strains. Serially passaged sCJD(MM2), sCJD(VV2), vCJD, and BSE prions all presented shorter incubation periods in Tg(BVPrP, M109) mice than have been observed in any human,[34,52] chimeric mouse/human,[32,33,53] or bovine[54,55] PrP-expressing lines. Upon secondary passage, sCJD(MM2) and sCJD(VV2) prions exhibited mean incubation periods of around 3 months, BSE of just over 2 months, and vCJD of just over 1 month.[19] These accelerated incubation periods were achieved within two passages, whereas M109 bank voles required three passages to fully adapt BSE prions to the BVPrP sequence.[23] However, a fundamental question remains as to whether these BVPrP-adapted strains exhibit the salient properties of the original prion isolates, which is a critical consideration when developing prion-targeting therapeutics.[35,56] Using a combination of biochemical analysis and histopathology, the fidelity of various BVPrP-adapted strains was assayed. Vole-passaged sCJD(MM1) likely represents the gold standard for fidelity of adapted strains as it maintained a virtually identical electrophoretic profile for PrP^{res}, an identical conformational stability, similar synaptic-like PrP^{Sc} deposition as seen in patients, and similar behavior to the original inoculum when retro-transmitted to transgenic mice expressing human PrP.[19,20] Although this level of exhaustive typing was not performed for every isolate, prion strain fidelity was found to be largely maintained upon propagation in Tg(BVPrP,M109) mice.[19,22] Interestingly, strain mutation was observed upon challenge of Tg(BVPrP,I109) mice with certain prion isolates, as judged by differences in the electrophoretic mobility of PrP^{res} species, which could be related to the concomitant spontaneous generation of prions that occurs in this line.[19,51]

3.2 Evidence from cultured cell studies

There are multiple lines of cultured cells that can be infected with various strains of mouse-adapted prions, but only a handful of lines that can be infected with prions from other species.[56] Despite the success in genetically modified mice, there have been relatively few attempts at leveraging BVPrP's unique properties to extend the range of prion strains that can be propagated in cultured cells. Early efforts included the successful transmission of bank vole-adapted BSE prions to BVPrP(M109)-expressing RK13 cells.[57] However, these cells, unlike M109 bank voles, were not directly susceptible to SS3 sheep scrapie prions.[23,57] This suggests that, at least in RK13 cells, expression of BVPrP on its own is not sufficient to confer a universal prion acceptor phenotype, perhaps indicating a role for other cellular factors. More recently, it has been demonstrated that CWD prions can be propagated in BVPrP-expressing CAD5 or MEF cell lines that had underwent gene editing to remove expression of their endogenous mouse PrP.[58] While there are drawbacks to this study, most notably that prion infection could only be detected using the ultrasensitive real-time quaking-induced conversion (RT-QuIC) technique, it provides proof-of-principal that BVPrP can function as a permissive substrate for cross-species prion replication in cultured cells.

However, it remains unclear how far BVPrP's enhanced prion strain susceptibility can be exploited in cultured cell paradigms. A recent study demonstrating that robust propagation of hamster prions can be achieved in gene-edited CAD5 cells expressing hamster PrP suggests that this cell line may be particularly useful for examining the promiscuity of BVPrP for replicating prions in cultured cells.[59] There is currently no immortalized cell line capable of replicating human prions, which has hindered the development of therapeutics for human prion diseases.[56] Given that bank voles and Tg(BVPrP) mice are highly susceptible to human prion strains and appear to maintain strain fidelity,[19,20,22,44] it seems worthwhile to explore whether this phenomenon can be recapitulated in cultured cells as it could lead to a high-throughput drug screening platform that could advance translational efforts significantly.

3.3 Evidence from PMCA studies

Protein misfolding cyclic amplification (PMCA) is a widely used *in vitro* cell-free prion conversion assay. It was developed by Claudio Soto's group as a means of detecting small quantities of PrP^{Sc} particles using iterative cycles of sonication and incubation in the presence of excess amounts of PrP^{C} substrate.[60] PMCA has been optimized for the detection of prions in a range of biological samples including blood,[61,62] urine,[63] nasal brushings,[64] lymph nodes,[65] CSF[66] and brain.[60] The ability of PMCA to generate *bona fide* infectious prions makes it a very attractive tool for studying prion biology.[67] The species barrier remains intact when employing standard PMCA reaction conditions, indicating that the underlying mechanism is similar to what occurs *in vivo*.[68] Therefore, PMCA represents a platform that can be used to investigate fundamental processes in prion biology such as the molecular basis of the species barrier, prion strain adaptation, and the effect of cofactors on prion replication.

Bank vole PrP has proven to be a highly flexible substrate in PMCA reactions because it appears to maintain its universal prion acceptor phenotype. Brain homogenate from M109 bank voles, which contains $BVPrP^{C}$, has been used in PMCA to amplify prions from species including humans, sheep, cattle, mouse, hamster, and deer.[69,70] The use of BVPrP in PMCA has improved the detection of otherwise difficult prion strains, including cases of FFI[64] and fCJD.[69] In this capacity, BVPrP offers great versatility in antemortem diagnostics, detecting a wide range of prion strains with a single substrate. This unique ability of BVPrP also translates to synthetic prion strains, as a non-infectious synthetic strain generated from recombinant mouse PrP was detected by PMCA using bank vole but not mouse brain homogenate.[71] This highlights the flexibility of BVPrP, which appears better equipped to replicate atypical prion strains than even host-matched substrates. Moreover, PMCA using BVPrP has been useful for demonstrating that the conformation of PrP^{Sc}, not the amino acid sequence of PrP^{C}, determines prion strain-specific glycosylation and cofactor preferences.[72]

3.4 Evidence from RT-QuIC studies

RT-QuIC relies on the ability of PrP^{Sc} to seed the multimerization of recombinant PrP monomers produced in *E. coli* into amyloid fibrils, which can be detected in real-time using Thioflavin T fluorescence.[73] Originally described as the amyloid seeding assay[74] and then the quaking-induced conversion (QuIC) assay,[75] RT-QuIC has been extensively optimized for the sensitive and specific detection of PrP^{Sc} seeds from a wide variety of tissues and biological samples.[76] In humans, RT-QuIC has been optimized for the detection of prions from various dermatomes of skin,[77] from the olfactory mucosa obtained from nasal brushings[78] and also from CSF.[79] When CSF seeding activity is combined with readings from the olfactory mucosa, the sensitivity and specificity of RT-QuIC for CJD diagnosis is near 100%. In other prion diseases like CWD, RT-QuIC has been able to detect seeding activity from a wide variety of biological samples, including urine and feces,[80] saliva,[81] lymph nodes,[82] olfactory mucosa,[83] CSF,[84] and third eyelid.[85] Collectively, these results suggest that naturally occurring prion diseases can be detected through RT-QuIC from a number of tissues, making it a very powerful diagnostic tool.

The single biggest challenge in RT-QuIC is determining which recombinant PrP substrate is capable of detecting your PrP^{Sc} species of interest. Since RT-QuIC does not always reproduce the species barrier effects observed in animals and generates amyloid end-products that are not particularly infectious,[86] a fundamentally distinct mechanism must be in operation. Recombinant BVPrP(M109) has been shown to function as a universal substrate for RT-QuIC reactions.[87] Following optimization of RT-QuIC parameters for the recombinant BVPrP substrate, 28 different prion strains from species including humans, sheep, hamsters, mice, cervids and cattle were able to be detected.[87] Atypical prion strains that were previously undetectable in the RT-QuIC assay were also able to be amplified using recombinant BVPrP, including the Nor98 scrapie strain and human GSS cases, all of which contain short PrP^{res} fragments. The ability of recombinant BVPrP to amplify these diverse prion strains also increases the utility of RT-QuIC for strain discrimination.[77,88] RT-QuIC has been optimized with recombinant hamster PrP substrate for the detection of common prion strains, including sCJD and classical scrapie, but hamster PrP is unable to replicate vCJD and atypical scrapie. Considering that recombinant BVPrP can amplify both classic and atypical prion cases, employing both hamster and bank vole substrates in RT-QuIC can help distinguish PrP^{Sc} strains.[77,87] Interestingly, recombinant BVPrP also displays a unique ability to detect synthetic prion strains generated *in vitro* in the presence or absence of specific cofactors, including strains that are non-pathogenic.[89]

Recombinant BVPrP's ability to detect prions is not only limited to the RT-QuIC format, and it has also been employed in simple QuIC experiments, where it is able to rapidly assay prion seeding activity from CJD and CWD samples.[90] In the context of CWD, recombinant BVPrP shows great promise for the screening of preclinical cervids, facilitating the surveillance of CWD in wild populations.[91,92] When paired with recombinant BVPrP, the ultrasensitive nature of RT-QuIC allows for the detection of infectivity that might otherwise be missed, either due to low prion levels, PK-sensitive species, or potential PrP^{Sc} intermediates. Therefore, it appears that most, if not all PrP^{Sc} seeds can be detected using RT-QuIC with recombinant BVPrP, making this paradigm a very powerful tool in both diagnostics and experimental biology.

4. Bank vole PrP as a tool for studying the spontaneous formation of prions

4.1 Transgenic mice expressing wild-type BVPrP

The vast majority of human prion disease cases do not arise from exposure to a pre-existing source of PrP^{Sc}, but instead from the spontaneous formation of prions in the brain. In sporadic CJD, the stochastic misfolding of PrP^{C} to PrP^{Sc} is thought to initiate the disease process. In genetic prion diseases such as fCJD, FFI, and GSS, mutations within the *PRNP* gene encode mutant PrPs that are believed to exhibit an enhanced propensity for forming PrP^{Sc}. Despite numerous attempts to model the spontaneous formation of prions using mice, little is known about the events governing PrP^{Sc} generation in the brain. Several lines of transgenic or knock-in mice that express mutant versions of PrP display certain aspects of the corresponding human prion disease, but typically fail to recapitulate key disease hallmarks such as the presence of PK-resistant PrP and the generation of prion infectivity.[93,94]

Unexpectedly, several lines of transgenic mice over-expressing wild-type BVPrP(I109) developed spontaneous signs of neurological illness (Table 2), including those consistent with a diagnosis of prion disease in mice such as kyphosis and ataxia.[51,96] This observation was robust, as 100% of mice examined eventually developed disease. Expression level of BVPrP(I109) had a significant impact on the disease incubation period, with higher-expressing lines developing spontaneous disease more rapidly than those with lower levels of expression.[51] Over-expression of BVPrP(I109) is likely necessary for this phenotype to manifest, as there is no evidence that I109 bank voles develop spontaneous disease as they age. Prion disease-specific neuropathological changes were found in the brains of spontaneously ill Tg(BVPrP,I109) mice, including spongiosis, gliosis, and deposition of misfolded PrP.[51] While stereotypical PrP27–30 similar to that observed in the brains of sCJD patients was not found in the brains of spontaneously ill Tg(BVPrP,I109) mice, they did exhibit an ~8 kDa PK-resistant PrP fragment similar to those seen in GSS-inoculated bank voles.[21,44,95] Thus, brains from spontaneously ill Tg(BVPrP,I109) mice exhibit the biochemical and neuropathological hallmarks of prion disease.

To assess whether prion infectivity is generated in spontaneously ill Tg(BVPrP,I109) mice, brain extracts were inoculated into young Tg(BVPrP,I109) mice. Rapid mean disease incubation periods of between 35 and 49 days were observed, indicating that *bona fide* transmissible prions are formed spontaneously in the brains of Tg(BVPrP,I109) mice.[51] Similar to the brains of spontaneously ill Tg(BVPrP,I109) mice, the brains of mice inoculated with the spontaneous prions also exhibited an ~8 kDa PrP^{res} fragment. The spontaneously formed BVPrP(I109) prions were also transmissible to transgenic mice that over-express mouse PrP, which do not themselves develop late-onset spontaneous illness,[97] indicative of authentic disease transmission and not simply acceleration of a disease intrinsic to the line. Like prototypical prion strains, brain extracts from spontaneously sick Tg(BVPrP,I109) mice could be serially diluted up to 10^{-6} and still cause disease in recipient animals.[95] These spontaneously formed prions were also infectious when administered *via* peripheral routes of inoculation.[95]

Table 2 BVPrP-based transgenic mouse models of spontaneous prion disease.

Line	BVPrP mutation	Corresponding human disease	BVPrP allotype	PrP level[a]	Spontaneous signs of neurologic illness (n/n_0)?	Mean age of disease onset (d)	Prion disease-specific neuropathology?	Highly PK-resistant PrP?[b]	Disease transmissible?	Reference(s)
Tg-Bv B495	WT	sCJD	M109	4 ×	Yes (19/20)	488	Yes	No	No	50
Tg3574	WT	sCJD	I109	3.6 ×	Yes (14/14)	340	Yes	Yes	Yes	51,95
Tg3581	WT	sCJD	I109	5.3 ×	Yes (19/19)	218	Yes	Yes	Yes	51,95
TgVole	WT	sCJD	I109	3–4 ×	Yes (10/10)	182	Yes	Yes	Yes	96
Tg15972	D178N	FFI	I109	0.4 ×	Yes (24/24)	240	Yes	Yes	Yes	95
Tg7253	E200K	fCJD	I109	1.7 ×	Yes (24/24)	268	Yes	Yes	Yes	95
Tg24600	ΔGPI	GSS	I109	0.5 ×	Yes (11/11)	415	Yes	Yes	Yes	95

[a]PrP expression level is noted as fold-change compared to PrP levels in non-transgenic animals.
[b]Defined as resistance to PK digestion using a PK concentration of at least 20 μg/mL.
WT, wild-type; n, number of animals that developed disease; n_0, number of animals inoculated.

While transgenic mice over-expressing BVPrP(I109) develop spontaneous prion disease, the situation is less clear for mice that express BVPrP(M109). Several lines of Tg(BVPrP,M109) mice remained healthy for at least 500 days, despite expressing BVPrP at levels that were sufficient to generate spontaneous disease in the Tg(BVPrP,I109) lines. In contrast, a different study found that another line of Tg(BVPrP,M109) mice developed spontaneous neurological illness, albeit with longer incubation times compared to Tg(BVPrP,I109) mice.[50] This line, which expresses BVPrP(M109) at approximately four times the normal level, exhibited a clinical attack rate of 95% with a mean disease incubation period of around 500 days[50] (Table 2). Prion disease-specific neuropathological changes including spongiform degeneration, astrocytic gliosis, and PrP deposition were observed in this line. However, there was no evidence for PK-resistant PrP, including the characteristic ~8 kDa PrP^{res} fragment observed in Tg(BVPrP,I109) lines, and brain extracts from spontaneously sick mice did not transmit disease to knock-in mice expressing BVPrP(M109), arguing against the formation of infectious prions. From these studies, it can be concluded that BVPrP(I109) is more prone to spontaneous misfolding than BVPrP(M109) and more readily forms aggregates that are infectious.

4.2 Transgenic mice expressing mutant BVPrP

Given the results obtained with Tg(BVPrP,I109) mice, it was next investigated whether BVPrP(I109) may serve as a superior backbone for developing mouse models of genetic prion diseases. Tg(BVPrP,I109,D178N) lines were designed to model FFI, Tg(BVPrP,I109,E200K) lines were generated to model fCJD, and the Tg(BVPrP,I109,R232X) line was created to model GSS by introducing a nonsense mutation to create a PrP variant lacking the GPI anchor, similar to the GSS-causing Y226X or Q227X mutations.[95] All lines expressing mutant BVPrP(I109) developed spontaneous disease (Table 2), and the same inverse relationship between BVPrP expression level and incubation period that was observed in mice expressing wild-type BVPrP(I109) was found in the mice expressing mutant BVPrP(I109).[51,95] However, each mutant line exhibited faster incubation periods for a given BVPrP expression level than their wild-type BVPrP-expressing counterparts, indicating that the mutations accelerated the spontaneous formation of prions in the brain. Additionally, all three mutations were found to significantly impact BVPrP steady-state levels. For a given transgene copy number, Tg(BVPrP,I109,D178N) mice produced approximately 10-fold less PrP than mice expressing wild-type BVPrP(I109), Tg(BVPrP,I109, E200K) mice approximately twofold less, and Tg(BVPrP,I109,R232X) mice approximately eightfold less.[95] The lower steady-state levels of mutant BVPrP were determined to result from an enhanced protein turnover rate, possibly indicative of PrP destabilization.[95]

Each mutation had a significant impact on neuropathology, implying that each mutation specified the formation of a distinct prion strain. In addition to astrocytic gliosis and spongiform degeneration, each line expressing mutant BVPrP(I109) developed prominent cerebral PrP deposition. Tg(BVPrP,I109,D178N) mice developed coarse and granular BVPrP aggregates, Tg(BVPrP,I109,E200K) lines developed small rounded BVPrP aggregates, and Tg(BVPrP,I109,R232X) mice developed larger amyloid plaques.[95] The Tg(BVPrP,I109,R232X) line most closely modeled the pathological features of the corresponding human disease, exhibiting larger GSS-like and Thioflavin S-positive aggregates as well as an 8 kDa PrP^{res} fragment. Despite the distinct neuropathological signatures, similar short PrP^{res} fragments were found in brains from spontaneously ill Tg(BVPrP, I109,D178N) and Tg(BVPrP,I109,E200K) mice, which did not match the much larger PrP27–30 species typically observed in patients with FFI and fCJD.

Injection of brain extracts from spontaneously ill, mutant BVPrP mouse brains into young mice expressing the same protein resulted in significant disease acceleration, as was seen with Tg(BVPrP,I109) mice.[51,95] Brain extracts from each line of mutant BVPrP(I109)-expressing mice could also be transmitted to transgenic mice expressing wild-type BVPrP(I109) and to transgenic mice over-expressing mouse PrP. In each case, an 8 kDa PrP^{res} fragment was observed in the brains of inoculated mice, and differences in the morphology of cerebral PrP deposits supported the contention that the spontaneously ill mice developed unique prion strains depending on which mutation was present.

Transgenic mice expressing wild-type or mutant BVPrP(I109) represented a major advancement toward modeling sporadic and heritable prion disease, respectively, arguably providing the most accurate mouse models available to date. The benefit of using BVPrP(I109) to model heritable prion disease becomes obvious when compared to genetically modified mice that have used either mouse or human PrP as the backbone for disease-causing mutations. For example, only 50% of Tg8015 mice that over-express GPI-anchorless mouse PrP at 1.7-fold above normal PrP levels developed clinical disease with a mean incubation period of approximately 600 days.[98]

Comparatively, an attack rate of 100% was observed in the Tg(BVPrP,I109, R232X) line as well as a considerably shorter mean incubation period, despite the fact that these mice express more than three times less PrP than Tg8015 mice.[95] Likewise, while clinical illness has been observed in transgenic mice expressing mouse PrP containing the mouse equivalent of the D178N mutation, there was no evidence of highly PK-resistant PrP species nor disease transmissibility.[99] Both of these features are present in Tg(BVPrP,I109,D178N) mice.[95] Transgenic mice expressing human PrP with prion disease-causing mutations fail to develop spontaneous disease altogether.[100] Although capable of propagating PrP27–30,[19] all spontaneously ill mice with the BVPrP(I109) backbone ultimately developed short ~8 kDa PrP^{res} species.[51,95] This may limit translatability of the Tg(BVPrP,I109,D178N), Tg(BVPrP,I109,E200K), and Tg(BVPrP,I109) models since FFI, fCJD, and sCJD are all associated with PrP27–30 in patients. However, all existing mouse models of spontaneous prion disease fail to generate PrP27–30,[93,98–104] suggesting a broader issue rather than a limitation of BVPrP(I109) itself.

4.3 *In vitro* studies

One study has found that recombinant BVPrP(M109) and BVPrP(I109) do not form aggregates after shaking for 90 h suggesting that, at least under the specific conditions tested, recombinant BVPrP is not especially prone to spontaneous polymerization.[105] Early attempts to use BVPrP(M109) as a substrate for PMCA reactions resulted in the occasional presence of PrP27–30 in unseeded reactions, potentially suggestive of spontaneous prion formation.[69] However, it was determined that cross-contamination was the most likely explanation for these results. More recently, it has been found that recombinant BVPrP(I109) can spontaneously form PK-resistant PrP aggregates when mixed with brain homogenate from PrP knockout mice and subjected to multiple rounds of PMCA.[106] When injected into I109 bank voles, spontaneously generated recombinant BVPrP prions induced disease with a mean incubation period of around 130 days and resulted in the production of an 8 kDa PrP^{res} fragment similar to ones observed in spontaneously ill transgenic mice expressing wild-type or mutant BVPrP(I109). These experiments provide hope that BVPrP may be useful for facilitating the production of "synthetic prions" that can be used to determine the high-resolution structure of PrP^{Sc}. Indeed, using a modified version of the PMCA protocol that involves shaking instead of sonication, large quantities of infectious recombinant BVPrP preparations that are suitable for structural studies have recently been produced.[107]

5. Structural features of the bank vole prion protein and the species barrier

5.1 The structural biology of BVPrP

Since the unique prion transmission properties of bank voles are enciphered within the sequence of BVPrP, it is reasonable to assert that a subset of BVPrP-specific amino acid residues governs this phenomenon. Connecting specific amino acid residues to the structural features that encode BVPrP's unique phenotype can help us understand the replication mechanism associated with cross-species prion conversion. An amino acid sequence alignment of BVPrP with PrPs from other animals commonly used for prion research reveals that the primary sequence of BVPrP is highly similar to those of other mammalian PrPs (Fig. 1). In fact, the sequence of mature BVPrP (following removal of N- and C-terminal signal sequences) differs from that of mouse PrP at only 8 positions and from that of hamster PrP at only 7 positions. Since neither mice nor hamsters are universal prion acceptors, this suggests that a small number of critical amino acids in BVPrP are responsible for encoding the unique properties of bank voles. There is only one truly unique amino acid residue in BVPrP that is not found in any other mammalian PrP sequences: glutamic acid at position 227 (E227) (Fig. 1). This residue is located near the GPI-anchor attachment site at the C-terminus of BVPrP, just beyond the third alpha-helix. Even this one unique residue represents only a modest sequence change, as aspartic acid, which is also negatively charged, is found in the corresponding position in mouse and hamster PrP. Most of the residues that differentiate BVPrP from mouse PrP are found in the sequence of hamster PrP, and most of the residues that differentiate BVPrP from hamster PrP are found in the sequence of mouse PrP. Thus, there is nothing particularly remarkable about the sequence of BVPrP that provides obvious clues toward its distinctive properties.

The structures of several recombinant PrP^{C}'s have been solved by NMR spectroscopy and reveal that all mammalian PrPs share a similar 3-dimensional fold, which consists of an intrinsically disordered N-terminal domain and an alpha-helical globular domain[108–117] (Figs. 1 and 2). The structure of

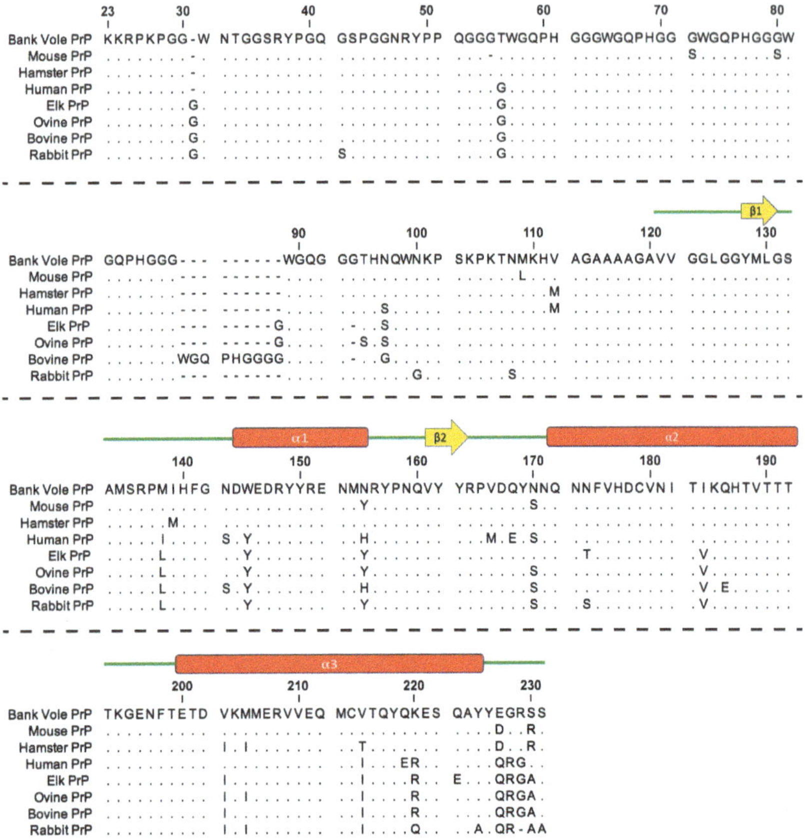

Fig. 1 Alignment of the amino acid sequences for a selection of mammalian PrPs. Only the sequence of the mature form of PrP (residues 23–231 of bank vole PrP), lacking the N- and C-terminal signal sequences, is shown. Elements of secondary structure in the C-terminal domain (residues 121–231), as determined from the NMR structure of recombinant PrP, are indicated above the alignment. The N-terminal domain of PrP is intrinsically disordered in solution. The numbering of the PrP sequences is based on the sequence of bank vole PrP.

recombinant BVPrP^{C} has also been solved and, not surprisingly, it bears a strong resemblance to the structure of other mammalian PrPs[118] (Fig. 2). Even rabbit PrP^{C} has a very similar structure to BVPrP^{C}, and rabbits are known to be relatively resistant to prion infection. The structure of recombinant BVPrP^{C} does contain a "rigid loop" that connects the second beta-strand with the second alpha-helix in PrP^{C} (the beta2-alpha2 loop). This structural change, which is mediated by the presence of asparagine at position 169 but is not unique to BVPrP, has been shown in other

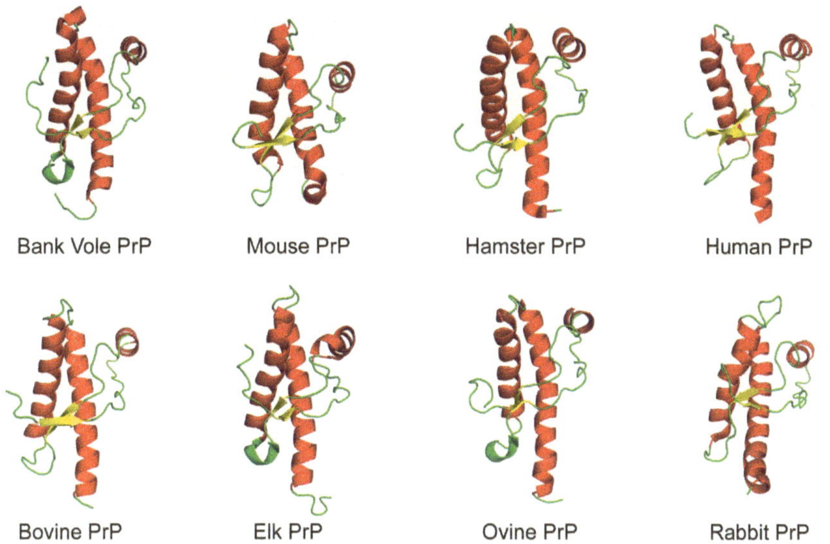

Fig. 2 The structure of the C-terminal domains of recombinant mammalian PrPs, as determined using solution NMR spectroscopy.[108–110,112,113,116–118] The alpha-helices are designated in red, and the beta-strands are highlighted in yellow. All mammalian PrPs share a similar overall structure, with three alpha-helices and two short beta-strands.

contexts to mediate spontaneous PrP misfolding[101,119,120] and the species barrier.[121] The ability of $BVPrP^{C}$ to become templated into $BVPrP^{Sc}$ by prions from a wide range of species, despite clear sequence mismatches, raises questions about the molecular basis of the species barrier. While prion susceptibility is clearly controlled by the sequence of PrP^{C}, species-specific differences in prion transmission efficiency are not mediated by alterations to the overall fold of PrP^{C}. The exact molecular mechanisms underlying BVPrP's anomalous behavior have not yet been fully elucidated.

5.2 Residues governing cross-species prion transmission

In an attempt to understand the residues driving BVPrP's prion strain and species promiscuity, a panel of phylogenetically similar rodents were challenged with sheep SS3 and cattle BSE prions.[23] The bank vole and field vole were compared to the wood mouse and C57BL/6 mouse, demonstrating strikingly divergent phenotypes. SS3 prions caused 100% clinical attack rates in both vole species, with the bank vole presenting a mean incubation period of around 200 days and the field vole showing a mean incubation period of approximately 260 days. Conversely, none of the

SS3-challenged wood mice or C57BL/6 mice developed clinical disease. The opposite results were obtained with the transmission of BSE prions: neither of the two inoculated vole species exhibited clinical symptoms, but 100% of wood mice and approximately 60% of C57BL/6 mice developed prion disease. The PrP sequence of oldfield mice contains vole-specific asparagine residues at positions 154 and 169 but is otherwise identical to the sequence of normal mouse PrP within the globular domain.[23] Oldfield mice exhibited a vole-like 100% attack rate and a mean incubation period of around 150 days when confronted with SS3, indicating a correlation between asparagine at positions 154/169 and prion susceptibility, at least for the SS3 scrapie strain. The same two residues were identified as being important in cell-free PrP^{Sc} conversion experiments.[25] Conversely, Mongolian gerbils and spiny mice, which have mouse PrP-specific residues at codons 154 and 169 but a similar sequence to BVPrP otherwise, were not very susceptible to SS3 prions.[23]

While these studies point to the importance of BVPrP residues 154 and 169 in defining its unique properties, they cannot tell the entire story. The presence of asparagine residues at positions 154 and 169 of PrP is clearly not sufficient to encode a universal prion acceptor phenotype since these residues are also present in the sequence of hamster PrP (Fig. 1). Thus, additional work is required to fully understand how individual amino acids within the BVPrP sequence contribute to its propensity to undergo spontaneous and templated misfolding. It has been suggested that residues at these positions are solvent exposed and may serve as a prion docking site. Therefore, molecular compatibility at these sites may influence the interaction between PrP^{Sc} and PrP^{C}, playing an important role in controlling cross-species prion conversion.[23,25]

At the other end of the prion susceptibility spectrum, rabbits and dogs are believed to be very resistant to prion conversion.[122] Thus, rabbit and dog PrP^{C} have been studied for their ability to resist seeding by various prion strains.[123–125] Certain residues and polymorphisms in canids have shown to be protective against prion conversion. One such residue is aspartic acid at position 163. When this amino acid is inserted into the corresponding position of BVPrP, it delays transmission of CWD prions.[96] Similarly, rabbit PrP demonstrates a drastic barrier to prion conversion, although rabbits are not absolutely resistant to prion disease.[126,127] Using PMCA, the ability of BVPrP, rabbit PrP and human PrP to undergo conversion by either mouse-adapted scrapie or elk CWD prions was investigated.[128] These studies utilized homogenates collected from RK13 cells expressing various chimeric PrP molecules as substrates for PMCA reactions, which rapidly facilitated the identification of key amino acid residues. Neither rabbit PrP nor human PrP could be converted by mouse or elk prions, whereas BVPrP was susceptible to both types of prions. Replacement of various BVPrP residues with the corresponding human PrP-specific residues blocked conversion, and residues in the beta2-alpha2 loop region appeared to have the greatest effect. Residue 169 of BVPrP is located within this region, again pointing to its importance in mediating the prion promiscuity of BVPrP.

Interestingly, replacement of various BVPrP residues with those from the sequence of rabbit PrP, revealed that a loss of asparagine/glutamine residues inhibits the conversion of BVPrP into prions.[128] One substitution, N100G, was particularly potent at blocking prion replication. Likewise, rabbit PrP with the G100N substitution could amplify both mouse and elk prions. Interestingly, this residue was previously highlighted for its inhibitory activity in leporids, both through PMCA[125] and in cell culture.[123] Through the systematic analysis of 40 residues, it was found that many of the key residues which facilitated prion conversion were asparagine residues.[128] When the asparagine/glutamine content of various PrPs were analyzed, the more prion-susceptible species like bank voles and hamsters demonstrated a high asparagine/glutamine content, whereas prion-resistant species like rabbits and dogs demonstrated a lower asparagine/glutamine content. This correlation between asparagine/glutamine residues and susceptibility to prion conversion is appealing, but how they contribute to cross-species prion conversion is not yet fully understood. For instance, hamster PrP has the same total number of asparagine and glutamine residues as BVPrP, yet hamsters are not universal prion acceptors.

The beta2-alpha2 loop of BVPrP (168-QYNNQNN-174) is enriched in asparagine and glutamine residues and has been proposed to participate in a strong hydrogen bond network.[128] Interestingly, the structure of a protofibril formed from a BVPrP peptide (residues 168–176), obtained using microED cryoelectron microscopy, also demonstrates this phenomenon, where the beta2-alpha2 loop behaves like a steric zipper with antiparallel beta-strands that pair front to back.[129] The asparagine and glutamine residues form a strong hydrogen bonding network in the core of the fibril, which is then stabilized by bulky aromatic residues. This hyperpolarized hydrogen bonding network of asparagines and glutamines serves to stabilize the prion protofibril and to present a nucleating surface for template-directed misfolding, potentially relaxing the requirement for sequence identity between PrP^{C} and PrP^{Sc}.

5.3 BVPrP residues influencing the spontaneous formation of prions

The methionine/isoleucine polymorphism at codon 109 of BVPrP has been shown to play a drastic role in regulating its spontaneous misfolding.[51,95] This residue is located in the intrinsically disordered N-terminal domain of PrP and how it influences spontaneous prion conversion is not known. *In vivo*, BVPrP(I109) has also been demonstrated as the fastest model of CWD replication, whereas BVPrP(M109) is not as rapid.[24] These differential prion susceptibilities, including the ability to spontaneously misfold, imply that the 109 residue carries structural importance in the process of PrP^{C} misfolding.

As mentioned previously, one line of transgenic mice expressing wild-type BVPrP(M109) developed spontaneous neurological illness with neuropathological changes consistent with prion disease.[50] In the same study, transgenic mice expressing mouse PrP containing two BVPrP-specific residues at the extreme C-terminus of the protein (E227 and S230) also developed spontaneous disease with a mean incubation period of nearly 700 days. This suggests that these two C-terminal residues may be sufficient for promoting the spontaneous misfolding of PrP into a disease-causing isoform. However, further studies are required to fully assess the contributions of various BVPrP domains to its propensity to misfold spontaneously. Interestingly, a single amino acid substitution at position 159 in BVPrP(I109) to the corresponding residue found in dogs (D159N) resulted in a substantial delay in the onset of spontaneous disease.[96] This suggests that while amino acids from naturally prion-resistant species are capable of interfering with spontaneous prion formation, they are not able to completely to override the misfolding propensity of BVPrP(I109).

6. Conclusions

The bank vole and its associated prion protein have had a significant impact on prion research, providing valuable experimental tools to explore many of the unanswered questions plaguing the field. The bank vole itself is a very flexible animal model since it can be used to bioassay a diverse set of prion strains. Further, the identification of BVPrP as the driver of the vole's unique susceptibility to prions has allowed this phenomenon to be adapted to systems that can be more widely implemented, such as genetically modified mice and completely *in vitro* systems of prion conversion like PMCA and RT-QuIC. The capacity of BVPrP to spontaneously misfold in whole organisms, with and without heritable disease-causing mutations, has also provided new models for sporadic and familial prion disease that can potentially be used for the identification and testing of drugs targeting the spontaneous formation of prions in the brain. Overall, the bank vole has provided new insights into prion transmission and the species barrier, as well as superior animal and *in vitro* models for studying and diagnosing prion disease. That being said, outstanding questions remain, including a detailed molecular explanation for the enhanced misfolding propensity of BVPrP. In addition, there may remain untapped potential for BVPrP, particularly with regards to creating improved cellular paradigms for studying prion transmission and propagation.

References

1. Watts JC, Balachandran A, Westaway D. The expanding universe of prion diseases. *PLoS Pathog*. 2006;2(3):e26.
2. Brown P, Preece M, Brandel JP, et al. Iatrogenic Creutzfeldt-Jakob disease at the millennium. *Neurology*. 2000;55(8):1075–1081.
3. Llewelyn CA, Hewitt PE, Knight RS, et al. Possible transmission of variant Creutzfeldt-Jakob disease by blood transfusion. *Lancet*. 2004;363(9407):417–421.
4. Colby DW, Prusiner SB. Prions. *Cold Spring Harb Perspect Biol*. 2011;3(1):a006833.
5. Collinge J. Mammalian prions and their wider relevance in neurodegenerative diseases. *Nature*. 2016;539(7628):217–226.
6. Scheckel C, Aguzzi A. Prions, prionoids and protein misfolding disorders. *Nat Rev Genet*. 2018;19(7):405–418.
7. Watts JC, Bourkas MEC, Arshad H. The function of the cellular prion protein in health and disease. *Acta Neuropathol*. 2018;135(2):159–178.
8. Prusiner SB. Novel proteinaceous infectious particles cause scrapie. *Science*. 1982; 216(4542):136–144.
9. Soto C, Pritzkow S. Protein misfolding, aggregation, and conformational strains in neurodegenerative diseases. *Nat Neurosci*. 2018;21(10):1332–1340.
10. Cali I, Castellani R, Yuan J, et al. Classification of sporadic Creutzfeldt-Jakob disease revisited. *Brain*. 2006;129(Pt 9):2266–2277.
11. Watts JC, Prusiner SB. Mouse models for studying the formation and propagation of prions. *J Biol Chem*. 2014;289(29):19841–19849.
12. Pattison IH. Experiments with scrapie with special reference to the nature of the agent and the pathology of the disease. In: Gajdusek DC, Gibbs Jr CJ, Alpers MP, eds. *Slow, Latent and Temperate Virus Infections, NINDB Monograph 2*. Washington, D.C.: U.S. Government Printing; 1965:249–257.
13. Lawson VA, Vella LJ, Stewart JD, et al. Mouse-adapted sporadic human Creutzfeldt-Jakob disease prions propagate in cell culture. *Int J Biochem Cell Biol*. 2008; 40(12):2793–2801.
14. Tamgüney G, Giles K, Bouzamondo-Bernstein E, et al. Transmission of elk and deer prions to transgenic mice. *J Virol*. 2006;80(18):9104–9114.
15. Telling GC, Scott M, Mastrianni J, et al. Prion propagation in mice expressing human and chimeric PrP transgenes implicates the interaction of cellular PrP with another protein. *Cell*. 1995;83(1):79–90.

16. Sanchez-Garcia J, Fernandez-Funez P. D159 and S167 are protective residues in the prion protein from dog and horse, two prion-resistant animals. *Neurobiol Dis*. 2018;119:1–12.
17. Chandler RL, Turfrey BA. Inoculation of voles, Chinese hamsters, gerbils and Guinea-pigs with scrapie brain material. *Res Vet Sci*. 1972;13(3):219–224.
18. Chandler RL. Experimental transmission of scrapie to voles and Chinese hamsters. *Lancet*. 1971;1(7692):232–233.
19. Watts JC, Giles K, Patel S, Oehler A, Dearmond SJ, Prusiner SB. Evidence that bank vole PrP is a universal acceptor for prions. *PLoS Pathog*. 2014;10(4):e1003990.
20. Nonno R, Di Bari MA, Cardone F, et al. Efficient transmission and characterization of Creutzfeldt-Jakob disease strains in bank voles. *PLoS Pathog*. 2006;2(2):e12.
21. Nonno R, Notari S, Di Bari MA, et al. Variable protease-sensitive prionopathy transmission to bank voles. *Emerg Infect Dis*. 2019;25(1):73–81.
22. Espinosa JC, Nonno R, Di Bari M, et al. PrPC governs susceptibility to prion strains in bank vole, while other host factors modulate strain features. *J Virol*. 2016;90(23): 10660–10669.
23. Agrimi U, Nonno R, Dell'Omo G, et al. Prion protein amino acid determinants of differential susceptibility and molecular feature of prion strains in mice and voles. *PLoS Pathog*. 2008;4(7):e1000113.
24. Di Bari MA, Nonno R, Castilla J, et al. Chronic wasting disease in bank voles: characterisation of the shortest incubation time model for prion diseases. *PLoS Pathog*. 2013;9(3):e1003219.
25. Piening N, Nonno R, Di Bari M, et al. Conversion efficiency of bank vole prion protein in vitro is determined by residues 155 and 170, but does not correlate with the high susceptibility of bank voles to sheep scrapie in vivo. *J Biol Chem*. 2006; 281(14):9373–9384.
26. Carlson CM, Schneider JR, Pedersen JA, Heisey DM, Johnson CJ. Experimental infection of meadow voles (Microtus pennsylvanicus) with sheep scrapie. *Can J Vet Res*. 2015;79(1):68–73.
27. Watts JC, Stohr J, Bhardwaj S, et al. Protease-resistant prions selectively decrease Shadoo protein. *PLoS Pathog*. 2011;7(11):e1002382.
28. Heisey DM, Mickelsen NA, Schneider JR, et al. Chronic wasting disease (CWD) susceptibility of several North American rodents that are sympatric with cervid CWD epidemics. *J Virol*. 2010;84(1):210–215.
29. Cartoni C, Schinina ME, Maras B, et al. Identification of the pathological prion protein allotypes in scrapie-infected heterozygous bank voles (*Clethrionomys glareolus*) by high-performance liquid chromatography-mass spectrometry. *J Chromatogr A*. 2005;1081(1):122–126.
30. Di Bari MA, Chianini F, Vaccari G, et al. The bank vole (*Myodes glareolus*) as a sensitive bioassay for sheep scrapie. *J Gen Virol*. 2008;89(Pt 12):2975–2985.
31. Béringue V, Bencsik A, Le Dur A, et al. Isolation from cattle of a prion strain distinct from that causing bovine spongiform encephalopathy. *PLoS Pathog*. 2006;2(10):e112.
32. Korth C, Kaneko K, Groth D, et al. Abbreviated incubation times for human prions in mice expressing a chimeric mouse—human prion protein transgene. *Proc Natl Acad Sci U S A*. 2003;100(8):4784–4789.
33. Taguchi Y, Mohri S, Ironside JW, Muramoto T, Kitamoto T. Humanized knock-in mice expressing chimeric prion protein showed varied susceptibility to different human prions. *Am J Pathol*. 2003;163(6):2585–2593.
34. Bishop MT, Will RG, Manson JC. Defining sporadic Creutzfeldt-Jakob disease strains and their transmission properties. *Proc Natl Acad Sci U S A*. 2010;107(26):12005–12010.
35. Berry DB, Lu D, Geva M, et al. Drug resistance confounding prion therapeutics. *Proc Natl Acad Sci U S A*. 2013;110(44):E4160–E4169.

36. Galeno R, Di Bari MA, Nonno R, et al. Prion strain characterization of a novel subtype of Creutzfeldt-Jakob disease. *J Virol*. 2017;91(11):e02390-16.
37. Rossi M, Saverioni D, Di Bari M, et al. Atypical Creutzfeldt-Jakob disease with PrP-amyloid plaques in white matter: molecular characterization and transmission to bank voles show the M1 strain signature. *Acta Neuropathol Commun*. 2017;5(1):87.
38. Piccardo P, Liepnieks JJ, William A, et al. Prion proteins with different conformations accumulate in Gerstmann-Sträussler-Scheinker disease caused by A117V and F198S mutations. *Am J Pathol*. 2001;158(6):2201–2207.
39. Parchi P, Chen SG, Brown P, et al. Different patterns of truncated prion protein fragments correlate with distinct phenotypes in P102L Gerstmann-Straussler-Scheinker disease. *Proc Natl Acad Sci U S A*. 1998;95(14):8322–8327.
40. Tateishi J, Kitamoto T, Hoque MZ, Furukawa H. Experimental transmission of Creutzfeldt-Jakob disease and related diseases to rodents. *Neurology*. 1996;46(2): 532–537.
41. Piccardo P, Manson JC, King D, Ghetti B, Barron RM. Accumulation of prion protein in the brain that is not associated with transmissible disease. *Proc Natl Acad Sci U S A*. 2007;104(11):4712–4717.
42. Asante EA, Linehan JM, Smidak M, et al. Inherited prion disease A117V is not simply a proteinopathy but produces prions transmissible to transgenic mice expressing homologous prion protein. *PLoS Pathog*. 2013;9(9):e1003643.
43. Manson JC, Jameison E, Baybutt H, et al. A single amino acid alteration (101L) introduced into murine PrP dramatically alters incubation time of transmissible spongiform encephalopathy. *EMBO J*. 1999;18(23):6855–6864.
44. Pirisinu L, Di Bari MA, D'Agostino C, et al. Gerstmann-Straussler-Scheinker disease subtypes efficiently transmit in bank voles as genuine prion diseases. *Sci Rep*. 2016;6:20443.
45. Gambetti P, Dong Z, Yuan J, et al. A novel human disease with abnormal prion protein sensitive to protease. *Ann Neurol*. 2008;63(6):697–708.
46. Zou WQ, Puoti G, Xiao X, et al. Variably protease-sensitive prionopathy: a new sporadic disease of the prion protein. *Ann Neurol*. 2010;68(2):162–172.
47. Notari S, Xiao X, Espinosa JC, et al. Transmission characteristics of variably protease-sensitive prionopathy. *Emerg Infect Dis*. 2014;20(12):2006–2014.
48. Diack AB, Ritchie DL, Peden AH, et al. Variably protease-sensitive prionopathy, a unique prion variant with inefficient transmission properties. *Emerg Infect Dis*. 2014;20(12):1969–1979.
49. Prusiner SB, Scott M, Foster D, et al. Transgenetic studies implicate interactions between homologous PrP isoforms in scrapie prion replication. *Cell*. 1990; 63(4):673–686.
50. Kobayashi A, Matsuura Y, Takeuchi A, et al. A domain responsible for spontaneous conversion of bank vole prion protein. *Brain Pathol*. 2019;29(2):155–163.
51. Watts JC, Giles K, Stöhr J, et al. Spontaneous generation of rapidly transmissible prions in transgenic mice expressing wild-type bank vole prion protein. *Proc Natl Acad Sci U S A*. 2012;109(9):3498–3503.
52. Asante EA, Linehan JM, Desbruslais M, et al. BSE prions propagate as either variant CJD-like or sporadic CJD-like prion strains in transgenic mice expressing human prion protein. *EMBO J*. 2002;21(23):6358–6366.
53. Giles K, Glidden DV, Patel S, et al. Human prion strain selection in transgenic mice. *Ann Neurol*. 2010;68(2):151–161.
54. Scott MR, Will R, Ironside J, et al. Compelling transgenetic evidence for transmission of bovine spongiform encephalopathy prions to humans. *Proc Natl Acad Sci U S A*. 1999;96(26):15137–15142.

55. Bishop MT, Hart P, Aitchison L, et al. Predicting susceptibility and incubation time of human-to-human transmission of vCJD. *Lancet Neurol*. 2006;5(5):393–398.
56. Krance SH, Luke R, Shenouda M, et al. Cellular models for discovering prion disease therapeutics: progress and challenges. *J Neurochem*. 2020;153(2):150–172.
57. Courageot MP, Daude N, Nonno R, et al. A cell line infectible by prion strains from different species. *J Gen Virol*. 2008;89(Pt 1):341–347.
58. Walia R, Ho CC, Lee C, Gilch S, Schatzl HM. Gene-edited murine cell lines for propagation of chronic wasting disease prions. *Sci Rep*. 2019;9(1):11151.
59. Bourkas MEC, Arshad H, Al-Azzawi ZAM, et al. Engineering a murine cell line for the stable propagation of hamster prions. *J Biol Chem*. 2019;294(13):4911–4923.
60. Saborio GP, Permanne B, Soto C. Sensitive detection of pathological prion protein by cyclic amplification of protein misfolding. *Nature*. 2001;411(6839):810–813.
61. Saa P, Castilla J, Soto C. Presymptomatic detection of prions in blood. *Science*. 2006;313(5783):92–94.
62. Castilla J, Saa P, Soto C. Detection of prions in blood. *Nat Med*. 2005;11(9):982–985.
63. Gonzalez-Romero D, Barria MA, Leon P, Morales R, Soto C. Detection of infectious prions in urine. *FEBS Lett*. 2008;582(21–22):3161–3166.
64. Redaelli V, Bistaffa E, Zanusso G, et al. Detection of prion seeding activity in the olfactory mucosa of patients with Fatal Familial Insomnia. *Sci Rep*. 2017;7:46269.
65. Shikiya RA, Langenfeld KA, Eckland TE, et al. PrPSc formation and clearance as determinants of prion tropism. *PLoS Pathog*. 2017;13(3):e1006298.
66. Barria MA, Lee A, Green AJ, Knight R, Head MW. Rapid amplification of prions from variant Creutzfeldt-Jakob disease cerebrospinal fluid. *J Pathol Clin Res*. 2018;4(2): 86–92.
67. Castilla J, Saa P, Hetz C, Soto C. *In vitro* generation of infectious scrapie prions. *Cell*. 2005;121(2):195–206.
68. Castilla J, Gonzalez-Romero D, Saa P, Morales R, De Castro J, Soto C. Crossing the species barrier by PrP^{Sc} replication in vitro generates unique infectious prions. *Cell*. 2008;134(5):757–768.
69. Cosseddu GM, Nonno R, Vaccari G, et al. Ultra-efficient PrP^{Sc} amplification highlights potentialities and pitfalls of PMCA technology. *PLoS Pathog*. 2011;7(11): e1002370.
70. Vanni I, Migliore S, Cosseddu GM, et al. Isolation of a defective prion mutant from natural scrapie. *PLoS Pathog*. 2016;12(11):e1006016.
71. Burke CM, Walsh DJ, Steele AD, et al. Full restoration of specific infectivity and strain properties from pure mammalian prion protein. *PLoS Pathog*. 2019;15(3): e1007662.
72. Burke CM, Walsh DJ, Mark KMK, et al. Cofactor and glycosylation preferences for in vitro prion conversion are predominantly determined by strain conformation. *PLoS Pathog*. 2020;16(4):e1008495.
73. Wilham JM, Orru CD, Bessen RA, et al. Rapid end-point quantitation of prion seeding activity with sensitivity comparable to bioassays. *PLoS Pathog*. 2010;6(12): e1001217.
74. Colby DW, Zhang Q, Wang S, et al. Prion detection by an amyloid seeding assay. *Proc Natl Acad Sci U S A*. 2007;104(52):20914–20919.
75. Atarashi R, Wilham JM, Christensen L, et al. Simplified ultrasensitive prion detection by recombinant PrP conversion with shaking. *Nat Methods*. 2008;5(3):211–212.
76. Orru CD, Hughson AG, Groveman BR, et al. Factors that improve RT-QuIC detection of prion seeding activity. *Viruses*. 2016;8(5):140.
77. Orru CD, Yuan J, Appleby BS, et al. Prion seeding activity and infectivity in skin samples from patients with sporadic Creutzfeldt-Jakob disease. *Sci Transl Med*. 2017;9(417):eaam7785.

78. Orru CD, Bongianni M, Tonoli G, et al. A test for Creutzfeldt-Jakob disease using nasal brushings. *N Engl J Med.* 2014;371(6):519–529.
79. Groveman BR, Orru CD, Hughson AG, et al. Extended and direct evaluation of RT-QuIC assays for Creutzfeldt-Jakob disease diagnosis. *Ann Clin Transl Neurol.* 2017;4(2):139–144.
80. John TR, Schatzl HM, Gilch S. Early detection of chronic wasting disease prions in urine of pre-symptomatic deer by real-time quaking-induced conversion assay. *Prion.* 2013;7(3):253–258.
81. Henderson DM, Manca M, Haley NJ, et al. Rapid antemortem detection of CWD prions in deer saliva. *PLoS One.* 2013;8(9):e74377.
82. Haley NJ, Carver S, Hoon-Hanks LL, et al. Detection of chronic wasting disease in the lymph nodes of free-ranging cervids by real-time quaking-induced conversion. *J Clin Microbiol.* 2014;52(9):3237–3243.
83. Haley NJ, Siepker C, Hoon-Hanks LL, et al. Seeded amplification of chronic wasting disease prions in nasal brushings and recto-anal mucosa-associated lymphoid tissues from elk by real-time quaking-induced conversion. *J Clin Microbiol.* 2016;54(4): 1117–1126.
84. Haley NJ, Van de Motter A, Carver S, et al. Prion-seeding activity in cerebrospinal fluid of deer with chronic wasting disease. *PLoS One.* 2013;8(11):e81488.
85. Cooper SK, Hoover CE, Henderson DM, Haley NJ, Mathiason CK, Hoover EA. Detection of CWD in cervids by RT-QuIC assay of third eyelids. *PLoS One.* 2019;14(8):e0221654.
86. Groveman BR, Raymond GJ, Campbell KJ, et al. Role of the central lysine cluster and scrapie templating in the transmissibility of synthetic prion protein aggregates. *PLoS Pathog.* 2017;13(9):e1006623.
87. Orru CD, Groveman BR, Raymond LD, et al. Bank vole prion protein as an apparently universal substrate for RT-QuIC-based detection and discrimination of prion strains. *PLoS Pathog.* 2015;11(6):e1004983.
88. Masujin K, Orru CD, Miyazawa K, et al. Detection of atypical H-type bovine spongiform encephalopathy and discrimination of bovine prion strains by real-time quaking-induced conversion. *J Clin Microbiol.* 2016;54(3):676–686.
89. Wang F, Wang X, Orru CD, et al. Self-propagating, protease-resistant, recombinant prion protein conformers with or without in vivo pathogenicity. *PLoS Pathog.* 2017;13(7):e1006491.
90. Kaelber N, Bett C, Asher DM, Gregori L. Quaking-induced conversion of prion protein on a thermal mixer accelerates detection in brains infected with transmissible spongiform encephalopathy agents. *PLoS One.* 2019;14(12):e0225904.
91. Bistaffa E, Vuong TT, Cazzaniga FA, et al. Use of different RT-QuIC substrates for detecting CWD prions in the brain of Norwegian cervids. *Sci Rep.* 2019;9(1):18595.
92. Hwang S, Greenlee JJ, Nicholson EM. Role of donor genotype in RT-QuIC seeding activity of chronic wasting disease prions using human and bank vole substrates. *PLoS One.* 2020;15(1):e0227487.
93. Watts JC, Prusiner SB. Experimental models of inherited PrP prion diseases. *Cold Spring Harb Perspect Med.* 2017;7(11):a027151.
94. Brandner S, Jaunmuktane Z. Prion disease: experimental models and reality. *Acta Neuropathol.* 2017;133(2):197–222.
95. Watts JC, Giles K, Bourkas ME, et al. Towards authentic transgenic mouse models of heritable PrP prion diseases. *Acta Neuropathol.* 2016;132(4):593–610.
96. Otero A, Hedman C, Fernandez-Borges N, et al. A single amino acid substitution, found in mammals with low susceptibility to prion diseases, delays propagation of two prion strains in highly susceptible transgenic mouse models. *Mol Neurobiol.* 2019;56(9):6501–6511.

97. Colby DW, Wain R, Baskakov IV, et al. Protease-sensitive synthetic prions. *PLoS Pathog*. 2010;6(1):e1000736.
98. Stohr J, Watts JC, Legname G, et al. Spontaneous generation of anchorless prions in transgenic mice. *Proc Natl Acad Sci U S A*. 2011;108(52):21223–21228.
99. Bouybayoune I, Mantovani S, Del Gallo F, et al. Transgenic fatal familial insomnia mice indicate prion infectivity-independent mechanisms of pathogenesis and phenotypic expression of disease. *PLoS Pathog*. 2015;11(4):e1004796.
100. Asante EA, Gowland I, Grimshaw A, et al. Absence of spontaneous disease and comparative prion susceptibility of transgenic mice expressing mutant human prion proteins. *J Gen Virol*. 2009;90(Pt 3):546–558.
101. Sigurdson CJ, Nilsson KP, Hornemann S, et al. *De novo* generation of a transmissible spongiform encephalopathy by mouse transgenesis. *Proc Natl Acad Sci U S A*. 2009; 106(1):304–309.
102. Telling GC, Haga T, Torchia M, Tremblay P, DeArmond SJ, Prusiner SB. Interactions between wild-type and mutant prion proteins modulate neurodegeneration in transgenic mice. *Genes Dev*. 1996;10(14):1736–1750.
103. Dossena S, Imeri L, Mangieri M, et al. Mutant prion protein expression causes motor and memory deficits and abnormal sleep patterns in a transgenic mouse model. *Neuron*. 2008;60(4):598–609.
104. Friedman-Levi Y, Meiner Z, Canello T, et al. Fatal prion disease in a mouse model of genetic E200K Creutzfeldt-Jakob disease. *PLoS Pathog*. 2011;7(11):e1002350.
105. Abskharon R, Dang J, Elfarash A, et al. Soluble polymorphic bank vole prion proteins induced by co-expression of quiescin sulfhydryl oxidase in E. coli and their aggregation behaviors. *Microb Cell Fact*. 2017;16(1):170.
106. Fernandez-Borges N, Di Bari MA, Erana H, et al. Cofactors influence the biological properties of infectious recombinant prions. *Acta Neuropathol*. 2018;135(2):179–199.
107. Erana H, Charco JM, Di Bari MA, et al. Development of a new largely scalable in vitro prion propagation method for the production of infectious recombinant prions for high resolution structural studies. *PLoS Pathog*. 2019;15(10):e1008117.
108. Riek R, Hornemann S, Wider G, Billeter M, Glockshuber R, Wuthrich K. NMR structure of the mouse prion protein domain PrP(121-231). *Nature*. 1996;382(6587): 180–182.
109. James TL, Liu H, Ulyanov NB, et al. Solution structure of a 142-residue recombinant prion protein corresponding to the infectious fragment of the scrapie isoform. *Proc Natl Acad Sci U S A*. 1997;94(19):10086–10091.
110. Zahn R, Liu A, Luhrs T, et al. NMR solution structure of the human prion protein. *Proc Natl Acad Sci U S A*. 2000;97(1):145–150.
111. Liu H, Farr-Jones S, Ulyanov NB, et al. Solution structure of Syrian hamster prion protein rPrP(90-231). *Biochemistry*. 1999;38(17):5362–5377.
112. Lysek DA, Schorn C, Nivon LG, et al. Prion protein NMR structures of cats, dogs, pigs, and sheep. *Proc Natl Acad Sci U S A*. 2005;102(3):640–645.
113. Lopez Garcia F, Zahn R, Riek R, Wuthrich K. NMR structure of the bovine prion protein. *Proc Natl Acad Sci U S A*. 2000;97(15):8334–8339.
114. Hornemann S, Schorn C, Wuthrich K. NMR structure of the bovine prion protein isolated from healthy calf brains. *EMBO Rep*. 2004;5(12):1159–1164.
115. Khan MQ, Sweeting B, Mulligan VK, et al. Prion disease susceptibility is affected by beta-structure folding propensity and local side-chain interactions in PrP. *Proc Natl Acad Sci U S A*. 2010;107(46):19808–19813.
116. Gossert AD, Bonjour S, Lysek DA, Fiorito F, Wuthrich K. Prion protein NMR structures of elk and of mouse/elk hybrids. *Proc Natl Acad Sci U S A*. 2005;102(3):646–650.
117. Wen Y, Li J, Yao W, et al. Unique structural characteristics of the rabbit prion protein. *J Biol Chem*. 2010;285(41):31682–31693.

118. Christen B, Perez DR, Hornemann S, Wüthrich K. NMR structure of the bank vole prion protein at 20 degrees C contains a structured loop of residues 165-171. *J Mol Biol.* 2008;383(2):306–312.
119. Kyle LM, John TR, Schätzl HM, Lewis RV. Introducing a rigid loop structure from deer into mouse prion protein increases its propensity for misfolding in vitro. *PLoS One.* 2013;8(6):e66715.
120. Sigurdson CJ, Joshi-Barr S, Bett C, et al. Spongiform encephalopathy in transgenic mice expressing a point mutation in the β2-α2 loop of the prion protein. *J Neurosci.* 2011;31(39):13840–13847.
121. Sigurdson CJ, Nilsson KP, Hornemann S, et al. A molecular switch controls interspecies prion disease transmission in mice. *J Clin Invest.* 2010;120(7):2590–2599.
122. Fernandez-Borges N, Parra B, Vidal E, et al. Unraveling the key to the resistance of canids to prion diseases. *PLoS Pathog.* 2017;13(11):e1006716.
123. Vorberg I, Groschup MH, Pfaff E, Priola SA. Multiple amino acid residues within the rabbit prion protein inhibit formation of its abnormal isoform. *J Virol.* 2003;77(3): 2003–2009.
124. Loftus B, Rogers M. Characterization of a prion protein (PrP) gene from rabbit; a species with apparent resistance to infection by prions. *Gene.* 1997;184(2):215–219.
125. Erana H, Fernandez-Borges N, Elezgarai SR, et al. In vitro approach to identify key amino acids in low susceptibility of rabbit prion protein to misfolding. *J Virol.* 2017;91(24):e01543-17.
126. Chianini F, Fernández-Borges N, Vidal E, et al. Rabbits are not resistant to prion infection. *Proc Natl Acad Sci U S A.* 2012;109(13):5080–5085.
127. Vidal E, Fernandez-Borges N, Pintado B, et al. Transgenic mouse bioassay: evidence that rabbits are susceptible to a variety of prion isolates. *PLoS Pathog.* 2015;11(8): e1004977.
128. Kurt TD, Aguilar-Calvo P, Jiang L, et al. Asparagine and glutamine ladders promote cross-species prion conversion. *J Biol Chem.* 2017;292(46):19076–19086.
129. Gallagher-Jones M, Glynn C, Boyer DR, et al. Sub-angstrom cryo-EM structure of a prion protofibril reveals a polar clasp. *Nat Struct Mol Biol.* 2018;25(2):131–134.

CHAPTER NINE

One or more β-amyloid(s)? New insights into the prion-like nature of Alzheimer's disease

Marcella Catania and Giuseppe Di Fede*
Neurology 5 / Neuropathology Unit, Fondazione IRCCS Istituto Neurologico Carlo Besta, Milan, Italy
*Corresponding author: e-mail address: giuseppe.difede@istituto-besta.it

Contents

Abstract

Misfolding and aggregation of proteins play a central role in the pathogenesis of several neurodegenerative diseases, including Alzheimer's disease (AD), Parkinson's and Lewy Body diseases, Frontotemporal Lobar Degeneration and prion diseases. Increasing evidence supports the view that Aβ and tau, which are the two main molecular players in AD, share with the prion protein several "prion-like" features that can be relevant for disease pathogenesis. These features essentially include structural/conformational/biochemical variations, resistance to degradation by endogenous proteases, seeding ability, attitude to form neurotoxic assemblies, spreading and propagation of toxic aggregates, transmissibility of tau- and Aβ-related pathology to animal models. Following this view, part of the recent scientific literature has generated a new reading frame for AD pathophysiology, based on the application of the prion paradigm to the amyloid cascade hypothesis in an attempt to definitely explain the key events causing the disease and inducing its occurrence under different clinical phenotypes.

Progress in Molecular Biology and Translational Science, Volume 175
ISSN 1877-1173
https://doi.org/10.1016/bs.pmbts.2020.07.003

1. Introduction

Neurodegenerative diseases are a group of devastating conditions, mainly caused by the gradual and irreversible loss of function and death of cells in the nervous system. A large subgroup of these disorders is due to misfolding and aggregation of proteins and has been classified as "prionoid disorders,"[1,2] which include—but are not limited to—*β-amiloidosis* such as Alzheimer's Diseases (AD), *tauopathies* like AD and Frontotemporal Lobar Degeneration (FTLD), *α-synucleinopathies* like Parkinson's Disease (PD), Multiple System Atrophy (MSA) and Lewy Body Disease (LBD), or other disorders such as Huntington's Disease (HD) and Amyotrophic Lateral Sclerosis (ALS). Although these diseases express different etiology and pathology, they share the degeneration of neurons and accompanied by progressive and irretrievable loss of functions.[3] However, the specific targeted neuronal pathways vary between diseases. The prion "nature" of the misfolding proteins involved in the pathogenesis of these illnesses is still under debate.[4–9] Despite extensive research, our understanding of the pathogenic mechanisms of prionoid diseases[2] is still limited and consequently no really effective treatments are available nowadays.[10] Hence, the urgent need to clarify the key molecular events leading from the initial structural/functional modifications of the disease-associated proteins to the generation of neuronal damage, neuropathological changes and clinical presentations of this group of neurological disorders.

The concept of prionoid disorders stems from the identification of prions by Stanley B. Prusiner.[11] He defined prion as a self-templating amyloidogenic state of a normal cellular protein (PrP) which serves as an infectious pathogenic agent (PrP^{Sc}),[12] whose structural and functional properties are determined not exclusively by its genetic code but mainly by the abnormal conformations it takes on and confers upon natively folded proteins. As anticipated, the distinction between prions and prionoid proteins has long been a controversial subject of academic discussion and yet no formal conclusion has been reached.[7,13] The term prionoid or "quasi-prion" to name misfolding proteins other than PrP is justified by the lack of definitive evidence that these peptides fully recapitulate the molecular behavior of prions but they can display some prion-like features such as the ability of template-based self-replication and cell-to-cell propagation.[14,15] This last phenomenon is common to prions and prionoids and is based on repeated nucleation-fragmentation cycles initiated by misfolding of proteins, induced by several triggers such as overexpression, mutations, stress or aging, and leading to aggregation into β-sheet-rich oligomers, which are considered the most toxic aggregates. The oligomers can then form higher-order structures, such as protofibrils or fibrils, by incorporating additional monomeric proteins. The fragmentation of higher-order structures produces new assemblies that can reinitiate the nucleation-fragmentation cycle and can disseminate the pathology over brain tissue.[15] This vicious cycle is most likely shared by virtually all misfolding peptides and represents a highly pathogenic loop that should be broken in order to prevent or stop the progression of the diseases associated with misfolding proteins, such as α-synuclein, tau, β-amyloid protein (Aβ), TDP-43, FUS protein, huntingtin, and many others.[1] However, infectivity of these aggregates, exemplified by serial transmissibility through consecutive hosts to prove unlimited self-replication of the agent, is not yet definitively demonstrated.[9,15–18] So, infectivity may be considered a sort of boundary line—at the moment not yet well defined—between prions and prionoids.[14] One of the most challenging and unsolved problems is how the aggregation of one given protein leads to distinct diseases. A structural-molecular explanation based on the ability of proteins such as α-synuclein, tau or Aβ to form assemblies that differ for their intrinsic architecture, stability, seeding capacity and surface properties has been proposed to account for distinct α-synucleinopathies and tauopathies,[19–21] following the model of prion diseases.[22] According to this intriguing hypothesis, the shape and surfaces of the aggregates generated by misfolding peptides may modulate at the same time their seeding capacity, interactome and tropism for defined neuronal cells or specific molecular pathways within the central nervous system, providing grounds to explain the phenotypic diversity of neurodegenerative diseases due to misfolding proteins.

2. Prion, quasi-prion or prion-like nature of Alzheimer's disease

AD is a prototypic neurodegenerative disorder due to two misfolding proteins—tau and Aβ—which show prion-like features. It is characterized by a striking clinical and neuropathological heterogeneity whose molecular bases are still largely unknown and have been at least in part ascribed to biochemical/structural diversity of the tau and/or Aβ aggregates.[20,23–29] However, this view is still questioned and needs further support from experimental data.[30]

AD is the most common form of dementia. It is defined clinically by a progressive decline in memory and other cognitive functions, and neuropathologically by brain atrophy and the accumulation of the two associated misfolded proteins Aβ and tau under the form of abundant extracellular deposits (Aβ plaques) and intraneuronal neurofibrillary lesions (tau tangles). Plaques and neurofibrillary lesions are made of abnormal filaments with the fine structure typical of amyloid.[22] Each amyloid filament exhibits peculiar dye-binding properties and consists of several protofilaments with β-sheet structures that are stabilized through hydrogen bonding. Aggregates formed by a given peptide or protein are polymorphic, in that they can adopt multiple molecular structures.[31]

The great number of studies produced in the field of AD during the last few decades promoted a progressive change in the nosography of this neurodegenerative disorder.[32,33] AD was initially defined as a clinicopathological entity with faint differentiation respect to other forms of dementia. It became early rather a neuropathological constellation based on a post-mortem pathobiology model that emphasized the pathogenic role of the two misfolding proteins involved in the molecular machinery leading to the disease. Then, it evolved towards an *antemortem* pathobiological model defined by neuropsychological, cerebrospinal fluid (CSF) and imaging biomarkers. Finally, it reached the identity of a complex disorder based on a strictly biological definition[34] and caused by an intricate cross-talking between tau and Aβ—with the contribution of many other molecules including ApoE, neuroinflammatory cytokines,[35] *etc.*—that leads to the occurrence of AD under distinct and quiet different disease phenotypes. In the last decade, a novel view of AD as a prionoid disorder has advanced based on similarities between the two disease associated peptides—tau and Aβ—and the pathogenic prions that cause spongiform transmissible encephalopathies.[1,36,37]

Indeed, similarly to neurodegenerative disorders caused by other proteopathic assemblies such as aggregates of α-synuclein, TDP43 and mutant superoxide dismutase,[38–40] in AD there is a long prodromal phase between the initial formation of pathogenic Aβ and tau seeds, during which neurodegenerative changes develop and eventually lead to brain damage and dysfunction.[34] The prolonged interval between the formation of the first protein assemblies and the clinical onset of disease potentially offers a therapeutic window, provided that sufficiently sensitive diagnostic tests can be developed.[22]

In this view, the most relevant questions to address by experimental approaches to AD pathophysiology are, in our opinion: *Does a prionoid nature of AD-associated proteins really exist? Can the pathological hallmarks of the disease be explained by prion-like properties of Aβ and tau? Can the phenotypic heterogeneity of AD be sustained at least in part by the prionoid nature of AD misfolding proteins? Can a structural/biochemical diversity of Aβ and tau aggregates—not sustained by a prion nature of the two peptides—be demonstrated and proposed as a cause of the occurrence of AD under different clinical phenotypes?*

During the last few years, several studies have been designed to solve these questions and provided results that are not yet definitive.

2.1 Does a prionoid nature of AD-associated proteins really exist?

Despite the first introduction of the term *amyloid* by Rudolph Virchow in 1854,[41] only in recent decades, thanks to developments in methods to create fibrils *in vitro* or to purify them from *ex vivo* material, as well as breakthroughs in cryo-electron microscopy (cryo-EM) and solid-state NMR spectroscopy (ssNMR), it was possible to analyze the fibril structures of different types of amyloid in near-atomic detail.[42] Amyloid fibrils have highly organized structures that consist of protofilaments in which individual monomeric subunits form usually parallel in-register β-strands.[43] These protofilaments then pack against each other, forming fibrils that are stabilized by dry steric interfaces. Further inter-fibril interaction leads to the formation of the plaques, inclusions or other types of deposits characteristic of amyloid diseases. Recent advances in cryo-EM and ssNMR have given insights into the structure of amyloid fibrils in unprecedented detail, revealing a variety of assemblies that conform to the cross-β amyloid fold.[29,30,44] Coupling these techniques with orthogonal data (for example, with other biophysical measurements) and data on disease type, presentation and progression helped to elucidate which amyloid formation pathways, amyloid intermediates and/or fibril structures are responsible for a specific disease.[42] However, what initiates the onset of an amyloid disease remains unclear. In the case of prion diseases, the formation of prions has been proposed to be driven by a direct interaction between the pathogenic prion protein (PrP^{Sc}) template and the endogenous cellular prion protein (PrP^{C}) substrate.[45] By contrast, in the genetic forms, the alteration in PrP^{C} conformation may be induced by a genetic mutation in *PRNP* gene encoding PrP^{C}.[46] Both Aβ and tau, as well as PrP^{C}, adopt a tertiary structure rich in β-sheets which in turn promotes the self-assembly of monomers into small oligomers, having neurotoxic properties, and fibrillary structures.[47] In genetically inherited forms of AD, it can be supposed that the initial trigger of aggregation is the presence of mutations which can deeply modify the structure of Aβ making it more prone to form oligomers and fibrils[48] or can cause overproduction of Aβ monomers.[23,49] Increasing the concentration of the monomeric precursor by mutations affecting the γ-secretase complex (PSEN1 and PSEN2 genes) or by duplication of the gene encoding amyloid precursor protein (APP) from which Aβ is derived can indeed cause disease, with gene duplication, triplication or even quadruplication associated with an early onset of AD.[50] Moreover, patients with trisomy of chromosome 21, which contains APP, have a higher risk of developing AD.[51] In addition, the failure of scavenger systems which are usually impaired by aging, with consequent break of the cellular homeostasis, may lead to the progressive accumulation of native Aβ and/or tau under the form of neurotoxic assemblies in brain tissue.[52] The last potentially pathogenic mechanisms do not ineludibly require a prionoid nature of the AD-associated misfolding proteins to induce the disease, but may coexist with prion or prion-like nature of Aβ and tau, reinforcing and amplifying the formation of oligomers, fibrils and amyloid aggregates.

While it is simple to hypothesize that a modification of the primary structure of a misfolding protein may deeply affect its trend toward polymerization, more difficult is to imagine how a wild type (wt) peptide can be responsible for progressive accumulation of intracerebral neurotoxic aggregates. Actually, many diseases due to misfolding proteins, including AD, are sporadic, strongly associated with aging and involve the aggregation of wt proteins.[53,54] Intrinsically disordered proteins (IDPs) like Aβ and tau are biologically active proteins without stable tertiary structure, characterized by the presence of disorder-promoting residues in their amino acid sequences.[55] They have a spontaneous higher propensity towards coacervation, which may explain why many of them are amyloidogenic.[56] Moreover, there is a number of factors that can modulate the amyloidogenicity of IDPs. They include post-translational modifications of a precursor's primary sequence (for example, truncations by the action of proteases or hyperphosphorylation) which can enhance or suppress amyloidogenicity, small molecules such as metabolites or metal ions, membranes and glycosaminoglycans[42] and the status of the cell itself (chaperone levels, rate of protein synthesis, *etc.*), which affect the rate of protein aggregation and the ability of the cell to respond to the formation of potentially toxic species.[42,56–58] More than one of these molecular players can cooperate in converting the native benign structure of an IDP into a pathogenic and toxic status. They can play their effects on native IDPs independently of the prion/prionoid nature of misfolding peptides.

Since the discovery that prion diseases can be transmitted to experimental animals by intracerebral injection of affected brain samples, the possibility that the brains of patients suffering from other neurodegenerative diseases might also harbor causative agents with prion-like transmissible properties has been proposed by several researchers.[59,60] A reasonable support to the hypothesis that the prion paradigm of nucleated protein self-assembly also applies to Aβ came mainly by the studies of Larry C. Walker's and Mathias Jucker's teams in Atlanta (USA) and Tübingen (Germany), respectively.[61] These research groups provided solid evidence that it is possible to induce Aβ deposition *in vivo* by injection of Aβ-containing extracts of autopsy-derived brain samples from AD patients into the hippocampal formation and/or neocortex of APP-transgenic mice before the age of initial amyloid deposition.[47,62,63] Importantly, *in vivo* Aβ seeding requires both a donor brain extract that contains Aβ seeds and a host that is capable of generating Aβ intracerebral deposits, since in transmission studies with AD brain extracts no Aβ deposition emerged in nontransgenic host mice, and brain extracts from non-AD control donors that lacked Aβ deposition did not induce significant Aβ-amyloidosis.[47,63] Moreover, denaturation of brain extracts with formic acid or the specific depletion of Aβ by antibodies or reagents that bind and/or remove amyloid proteins from the extract, abolished the seeding effect, indicating that the seed responsible for the effects of AD brain extract transmission to animal models is actually Aβ.[63]

It was also demonstrated that—similarly with prions—pure, aggregated, synthetic Aβ is able to induce deposition in mice, provided that a robust dose of aggregated Aβ and a long incubation time are set in the transmission experiments.[64] Aβ-seeding experiments were undertaken in animal models that do not generate spontaneously amyloid deposits within their normal life spans. In both instances, Aβ deposition was induced in the brain after a suitable incubation period.[47,65,66] On the other hand, it seems that Aβ in CSF of AD patients and aged APP-transgenic donor mice is devoid of significant *in vivo* seeding activity, even at levels of Aβ exceeding those in the most concentrated brain extracts.[67] The reasons for the inability of Aβ in the CSF to seed *in vivo* remain uncertain.[47]

Recent observations from human pathology seems to support the transmissibility of AD pathology—or at least of some features of AD pathology—even among humans, after a long period of incubation,[68–70] a phenomenon that may parallel the iatrogenic transmission of prion diseases after treatment with growth hormone that had been isolated from cadaveric human pituitary glands contaminated with prions.[71] However, the iatrogenic transmission of the full AD pathology is still questioned for several reasons,[72,73] including the lack of tau-related neuropathological changes in the brains of the recipient individuals.

Aβ seeds actually share with prions several other properties.[47] It has been proposed that they may exist as variant structural *strains* that can adopt different molecular conformations modulating their functionality *in vivo*.[74–77] The resistance to the inactivation by conventional chemical agents (*i.e.*, formaldehyde)—analogously to prions—was also attributed to Aβ seeds, following the observation that formaldehyde-fixed brain samples from AD patients and APP-transgenic mice retain their seeding potential, even after 2 years in formaldehyde.[78]

Small, soluble oligomers of Aβ assemblies, like small PrP prions, are the Aβ species with the most effective seeding properties in the brain extracts.[79] Moreover, given their limited dimensions, it is conceivable that these soluble Aβ seeds traffic through the brain more readily than do larger fibrils and so they can be responsible of the spreading of AD pathology.[47] The transmission studies from the Tübingen research group clearly demonstrated that Aβ injected into the hippocampus diffuses away from the injection site and, over time, Aβ deposits appear in other brain regions, especially the entorhinal cortex.[16,63,80] These data, together with experimental evidence from other groups,[81] suggest that Aβ aggregates most likely traffic from cell to cell, even if at the moment the exact pathway used by Aβ assemblies for cerebral spreading remains unknown.

The main analogies between prions and Aβ seeds have been well summarized by Walker et al.,[47,82] which emphasized, among others, the capacity of adopting β-sheet-enriched conformations and of assembling into neurotoxic aggregates, the seeding abilities, the strain variation, the stability and resistance to chemical inactivation and enzymatic degradation, the spreading ability within and to the brain, the transmissibility to animal models and humans. Even if not all these features are definitively demonstrated for Aβ, they worth further investigations for an uncontroversial validation.

2.2 Can the key events in AD pathogenesis be explained by prion-like properties of Aβ and tau?

Aβ and tau assemblies deposit outside and inside the cells respectively and both can cause cellular dysfunction and death. In general, the aggregates generated by different protein precursors may have different and not always 'negative' effects on living cells, but deciphering the toxic mechanisms of an individual protein and its ensemble of misfolded or aggregated states (misfolded monomers, oligomers, fibrils or plaques) remains a challenge.[42] For example, we have still poor knowledge of the trafficking of these assemblies. Small-size assemblies can be taken up into a cell *via* endocytosis, but this can be perturbed by preventing binding to certain cell surface receptors. Within endosomes and lysosomes, fibrils can release toxic oligomers and can disrupt the endosomal and lysosomal function.[83,84] Fibrils can also access the intracellular space following release from cells, thus spreading disease by uptake into adjacent cells. Other effects of aggregates within cells include disruption of endoplasmic reticulum (ER) dynamics,[85] release of reactive oxygen species from mitochondria and induction of stress responses.[86] All these events may explain the most relevant features of AD pathology as well as of other neurodegenerative disorders due to misfolding proteins.[87] In AD, they are triggered by aggregation of Aβ and tau monomers into more complex assemblies and by their spreading throughout the brain,[88–90] but whether or not the primary trigger is an upstream prion-like change in the protein conformation and in the chemo-physical properties of the two peptides as a result of the templated corruption of like proteins,[82,91] is still an open question.

One of the most important milestones supporting the "amyloid cascade hypothesis" as the most endorsed theory for the interpretation of AD pathogenesis is the link between the two main molecular players: Aβ and tau. In the past, this link was substantiated in the induction of hyperphosphorylation and loss of microtubule binding of tau by Aβ oligomers.[92] The Aβ-tau connections are mostly reported as indirect, such as Aβ small oligomers induce activation of protein kinases, which then catalyze abnormal tau phosphorylation.[92,93] However, several studies supported the view that a direct contact occurs between the two peptides and that misfolded Aβ is able to induce tau misfolding *in vitro* through direct intermolecular interaction suggesting even the intriguing possibility that Aβ acts as a template for tau misfolding *in vivo*.[94] Indeed, in the absence of any other peptides, Aβ can bind to tau, and tau monomers can be induced to oligomerize *in vitro* after exposure to low levels of Aβ oligomers.[95,96] These findings raise the obvious possibility that, *in vivo*, Aβ oligomers seed the initial formation of tau oligomers, which can then self-propagate in the absence of additional input from Aβ. If such a phenomenon were to occur *in vivo*, it would represent a prion-like pivotal step in AD pathogenesis.[94] Recent data seem to support this hypothesis by demonstration that heterotypic seeding of tau by pre-aggregated Aβ provides efficient seeds for induction and propagation of tau pathology *in vivo*.[97] Prion-like, heterotypic seeding of tau fibrillization by Aβ supports a model of propagation of AD pathology based on the direct interaction between the two main molecular players of AD, and opens the way to a more general re-thinking on the specific value of cross-seeding phenomena in the neurodegenerative disorders due to misfolding proteins.

Increasing data from the recent scientific literature brace the conclusion that both Aβ and tau can aggregate and spread in the brain of AD patients and AD animal models by a prion-like mechanism.[36] Indeed, the ability of protein particles, deriving from misfolding and aggregation of Aβ and tau, to transfer from one cell to another, similar to misfolded PrP, accounts for the widespread pathophysiology seen in AD and other tauopathies.[98] Thus, the concept of intercellular transfer of protein inclusions[99] represents an important ground for the definition of AD as a "prion-like" disorder.[100]

The fibrillar assemblies derived from proteins associated with AD and prion diseases differ by their surfaces because of diversity in their primary structures. These differences give rise to definition of the term *strains* that was first applied to prions[101] and only recently extended to α-synuclein, Aβ and tau.[25,102–104] The strain concept, as a consequence, implicates different biophysical properties of a given misfolded peptide and assumes the existence of different interactomes for α-synuclein, Aβ and tau, which may generate strain-dependent pathologies among neurodegenerative disorders due to misfolding proteins.[100] In this context, the distinct surface properties of pathogenic protein assemblies regulate the different affinity for molecules at the cell surface. A number of peptidic and lipidic components of the plasma membrane and extracellular matrix, including chondroitin sulfate proteoglycans, extracellular glycoproteins, heparan sulfate proteoglycans, cholesterol and the gangliosides, have been reported to affect the interaction with AD-associated misfolded proteins, thus contributing to their binding, aggregation into toxic assemblies and cell-to-cell propagation.[105–108]

As for prion diseases,[109] this scenario is further complicated by the still unclear role of chaperones in mediating the interactions between misfolding proteins and plasma membrane or extracellular matrix components. Huge evidence from scientific literature suggest that cell chaperones can directly interact with the hydrophobic surfaces exposed by the oligomers/fibrils that can mediate aberrant interactions with various targets. Moreover, some chaperones are able to convert the oligomers into large and innocuous species by clearance mechanism, such as autophagy and so, they can play a protective role against neurodegeneration induced by Aβ and tau in AD, whose detailed mechanisms remain an open issue that deserves further investigation.[100,110] The prion-like spreading of misfolding peptides may also be evoked to explain the selective vulnerability of distinct neuronal pathways and brain anatomical regions in different neurodegenerative disorders.[111,112] It is likely that not only Aβ but tau also may form prion-like strains or diverse conformational structures that may underlie some of the differences in clinical pictures and pathology among neurodegenerative tauopathies.[113] Overall, the shape and surface properties of distinct pathogenic protein assemblies appear to tightly regulate their respective interactomes. Their affinity and ability to interact with distinct molecular partners may drive their tropism for defined cell populations within the central nervous system and may account for the spatial distribution of the lesions that characterize different neurodegenerative diseases.[1,19] This general paradigm may be reasonably applied to AD too. Differential diffusion of tau pathology was demonstrated in AD and other tauopathies, where the tau lesions appear to spread throughout distinct brain areas following a cell-to-cell propagation of tau aggregates that is driven by differences in isoform composition and morphology of tau filaments among tauopathies suggesting the existence of conformationally diverse tau strains responsible for the spreading of tau pathology.[114,115]

Taking into account all these data, AD pathogenesis could be recapitulated in a *prion-like reading frame* as follows: Aβ changes its conformation through a template-mediated protein misfolding and form toxic oligomers that are able to propagate from cell to cell by a prion-like mechanism; Aβ oligomers activate tau kinases triggering the abnormal phosphorylation of tau and also serve as prion-like seeds that induce tau to oligomerize; tau oligomers also self-propagate according to the prion model, and along with hyperphosphorylated monomeric tau and Aβ oligomers, drive the degeneration and death of neurons involved in memory and cognition; the template-mediated conformational changes of the two main peptides also confer the ability to interact with molecular partners (chaperons, components of plasma membrane and extracellular matrix, *etc.*...), which may favor their stabilization under the form of toxic small-size assemblies, or may tag Aβ and tau aggregates promoting their elimination or sequestering them into nontoxic large-size assemblies or targeting them towards inappropriate subcellular organelles or cell compartments where the neurotoxic effects can take place.

2.3 Can the phenotypic heterogeneity of AD be sustained at least in part by the prionoid nature of AD misfolding proteins?

AD is increasingly recognized as a heterogeneous disorder comprising different phenotypes that are characterized by diverse clinical presentations, neuroanatomical involvement and neuropathological profiles. Although this heterogeneity is most striking in the inherited forms of AD, it is also well recognized in sporadic cases[116–122] where it results in relevant differences in the prognosis of the disease and most likely affects the resistance to treatments in AD patients with different clinical phenotypes.[23,123,124] The molecular basis of phenotypic variations in AD is largely unknown and most likely encompasses a complex series of molecular players beyond the two main actors of amyloid cascade hypothesis: Aβ and tau. However, some prion-like properties of Aβ and tau could play a pivotal role in driving the clinical and neuropathological diversity of AD cases and the presentation of AD under distinct clinicopathological phenotypes.

Several studies from scientific literature have indeed provided proofs on the existence of structural diversities of Aβ or tau assemblies which can parallel different presentations of the disease. In a general view, although the fibrils generated by misfolding proteins adopt the canonical cross-β structure of amyloid—in which the β-strands within each protofilament align perpendicular to the long axis of the fibril—these fibrils are more complex and elaborate than suspected previously, suggesting that a wide variety of amyloid structures exists.[42] Such diversity may explain why amyloid diseases are difficult to understand and have polymorphic clinical pictures, and why their treatment is still a challenge.

Coming back to prion diseases, one of the most fascinating features of prions is their occurrence under distinct strains.[24,125,126] Prion strains are characterized by infecting identical host animals and characterizing their distinct pathogenic phenotypes based on disease incubation time, deposition pattern in the brain, and the physicochemical features of the PrP aggregates which are solely enciphered in the secondary, tertiary and quaternary structure of the misfolded PrP.[127,128] These characteristics are stable upon repeated, serial transmission in animals, a requirement for prion strain classification. Even if a rigorous demonstration of these features has not yet reached, several attempts from many scientific groups to apply the prion strain paradigm to other neurodegenerative diseases due to misfolding proteins are ongoing. Both Aβ and tau have the capacity to aggregate into assemblies with multiple conformations *in vitro* generating structural heterogeneity of the resulting fibrils as shown by electron microscopy.[129–132] This attitude potentially confers to the two misfolding peptides the molecular basis for the manifestation of prion-like properties *in vivo*. However, isolation and characterization of fibrils from AD patients have generated until now unclear or contradictory results. Some studies supported the view of an inter-individual variability of fibril morphology. Seeding fibril formation of Aβ40/42 using plaques extracted from patients who displayed different disease phenotypes and progression rates resulted in different ssNMR spectra, suggesting that a different fibril polymorph was predominant in each patient.[133] A further complication is due to the observation that all fibrils formed *in vivo* are not necessarily monomorphic.[42] Indeed, multiple polymorphs of tau were isolated from a single patient-derived sample,[29] and various Aβ polymorphs have been identified within the same patients using X-ray diffraction.[134] More recent studies showed that there is no significant variation in tau filament structures among individuals with AD.[30] In other studies, a correlation between clinical subgroups of AD and structural variations of Aβ fibrils was advanced and structural peculiarities were proposed as key determinants of clinical heterogeneity of the disease.[24,75] The analysis of fibrils seeded from *ex vivo* material from patients with different clinical presentations of AD—posterior cortical atrophy form of AD (PCA-AD), rapid-AD (r-AD) or typical AD (t-AD) revealed a single Aβ40 polymorph that is especially represented in PCA-AD and t-AD brains, whereas the r-AD samples contained a higher proportion of additional structures.[75] Some studies headed by Stanley B. Prusiner pointed to a view where structural variations of Aβ assemblies exist and are due to prion-like properties of Aβ.[77,135] Among them, Stohr and colleagues showed that two distinct synthetic Aβ prion strains are able to produce different amyloid deposits in bigenic mice and concluded that the composition of Aβ deposits in the brains of injected animals depends on the conformation of the inoculated Aβ polymers, suggesting that these Aβ polymers represent distinct Aβ prion strains.[136] Condello et al. demonstrated that distinct strains of Aβ can be discerned in different disease subgroups of AD, or in different brain compartments within a given patient.[135] So, the diversity of amyloid polymorphs could explain the different disease phenotypes observed in patients in which the same misfolding peptide aggregates.[42] Moreover, it was demonstrated that Aβ fibril in transgenic mouse models undergo slow structural rearrangements over time,[137] and more in general it was reported that fibril structures in the brains of patients may vary during disease progression.[42] These last findings constitute remarkable limitations for more in-depth studies to confirm definitively the causal link between structural variation of misfolding protein aggregates and disease phenotypes.

The role of cross-seeding among different misfolding proteins in the generation of distinct AD clinical subgroups also deserves further in-depth analysis.[138] It is well known from several neuropathological studies the overlapping among Aβ-, tau-, α-synuclein, PrP and TDP43-dependent pathology in the brains of patients with neurodegenerative diseases like AD, PD, LBD, Creutzfeldt-Jakob disease (CJD) and FTLD.[23,121,139–146] How Aβ and tau interact has not been fully characterized. Recently, it was shown that different Aβ self-assemblies have distinct effects on tau aggregation,[147] suggesting that the cross-talking among different misfolding peptides may be driven by difference in the biochemical composition and structural properties of interacting aggregates. Moreover, *in vitro* and *in vivo* studies have demonstrated that tau, a-synuclein and prion proteins can interact with Aβ and modulate the onset and course of AD.[95,96,148,149] It has also been demonstrated that Aβ can trigger α-synuclein polymerization *in vitro* and overexpression of human APP promotes the accumulation of α-synuclein inducing the associated disease phenotype.[150] However, the detailed effects of the cross-seeding network of misfolding proteins are still largely unknown and it is not possible to rule out the possibility that some interactions are detrimental and some others are protective. For example, injection of APP-tg mice with α-synuclein-derived extracts abolishes their seeding effect, suggesting that α-synuclein is not able to cross-seed Aβ plaques *in vivo*, and α-synuclein even hampers amyloid plaque formation in APP-tg mice.[150] Finally, brains of APP-tg mice lacking α-synuclein exhibited a significant increase in Aβ burden, suggesting a suppressive role of α-synuclein on amyloidogenesis in AD.[151]

2.4 Can a structural/biochemical diversity of Aβ and tau aggregates—not sustained by a prion nature of the two peptides—be demonstrated and proposed as a cause of the occurrence of AD under different clinical phenotypes?

A number of recent studies has been focused on the dissection of the structural/conformational properties of Aβ and tau assemblies and on their biochemical composition as well as on the correlation between the structural/biochemical diversity of such aggregates and the occurrence of AD under distinct phenotypes. Following the prion concept, it is hypothesized that a prion mechanism may be the cause of the diverse clinicopathologic phenotypes observed across the spectrum of AD patients. However, the variability of the physicochemical properties of Aβ and tau assemblies does not necessarily stem from a prion or prion-like nature of the two peptides.

Recently, tau filaments from patients with Pick's disease were revealed to contain two residues of 3R tau, which are folded differently from the tau filaments in AD, providing a proof of the existence of different conformers of assembled tau.[152] The observed tau fold in the filaments of patients with Pick's disease may explain the selective incorporation of 3R tau in Pick bodies and the differences in phosphorylation of tau between AD and other tauopathies. Growing evidence supports the hypothesis that the formation and propagation of distinct conformers of assembled tau occurs in different neurodegenerative diseases.[153]

In addition, increasing evidence supports the hypothesis that polymorphisms of Aβ fibrils correlate with variations in clinical and pathological features of AD. Structural variants of Aβ40 fibrils show different levels of toxicity in primary neuronal cell cultures.[31] Some features of amyloid deposits in transgenic mice, induced by exogenous amyloid-containing biological material or by synthetic Aβ, vary with the source of the injected material.[25,154] Differences in plaque morphology have been described in AD brains at different disease stages.[155,156] However, it remains unknown whether a distinct Aβ conformation, or ensemble of conformers, in a given patient contributes to the specific pattern of amyloid deposition and associated neuronal injury.[128] Following the hypothesis that the phenotypic variability of AD results from variations in the molecular architecture of aggregated Aβ, Rasmussen et al. showed that Aβ-amyloid plaques in the brain cluster as clouds of conformational variants that differ among certain subtypes of AD.[28] The conformational features of AD plaques can be resolved by using luminescent conjugated oligothiophenes that bind to the repetitive cross–β-sheet structures of amyloid fibrils and display spectral differences based on the twisting of the flexible dye backbone.[157] These conformational properties of Aβ aggregates were partially transmissible to transgenic mice.

As already mentioned above, structural differences in amyloid fibrils from distinct AD clinical subtypes have been shown by Tycko and coworkers.[153,158] These observations together with many others, support the view that a structural heterogeneity of Aβ and tau exists and may be a relevant source of variation in the clinical/neuropathological presentations of AD.

Additional data from scientific literature of the recent past provided further insights into the striking heterogeneity of Aβ assemblies accumulating in AD brains. These studies highlighted the presence of N- and C-terminally truncated Aβ isoforms, in addition to the most represented Aβ1–42 and Aβ1–40 peptides and were performed by using several approaches and methodologies, including mass spectrometry and immunoprecipitation with anti-Aβ antibodies raised against truncated Aβ isoforms. Site-specific Aβ autocleavage during the early phases of aggregation, producing a typical Aβ fragment signature, was demonstrated by Rudinskiy and coworkers.[159] These authors also showed that truncated Aβ peptides can form stable oligomeric complexes with full-length Aβ. Antibody-enabled screening of Aβ aggregates in brain tissue and CSF of AD patients suggest that aggregation-associated Aβ cleavage may play a relevant role in the generation of structural and biochemical diversity of Aβ assemblies and, at least potentially, in the generation of clinical phenotypes of AD. It was also remarked that some truncated and post-translationally modified Aβ fragments seem to be particularly involved in the generation of neurotoxic Aβ assemblies.[160–162]

The pathogenic role of Aβ(pE3) (N-terminal truncated Aβ starting with pyroglutamate) in AD has been proposed after the discovery that this peptide represents a dominant fraction of Aβ peptides in amyloid deposits of AD brains.[163] The involvement of Aβ(pE3) in AD pathogenesis is suggested by its stability, higher aggregation propensity and increased toxicity compared to full-length Aβ.[164,165] Moreover, intraneuronal accumulation of Aβ(pE3) peptides induces severe neuronal loss and a novel neurological phenotype in a mouse model of AD.[166] Nevertheless, a specific role of Aβ species in determining the phenotypic diversity of AD is not yet fully explored, and overall these potential performers of phenotypic heterogeneity of AD could play their role independently of any prion-like features of Aβ.

In 2018, we found that: (i) β-amyloid fraction isolated from AD brains is generated starting from different mixtures of Aβ peptides including N- and C-truncated Aβ fragments; (ii) different subtypes of amyloid might undergo distinct aggregation pathways, generating Aβ assemblies that display different toxic properties, depending on the time-course of the lag and growth phases in their aggregation kinetics; (iii) amyloid subtypes also show variability in their seeding activity on synthetic monomers of Aβ1–42, which depends on the seed-substrate affinity during the polymerization process; (iv) aggregates from different β-amyloid subtypes are differently resistant to degradation by proteases, giving rise to assemblies that can be variably stable and toxic *in vivo*; (v) amyloid subtypes have differential ability toward anticipating or accelerating amyloidogenesis in animal models of AD and generate amyloidosis showing distinct features in mice injected with human brain extracts.[20] These findings suggest that more than one type of β-amyloid accumulates in the brain tissue of AD patients and that the biochemical composition of the molecular assemblies of Aβ—essentially the relative amount of Aβ isoforms participating in the formation of aggregates- may contribute to the generation of distinct disease phenotypes. Moreover, together with the results of other studies mentioned above, these data offer, in our opinion, grounds for a novel classification of AD based on the variations characterizing the assemblies of the two disease-associated misfolding proteins Aβ and tau.

We also found that the preferential accumulation of some Aβ fragments in amyloid was paralleled by a reduction of the very same fragments in the CSF of AD cases.[167] This provides the rationale for the detection of distinct AD subgroups by the analysis of Aβ profile in CSF of AD patients.

3. Concluding remarks

The recent investigations on neurodegenerative disorders due to misfolding of Aβ, tau, α-synuclein and others suggest that at least some of the structural/biochemical features of the pathogenic aggregates generated by these peptides have close similarities with prions. Although these pivotal findings are exciting and may constitute a *proof-of-principle* of the prion-like nature of misfolding proteins linked to AD, PD, FTLD, LBD and other neurological disorders, and even though they provided an new reading frame for intricate mechanisms such as brain propagation or selective vulnerability to pathology, whether Aβ, tau, and α-synuclein really misfold and polymerize following the template-driven self-aggregation model of prions remains to be confirmed. Whether or not the prion-like mechanisms of misfolding, aggregation and propagation of Aβ and tau are actually responsible for the phenotypic variability of AD remains to be established too. So, although a prion nature of these two peptides is suggested by growing evidence from experimental studies, many issues concerning the existence of a misfolding phase preceding aggregation, the connection between seeding capacity and structural features of Aβ and tau assemblies and their neurotoxic ability, the relationship between Aβ isoform content in Aβ aggregates and their pathobiology, the role of cross-seeding phenomena, the specific weight of interactome and how is it modulated by different types of aggregates, the real existence of Aβ and tau strains fully matching the prion strain concept, require further explanation. The upcoming efforts to resolve overall these issues are very important because they may pave the way to the discovery of new molecular tools to dissect the heterogeneity of AD and may lead to the development of more informative diagnostics and personalized therapeutics.

References

1. Wells C, Brennan SE, Keon M, Saksena NK. Prionoid proteins in the pathogenesis of neurodegenerative diseases. *Front Mol Neurosci*. 2019;12:271.
2. Aguzzi A. Cell biology: beyond the prion principle. In: *Nature*. England: Springer Nature; 2009:924–925. vol 459.

3. Dar KB, Bhat AH, Amin S, et al. Elucidating critical proteinopathic mechanisms and potential drug targets in neurodegeneration. *Cell Mol Neurobiol.* 2019;40:313–345.
4. O'Carroll A, Coyle J, Gambin Y. Prions and prion-like assemblies in neurodegeneration and immunity: the emergence of universal mechanisms across health and disease. *Semin Cell Dev Biol.* 2019;99:115–130.
5. Nonaka T, Hasegawa M. Prion-like properties of assembled TDP-43. *Curr Opin Neurobiol.* 2019;61:23–28.
6. Jaunmuktane Z, Brandner S. The role of prion-like mechanisms in neurodegenerative diseases. *Neuropathol Appl Neurobiol.* 2019.
7. Soto C, Pritzkow S. Protein misfolding, aggregation, and conformational strains in neurodegenerative diseases. *Nat Neurosci.* 2018;21(10):1332–1340.
8. Clavaguera F, Duyckaerts C, Haik S. Prion-like properties of tau assemblies. *Curr Opin Neurobiol.* 2020;61:49–57.
9. Jucker M, Christen Y. *Proteopathic Seeds and Neurodegenerative Diseases.* Heidelberg: Springer; 2013.
10. McAlary L, Plotkin SS, Cashman NR. Emerging developments in targeting proteotoxicity in neurodegenerative diseases. *CNS Drugs.* 2019;33(9):883–904.
11. Prusiner SB. Novel proteinaceous infectious particles cause scrapie. *Science.* 1982;216(4542):136–144.
12. Prusiner SB, Scott MR, DeArmond SJ, Cohen FE. Prion protein biology. *Cell.* 1998; 93(3):337–348.
13. Erana H. The prion 2018 round tables (II): Abeta, tau, alpha-synuclein... are they prions, prion-like proteins, or what? *Prion.* 2019;13(1):41–45.
14. Harbi D, Harrison PM. Classifying prion and prion-like phenomena. *Prion.* 2014;8(2):161–165.
15. Scheckel C, Aguzzi A. Prions, prionoids and protein misfolding disorders. *Nat Rev Genet.* 2018;19(7):405–418.
16. Eisele YS, Obermüller U, Heilbronner G, et al. Peripherally applied Abeta-containing inoculates induce cerebral beta-amyloidosis. *Science.* 2010;330(6006):980–982.
17. Clavaguera F, Bolmont T, Crowther RA, et al. Transmission and spreading of tauopathy in transgenic mouse brain. *Nat Cell Biol.* 2009;11(7):909–913.
18. Jucker M, Walker LC. Self-propagation of pathogenic protein aggregates in neurodegenerative diseases. *Nature.* 2013;501(7465):45–51.
19. Melki R. How the shapes of seeds can influence pathology. *Neurobiol Dis.* 2018; 109(pt B):201–208.
20. Di Fede G, Catania M, Maderna E, et al. Molecular subtypes of Alzheimer's disease. *Sci Rep.* 2018;8(1):3269.
21. Lim KH. Diverse misfolded conformational strains and cross-seeding of misfolded proteins implicated in neurodegenerative diseases. *Front Mol Neurosci.* 2019;12:158.
22. Goedert M. Neurodegeneration. Alzheimer's and Parkinson's diseases: the prion concept in relation to assembled Abeta, tau, and alpha-synuclein. *Science.* 2015;349(6248): 1255555.
23. Di Fede G, Giaccone G, Salmona M, Tagliavini F. Translational research in Alzheimer's and prion diseases. *J Alzheimers Dis.* 2018;62(3):1247–1259.
24. Cohen M, Appleby B, Safar JG. Distinct prion-like strains of amyloid beta implicated in phenotypic diversity of Alzheimer disease. *Prion.* 2016;10:9–17.
25. Stöhr J, Condello C, Watts JC, et al. Distinct synthetic Aβ prion strains producing different amyloid deposits in bigenic mice. *Proc Natl Acad Sci USA.* 2014;111(28): 10329–10334.
26. Watts JC, Condello C, Stöhr J, et al. Serial propagation of distinct strains of Aβ prions from Alzheimer's disease patients. *Proc Natl Acad Sci USA.* 2014;111(28):10323–10328.

27. Heilbronner G, Eisele YS, Langer F, et al. Seeded strain-like transmission of β-amyloid morphotypes in APP transgenic mice. *EMBO Rep*. 2013;14(11):1017–1022.
28. Rasmussen J, Mahler J, Beschorner N, et al. Amyloid polymorphisms constitute distinct clouds of conformational variants in different etiological subtypes of Alzheimer's disease. *Proc Natl Acad Sci USA*. 2017;114(49):13018–13023.
29. Fitzpatrick AWP, Falcon B, He S, et al. Cryo-EM structures of tau filaments from Alzheimer's disease. *Nature*. 2017;547(7662):185–190.
30. Falcon B, Zhang W, Schweighauser M, et al. Tau filaments from multiple cases of sporadic and inherited Alzheimer's disease adopt a common fold. *Acta Neuropathol*. 2018;136(5):699–708.
31. Petkova AT, Leapman RD, Guo Z, Yau WM, Mattson MP, Tycko R. Self-propagating, molecular-level polymorphism in Alzheimer's beta-amyloid fibrils. *Science*. 2005;307(5707):262–265.
32. Knopman DS, Petersen RC, Jack Jr CR. A brief history of "Alzheimer disease": multiple meanings separated by a common name. *Neurology*. 2019;92(22):1053–1059.
33. Goedert M, Spillantini MG. A century of Alzheimer's disease. *Science*. 2006;314(5800): 777–781.
34. Jack Jr CR, Bennett DA, Blennow K, et al. NIA-AA research framework: toward a biological definition of Alzheimer's disease. *Alzheimers Dement*. 2018;14(4):535–562.
35. Penke B, Bogar F, Fulop L. Beta-amyloid and the pathomechanisms of Alzheimer's disease: a comprehensive view. *Molecules*. 2017;22(10):1692.
36. Walker LC. Prion-like mechanisms in Alzheimer disease. *Handb Clin Neurol*. 2018;153:303–319.
37. Walker LC, Diamond MI, Duff KE, Hyman BT. Mechanisms of protein seeding in neurodegenerative diseases. *JAMA Neurol*. 2013;70(3):304–310.
38. Munch C, O'Brien J, Bertolotti A. Prion-like propagation of mutant superoxide dismutase-1 misfolding in neuronal cells. *Proc Natl Acad Sci USA*. 2011;108(9): 3548–3553.
39. Nonaka T, Masuda-Suzukake M, Arai T, et al. Prion-like properties of pathological TDP-43 aggregates from diseased brains. *Cell Rep*. 2013;4(1):124–134.
40. Goedert M, Masuda-Suzukake M, Falcon B. Like prions: the propagation of aggregated tau and alpha-synuclein in neurodegeneration. *Brain*. 2017;140(2):266–278.
41. Sipe JD, Cohen AS. Review: history of the amyloid fibril. *J Struct Biol*. 2000; 130(2–3):88–98.
42. Iadanza MG, Jackson MP, Hewitt EW, Ranson NA, Radford SE. A new era for understanding amyloid structures and disease. *Nat Rev Mol Cell Biol*. 2018;19(12):755–773.
43. Luhrs T, Ritter C, Adrian M, et al. 3D structure of Alzheimer's amyloid-beta(1-42) fibrils. *Proc Natl Acad Sci USA*. 2005;102(48):17342–17347.
44. Qiang W, Yau WM, Lu JX, Collinge J, Tycko R. Structural variation in amyloid-beta fibrils from Alzheimer's disease clinical subtypes. *Nature*. 2017;541(7636):217–221.
45. Prusiner SB. The prion diseases. *Brain Pathol*. 1998;8(3):499–513.
46. Morinet F. Prions: a model of conformational disease? *Pathol Biol*. 2014;62(2):96–99.
47. Walker LC, Schelle J, Jucker M. The prion-like properties of amyloid-beta assemblies: implications for Alzheimer's disease. *Cold Spring Harb Perspect Med*. 2016;6(7):a024398.
48. Krone MG, Baumketner A, Bernstein SL, et al. Effects of familial Alzheimer's disease mutations on the folding nucleation of the amyloid beta-protein. *J Mol Biol*. 2008;381(1):221–228.
49. Rosenberg RN, Lambracht-Washington D, Yu G, Xia W. Genomics of Alzheimer disease: a review. *JAMA Neurol*. 2016;73(7):867–874.
50. Hardy J. The discovery of Alzheimer-causing mutations in the APP gene and the formulation of the "amyloid cascade hypothesis". *FEBS J*. 2017;284(7):1040–1044.

51. Gomez W, Morales R, Maracaja-Coutinho V, Parra V, Nassif M. Down syndrome and Alzheimer's disease: common molecular traits beyond the amyloid precursor protein. *Aging*. 2020;12(1):1011–1033.
52. Cornejo F, von Bernhardi R. Role of scavenger receptors in glia-mediated neuroinflammatory response associated with Alzheimer's disease. *Mediators Inflamm*. 2013; 2013:895651.
53. Drummond E, Wisniewski T. Alzheimer's disease: experimental models and reality. *Acta Neuropathol*. 2017;133(2):155–175.
54. Zetterberg H, Mattsson N. Understanding the cause of sporadic Alzheimer's disease. *Expert Rev Neurother*. 2014;14(6):621–630.
55. Uversky VN. Intrinsic disorder, protein-protein interactions, and disease. *Adv Protein Chem Struct Biol*. 2018;110:85–121.
56. Xiang S, Kato M, Wu LC, et al. The LC domain of hnRNPA2 adopts similar conformations in hydrogel polymers, liquid-like droplets, and nuclei. *Cell*. 2015;163(4): 829–839.
57. Woerner AC, Frottin F, Hornburg D, et al. Cytoplasmic protein aggregates interfere with nucleocytoplasmic transport of protein and RNA. *Science*. 2016;351(6269): 173–176.
58. Olzscha H, Schermann SM, Woerner AC, et al. Amyloid-like aggregates sequester numerous metastable proteins with essential cellular functions. *Cell*. 2011;144(1): 67–78.
59. Prusiner SB. Some speculations about prions, amyloid, and Alzheimer's disease. *N Engl J Med*. 1984;310(10):661–663.
60. Gajdusek DC. Spontaneous generation of infectious nucleating amyloids in the transmissible and nontransmissible cerebral amyloidoses. *Mol Neurobiol*. 1994;8(1):1–13.
61. Walker LC, Jucker M. Neurodegenerative diseases: expanding the prion concept. *Annu Rev Neurosci*. 2015;38:87–103.
62. Kane MD, Lipinski WJ, Callahan MJ, et al. Evidence for seeding of beta -amyloid by intracerebral infusion of Alzheimer brain extracts in beta -amyloid precursor protein-transgenic mice. *J Neurosci*. 2000;20(10):3606–3611.
63. Meyer-Luehmann M, Coomaraswamy J, Bolmont T, et al. Exogenous induction of cerebral beta-amyloidogenesis is governed by agent and host. *Science*. 2006;313(5794): 1781–1784.
64. Stohr J, Watts JC, Mensinger ZL, et al. Purified and synthetic Alzheimer's amyloid beta (Abeta) prions. *Proc Natl Acad Sci USA*. 2012;109(27):11025–11030.
65. Morales R, Duran-Aniotz C, Castilla J, Estrada LD, Soto C. De novo induction of amyloid-beta deposition in vivo. *Mol Psychiatry*. 2012;17(12):1347–1353.
66. Rosen RF, Fritz JJ, Dooyema J, et al. Exogenous seeding of cerebral beta-amyloid deposition in betaAPP-transgenic rats. *J Neurochem*. 2012;120(5):660–666.
67. Fritschi SK, Langer F, Kaeser SA, et al. Highly potent soluble amyloid-beta seeds in human Alzheimer brain but not cerebrospinal fluid. *Brain*. 2014;137(pt 11):2909–2915.
68. Jaunmuktane Z, Mead S, Ellis M, et al. Evidence for human transmission of amyloid-beta pathology and cerebral amyloid angiopathy. *Nature*. 2015;525(7568):247–250.
69. Frontzek K, Lutz MI, Aguzzi A, Kovacs GG, Budka H. Amyloid-beta pathology and cerebral amyloid angiopathy are frequent in iatrogenic Creutzfeldt-Jakob disease after dural grafting. *Swiss Med Wkly*. 2016;146:w14287.
70. Giaccone G, Maderna E, Marucci G, et al. Iatrogenic early onset cerebral amyloid angiopathy 30 years after cerebral trauma with neurosurgery: vascular amyloid deposits are made up of both Abeta40 and Abeta42. *Acta Neuropathol Commun*. 2019;7(1):70.
71. Rudge P, Jaunmuktane Z, Adlard P, et al. Iatrogenic CJD due to pituitary-derived growth hormone with genetically determined incubation times of up to 40 years. *Brain*. 2015;138(pt 11):3386–3399.

72. Souza IN, De-Souza EA. Commentary: Evidence for human transmission of amyloid-beta pathology and cerebral amyloid angiopathy. *Front Aging Neurosci*. 2016;8:5.
73. Irwin DJ, Abrams JY, Schonberger LB, et al. Evaluation of potential infectivity of Alzheimer and Parkinson disease proteins in recipients of cadaver-derived human growth hormone. *JAMA Neurol*. 2013;70(4):462–468.
74. Eisenberg D, Jucker M. The amyloid state of proteins in human diseases. *Cell*. 2012; 148(6):1188–1203.
75. Cohen ML, Kim C, Haldiman T, et al. Rapidly progressive Alzheimer's disease features distinct structures of amyloid-beta. *Brain*. 2015;138(Pt 4):1009–1022.
76. Hatami A, Albay 3rd R, Monjazeb S, Milton S, Glabe C. Monoclonal antibodies against Abeta42 fibrils distinguish multiple aggregation state polymorphisms in vitro and in Alzheimer disease brain. *J Biol Chem*. 2014;289(46):32131–32143.
77. Watts JC, Condello C, Stohr J, et al. Serial propagation of distinct strains of Abeta prions from Alzheimer's disease patients. *Proc Natl Acad Sci USA*. 2014;111(28):10323–10328.
78. Fritschi SK, Cintron A, Ye L, et al. Abeta seeds resist inactivation by formaldehyde. *Acta Neuropathol*. 2014;128(4):477–484.
79. Langer F, Eisele YS, Fritschi SK, Staufenbiel M, Walker LC, Jucker M. Soluble Abeta seeds are potent inducers of cerebral beta-amyloid deposition. *J Neurosci*. 2011; 31(41):14488–14495.
80. Eisele YS, Bolmont T, Heikenwalder M, et al. Induction of cerebral beta-amyloidosis: intracerebral versus systemic Abeta inoculation. *Proc Natl Acad Sci USA*. 2009;106(31): 12926–12931.
81. Ronnback A, Sagelius H, Bergstedt KD, et al. Amyloid neuropathology in the single arctic APP transgenic model affects interconnected brain regions. *Neurobiol Aging*. 2012;33(4):831.e11–831.e19.
82. Rasmussen J, Jucker M, Walker LC. Abeta seeds and prions: how close the fit? *Prion*. 2017;11(4):215–225.
83. Tipping KW, Karamanos TK, Jakhria T, et al. pH-induced molecular shedding drives the formation of amyloid fibril-derived oligomers. *Proc Natl Acad Sci USA*. 2015; 112(18):5691–5696.
84. Jakhria T, Hellewell AL, Porter MY, et al. Beta2-microglobulin amyloid fibrils are nanoparticles that disrupt lysosomal membrane protein trafficking and inhibit protein degradation by lysosomes. *J Biol Chem*. 2014;289(52):35781–35794.
85. Bauerlein FJB, Saha I, Mishra A, et al. In situ architecture and cellular interactions of polyQ inclusions. *Cell*. 2017;171(1):179–187.e110.
86. Uttara B, Singh AV, Zamboni P, Mahajan RT. Oxidative stress and neurodegenerative diseases: a review of upstream and downstream antioxidant therapeutic options. *Curr Neuropharmacol*. 2009;7(1):65–74.
87. Bucciantini M, Giannoni E, Chiti F, et al. Inherent toxicity of aggregates implies a common mechanism for protein misfolding diseases. *Nature*. 2002;416(6880): 507–511.
88. Serra-Batiste M, Ninot-Pedrosa M, Bayoumi M, Gairi M, Maglia G, Carulla N. Abeta42 assembles into specific beta-barrel pore-forming oligomers in membrane-mimicking environments. *Proc Natl Acad Sci USA*. 2016;113(39):10866–10871.
89. Shankar GM, Li S, Mehta TH, et al. Amyloid-beta protein dimers isolated directly from Alzheimer's brains impair synaptic plasticity and memory. *Nat Med*. 2008;14(8): 837–842.
90. Nelson PT, Alafuzoff I, Bigio EH, et al. Correlation of Alzheimer disease neuropathologic changes with cognitive status: A review of the literature. *J Neuropathol Exp Neurol*. 2012;71(5):362–381.
91. Prusiner SB. Biology and genetics of prions causing neurodegeneration. *Annu Rev Genet*. 2013;47:601–623.

92. Jin M, Shepardson N, Yang T, Chen G, Walsh D, Selkoe DJ. Soluble amyloid beta-protein dimers isolated from Alzheimer cortex directly induce tau hyperphosphorylation and neuritic degeneration. *Proc Natl Acad Sci USA*. 2011;108(14): 5819–5824.
93. Busciglio J, Lorenzo A, Yeh J, Yankner BA. Beta-amyloid fibrils induce tau phosphorylation and loss of microtubule binding. *Neuron*. 1995;14(4):879–888.
94. Nussbaum JM, Seward ME, Bloom GS. Alzheimer disease: a tale of two prions. *Prion*. 2013;7(1):14–19.
95. Guo JP, Arai T, Miklossy J, McGeer PL. Abeta and tau form soluble complexes that may promote self aggregation of both into the insoluble forms observed in Alzheimer's disease. *Proc Natl Acad Sci USA*. 2006;103(6):1953–1958.
96. Lasagna-Reeves CA, Castillo-Carranza DL, Guerrero-Muoz MJ, Jackson GR, Kayed R. Preparation and characterization of neurotoxic tau oligomers. *Biochemistry*. 2010;49(47):10039–10041.
97. Vasconcelos B, Stancu IC, Buist A, et al. Heterotypic seeding of tau fibrillization by pre-aggregated Abeta provides potent seeds for prion-like seeding and propagation of tau-pathology in vivo. *Acta Neuropathol*. 2016;131(4):549–569.
98. Shrivastava AN, Aperia A, Melki R, Triller A. Physico-pathologic mechanisms involved in neurodegeneration: misfolded protein-plasma membrane interactions. *Neuron*. 2017;95(1):33–50.
99. Goedert M, Clavaguera F, Tolnay M. The propagation of prion-like protein inclusions in neurodegenerative diseases. *Trends Neurosci*. 2010;33(7):317–325.
100. Sarnataro D. Attempt to untangle the prion-like misfolding mechanism for neurodegenerative diseases. *Int J Mol Sci*. 2018;19(10):3081.
101. Collinge J, Clarke AR. A general model of prion strains and their pathogenicity. *Science*. 2007;318(5852):930–936.
102. Melki R. Role of different alpha-synuclein strains in synucleinopathies, similarities with other neurodegenerative diseases. *J Parkinsons Dis*. 2015;5(2):217–227.
103. Kaufman SK, Sanders DW, Thomas TL, et al. Tau prion strains dictate patterns of cell pathology, progression rate, and regional vulnerability in vivo. *Neuron*. 2016;92(4): 796–812.
104. Hasegawa M. Molecular mechanisms in the pathogenesis of Alzheimer's disease and tauopathies-prion-like seeded aggregation and phosphorylation. *Biomolecules*. 2016; 6(2):24.
105. Suttkus A, Holzer M, Morawski M, Arendt T. The neuronal extracellular matrix restricts distribution and internalization of aggregated tau-protein. *Neuroscience*. 2016;313:225–235.
106. Morawski M, Bruckner G, Jager C, Seeger G, Arendt T. Neurons associated with aggrecan-based perineuronal nets are protected against tau pathology in subcortical regions in Alzheimer's disease. *Neuroscience*. 2010;169(3):1347–1363.
107. Holmes BB, DeVos SL, Kfoury N, et al. Heparan sulfate proteoglycans mediate internalization and propagation of specific proteopathic seeds. *Proc Natl Acad Sci USA*. 2013;110(33):E3138–E3147.
108. Kanekiyo T, Zhang J, Liu Q, Liu CC, Zhang L, Bu G. Heparan sulphate proteoglycan and the low-density lipoprotein receptor-related protein 1 constitute major pathways for neuronal amyloid-beta uptake. *J Neurosci*. 2011;31(5):1644–1651.
109. Sarnataro D, Pepe A, Zurzolo C. Cell biology of prion protein. *Prog Mol Biol Transl Sci*. 2017;150:57–82.
110. Mannini B, Chiti F. Chaperones as suppressors of protein misfolded oligomer toxicity. *Front Mol Neurosci*. 2017;10:98.
111. Jackson WS. Selective vulnerability to neurodegenerative disease: the curious case of prion protein. *Dis Model Mech*. 2014;7(1):21–29.

112. Costanzo M, Zurzolo C. The cell biology of prion-like spread of protein aggregates: mechanisms and implication in neurodegeneration. *Biochem J.* 2013;452(1):1–17.
113. Gerson JE, Mudher A, Kayed R. Potential mechanisms and implications for the formation of tau oligomeric strains. *Crit Rev Biochem Mol Biol.* 2016;51(6):482–496.
114. Clavaguera F, Hench J, Goedert M, Tolnay M. Invited review: prion-like transmission and spreading of tau pathology. *Neuropathol Appl Neurobiol.* 2015;41(1):47–58.
115. Kundel F, Hong L, Falcon B, et al. Measurement of tau filament fragmentation provides insights into prion-like spreading. *ACS Chem Neurosci.* 2018;9(6):1276–1282.
116. McKhann GM, Knopman DS, Chertkow H, et al. The diagnosis of dementia due to Alzheimer's disease: recommendations from the National Institute on Aging-Alzheimer's Association workgroups on diagnostic guidelines for Alzheimer's disease. *Alzheimers Dement.* 2011;7(3):263–269.
117. Warren JD, Fletcher PD, Golden HL. The paradox of syndromic diversity in Alzheimer disease. *Nat Rev Neurol.* 2012;8(8):451–464.
118. Maarouf CL, Daugs ID, Spina S, et al. Histopathological and molecular heterogeneity among individuals with dementia associated with presenilin mutations. *Mol Neurodegener.* 2008;3:20.
119. Suarez-Gonzalez A, Henley SM, Walton J, Crutch SJ. Posterior cortical atrophy: an atypical variant of Alzheimer disease. *Psychiatr Clin North Am.* 2015;38(2):211–220.
120. Larner AJ. "Frontal variant Alzheimer's disease": a reappraisal. *Clin Neurol Neurosurg.* 2006;108(7):705–708.
121. Villain N, Dubois B. Alzheimer's disease including focal presentations. *Semin Neurol.* 2019;39(2):213–226.
122. Di Fede G, Catania M, Morbin M, et al. Good gene, bad gene: new APP variant may be both. *Prog Neurobiol.* 2012;99(3):281–292.
123. Catania M, Giaccone G, Salmona M, Tagliavini F, Di Fede G. Dreaming of a New World where Alzheimer's is a treatable disorder. *Front Aging Neurosci.* 2019;11:317.
124. Marciani DJ. Promising results from Alzheimer's disease passive immunotherapy support the development of a preventive vaccine. *Research.* 2019;5341375:1–14.
125. Colby DW, Prusiner SB. De novo generation of prion strains. *Nat Rev Microbiol.* 2011;9(11):771–777.
126. Scott MR, Groth D, Tatzelt J, et al. Propagation of prion strains through specific conformers of the prion protein. *J Virol.* 1997;71(12):9032–9044.
127. Safar J, Wille H, Itri V, et al. Eight prion strains have PrP(Sc) molecules with different conformations. *Nat Med.* 1998;4(10):1157–1165.
128. Condello C, Stoehr J. Abeta propagation and strains: implications for the phenotypic diversity in Alzheimer's disease. *Neurobiol Dis.* 2018;109(pt B):191–200.
129. Meinhardt J, Sachse C, Hortschansky P, Grigorieff N, Fandrich M. Abeta(1-40) fibril polymorphism implies diverse interaction patterns in amyloid fibrils. *J Mol Biol.* 2009;386(3):869–877.
130. Elkins MR, Wang T, Nick M, et al. Structural polymorphism of Alzheimer's beta-amyloid fibrils as controlled by an E22 switch: a solid-state NMR study. *J Am Chem Soc.* 2016;138(31):9840–9852.
131. Clavaguera F, Lavenir I, Falcon B, Frank S, Goedert M, Tolnay M. "Prion-like" templated misfolding in tauopathies. *Brain Pathol.* 2013;23(3):342–349.
132. Frost B, Ollesch J, Wille H, Diamond MI. Conformational diversity of wild-type tau fibrils specified by templated conformation change. *J Biol Chem.* 2009;284(6): 3546–3551.
133. Lu JX, Qiang W, Yau WM, Schwieters CD, Meredith SC, Tycko R. Molecular structure of β-amyloid fibrils in Alzheimer's disease brain tissue. *Cell.* 2013;154(6):1257–1268.
134. Liu J, Costantino I, Venugopalan N, et al. Amyloid structure exhibits polymorphism on multiple length scales in human brain tissue. *Sci Rep.* 2016;6:33079.

135. Condello C, Lemmin T, Stohr J, et al. Structural heterogeneity and intersubject variability of Abeta in familial and sporadic Alzheimer's disease. *Proc Natl Acad Sci USA*. 2018;115(4):E782–e791.
136. Stohr J, Condello C, Watts JC, et al. Distinct synthetic Abeta prion strains producing different amyloid deposits in bigenic mice. *Proc Natl Acad Sci USA*. 2014;111(28): 10329–10334.
137. Nystrom S, Psonka-Antonczyk KM, Ellingsen PG, et al. Evidence for age-dependent in vivo conformational rearrangement within Abeta amyloid deposits. *ACS Chem Biol*. 2013;8(6):1128–1133.
138. Friesen M, Meyer-Luehmann M. Abeta seeding as a tool to study cerebral amyloidosis and associated pathology. *Front Mol Neurosci*. 2019;12:233.
139. McKeith IG. Consensus guidelines for the clinical and pathologic diagnosis of dementia with Lewy bodies (DLB): Report of the consortium on DLB international workshop. *J Alzheimers Dis*. 2006;9(3 suppl):417–423.
140. Adler CH, Beach TG, Zhang N, et al. Unified staging system for Lewy body disorders: clinicopathologic correlations and comparison to Braak staging. *J Neuropathol Exp Neurol*. 2019;78(10):891–899.
141. Hyman BT, Phelps CH, Beach TG, et al. National Institute on Aging-Alzheimer's association guidelines for the neuropathologic assessment of Alzheimer's disease. *Alzheimers Dement*. 2012;8(1):1–13.
142. Cairns NJ, Bigio EH, Mackenzie IR, et al. Neuropathologic diagnostic and nosologic criteria for frontotemporal lobar degeneration: consensus of the consortium for frontotemporal lobar degeneration. *Acta Neuropathol*. 2007;114(1):5–22.
143. Haraguchi T, Terada S, Ishizu H, et al. Coexistence of Creutzfeldt-Jakob disease, Lewy body disease, and Alzheimer's disease pathology: an autopsy case showing typical clinical features of Creutzfeldt-Jakob disease. *Neuropathology*. 2009;29(4):454–459.
144. Bugiani O, Giaccone G, Piccardo P, Morbin M, Tagliavini F, Ghetti B. Neuropathology of Gerstmann-Straussler-Scheinker disease. *Microsc Res Tech*. 2000;50(1):10–15.
145. Colom-Cadena M, Gelpi E, Charif S, et al. Confluence of alpha-synuclein, tau, and beta-amyloid pathologies in dementia with Lewy bodies. *J Neuropathol Exp Neurol*. 2013;72(12):1203–1212.
146. Di Fede G, Giaccone G, Tagliavini F. Hereditary and sporadic beta-amyloidoses. *Front Biosci*. 2013;18:1202–1226.
147. Shin WS, Di J, Murray KA, et al. Different amyloid-beta self-assemblies have distinct effects on intracellular tau aggregation. *Front Mol Neurosci*. 2019;12:268.
148. Götz J, Chen F, van Dorpe J, Nitsch RM. Formation of neurofibrillary tangles in P301l tau transgenic mice induced by Abeta 42 fibrils. *Science*. 2001;293(5534):1491–1495.
149. Tsigelny IF, Crews L, Desplats P, et al. Mechanisms of hybrid oligomer formation in the pathogenesis of combined Alzheimer's and Parkinson's diseases. *PLoS One*. 2008;3(9):e3135.
150. Masliah E, Rockenstein E, Veinbergs I, et al. Beta-amyloid peptides enhance alpha-synuclein accumulation and neuronal deficits in a transgenic mouse model linking Alzheimer's disease and Parkinson's disease. *Proc Natl Acad Sci USA*. 2001;98(21): 12245–12250.
151. Kallhoff V, Peethumnongsin E, Zheng H. Lack of alpha-synuclein increases amyloid plaque accumulation in a transgenic mouse model of Alzheimer's disease. *Mol Neurodegener*. 2007;2:6.
152. Falcon B, Zhang W, Murzin AG, et al. Structures of filaments from Pick's disease reveal a novel tau protein fold. *Nature*. 2018;561(7721):137–140.
153. Lu JX, Qiang W, Yau WM, Schwieters CD, Meredith SC, Tycko R. Molecular structure of beta-amyloid fibrils in Alzheimer's disease brain tissue. *Cell*. 2013;154(6): 1257–1268.

154. Langer F, Eisele YS, Fritschi SK, Staufenbiel M, Walker LC, Jucker M. Soluble Aβ seeds are potent inducers of cerebral β-amyloid deposition. *J Neurosci*. 2011;31(41): 14488–14495.
155. Dickson TC, Vickers JC. The morphological phenotype of beta-amyloid plaques and associated neuritic changes in Alzheimer's disease. *Neuroscience*. 2001;105(1):99–107.
156. Bugiani O, Tagliavini F, Giaccone G. Preamyloid deposits, amyloid deposits, and senile plaques in Alzheimer's disease, down syndrome, and aging. *Ann N Y Acad Sci*. 1991;640:122–128.
157. Aslund A, Sigurdson CJ, Klingstedt T, et al. Novel pentameric thiophene derivatives for in vitro and in vivo optical imaging of a plethora of protein aggregates in cerebral amyloidoses. *ACS Chem Biol*. 2009;4(8):673–684.
158. Qiang W, Yau WM, Lu JX, Collinge J, Tycko R. Structural variation in amyloid-β fibrils from Alzheimer's disease clinical subtypes. *Nature*. 2017;541(7636):217–221.
159. Rudinskiy N, Fuerer C, Demurtas D, et al. Amyloid-beta oligomerization is associated with the generation of a typical peptide fragment fingerprint. *Alzheimers Dement*. 2016;12(9):996–1013.
160. Portelius E, Lashley T, Westerlund A, et al. Brain amyloid-beta fragment signatures in pathological ageing and Alzheimer's disease by hybrid immunoprecipitation mass spectrometry. *Neurodegener Dis*. 2015;15(1):50–57.
161. Kubo T, Kumagae Y, Miller CA, Kaneko I. Beta-amyloid racemized at the Ser26 residue in the brains of patients with Alzheimer disease: implications in the pathogenesis of Alzheimer disease. *J Neuropathol Exp Neurol*. 2003;62(3):248–259.
162. Kumar S, Wirths O, Stuber K, et al. Phosphorylation of the amyloid beta-peptide at Ser26 stabilizes oligomeric assembly and increases neurotoxicity. *Acta Neuropathol*. 2016;131(4):525–537.
163. Kuo YM, Emmerling MR, Woods AS, Cotter RJ, Roher AE. Isolation, chemical characterization, and quantitation of a beta 3-pyroglutamyl peptide from neuritic plaques and vascular amyloid deposits. *Biochem Biophys Res Commun*. 1997;237(1): 188–191.
164. Kuo YM, Kokjohn TA, Beach TG, et al. Comparative analysis of amyloid-beta chemical structure and amyloid plaque morphology of transgenic mouse and Alzheimer's disease brains. *J Biol Chem*. 2001;276(16):12991–12998.
165. Russo C, Violani E, Salis S, et al. Pyroglutamate-modified amyloid beta-peptides—AbetaN3(pE)—strongly affect cultured neuron and astrocyte survival. *J Neurochem*. 2002;82(6):1480–1489.
166. Wirths O, Bethge T, Marcello A, et al. Pyroglutamate Abeta pathology in APP/PS1KI mice, sporadic and familial Alzheimer's disease cases. *J Neural Transm (Vienna)*. 2010;117(1):85–96.
167. Catania M, Di Fede G, Tonoli E, et al. Mirror image of the amyloid-beta species in cerebrospinal fluid and cerebral amyloid in Alzheimer's disease. *J Alzheimers Dis*. 2015;47(4):877–881.

CHAPTER TEN

Tau proteinopathies and the prion concept

Michel Goedert*
MRC Laboratory Molecular Biology, Cambridge, United Kingdom
*Corresponding author: e-mail address: mg@mrc-lmb.cam.ac.uk

Contents

Abstract

The ordered assembly of a small number of proteins into amyloid filaments is central to age-related neurodegenerative diseases. Tau is the most commonly affected of these proteins. In sporadic diseases, assemblies of tau form in a stochastic manner in certain brain regions, from where they appear to spread in a deterministic way, giving rise to disease symptoms. Over the past decade, multiple lines of evidence have shown that assembled tau behaves like a prion. More recently, electron cryo-microscopy of tau filaments has shown that distinct conformers are present in different diseases, with no inter-individual variation for a given disease.

1. Introduction

Ordered assembly of a small number of proteins into filaments characterizes the majority of cases of age-related neurodegenerative diseases, including Alzheimer's and Parkinson's. Most cases are sporadic, but a small

Progress in Molecular Biology and Translational Science, Volume 175
ISSN 1877-1173
https://doi.org/10.1016/bs.pmbts.2020.08.003

number is inherited in a dominant manner. Huntington's disease and other glutamine repeat disorders are always inherited, whereas chronic traumatic encephalopathy is probably environmentally induced. Study of dominantly inherited forms of disease has established a causative role for ordered assembly. By extrapolation, it appears likely that inclusion formation is central to neurodegeneration in all cases of disease.

Neurodegenerative diseases that are characterized by the ordered assembly of tau protein into insoluble filaments are the most common proteinopathies of the human nervous system.[1] They include Alzheimer's disease (AD), Pick's disease (PiD), chronic traumatic encephalopathy (CTE), tangle-only dementia, progressive supranuclear palsy (PSP), corticobasal degeneration (CBD), argyrophilic grain disease (AGD), globular glial tauopathy (GGT) and several rarer diseases.

2. Tau protein and its isoforms

Tau is an intrinsically disordered protein, with many potential interaction partners. It can be divided into an N-terminal domain, a proline-rich region, the repeat domain and a C-terminal region (Fig. 1). The N-terminal domain projects from the microtubule surface and is believed to interact with components of the plasma membrane. It contains a primate-specific sequence between residues 19 and 29 of human tau (GLGDRKDQGGY). This sequence lends itself to the production of antibodies that distinguish between human and rodent tau. The PXXP motifs in the proline-rich region are recognized by SH3 domain-containing proteins of the Src family of nonreceptor tyrosine kinases, such as Fyn.[2]

The repeat region and some adjacent sequences mediate interactions between tau and microtubules. By electron cryo-microscopy (cryo-EM), each tau repeat binds to the outer microtubule surface and adopts an extended structure along protofilaments, interacting with α- and β-tubulins.[3,4] Single-molecule tracking revealed a kiss-and-hop mechanism, with a dwell time of tau on individual microtubules of approximately 40 ms.[5,6] Isoform differences did not influence this interaction. Despite these rapid dynamics, tau promoted microtubule assembly. It remains to be seen if microtubules were also stabilized. Tau is most abundant in the labile domain of microtubules, which has led to the suggestion that it may not stabilize microtubules, but enable them to have long labile domains.[7,8] Less is known about the function of the C-terminal region, which may inhibit assembly into filaments.

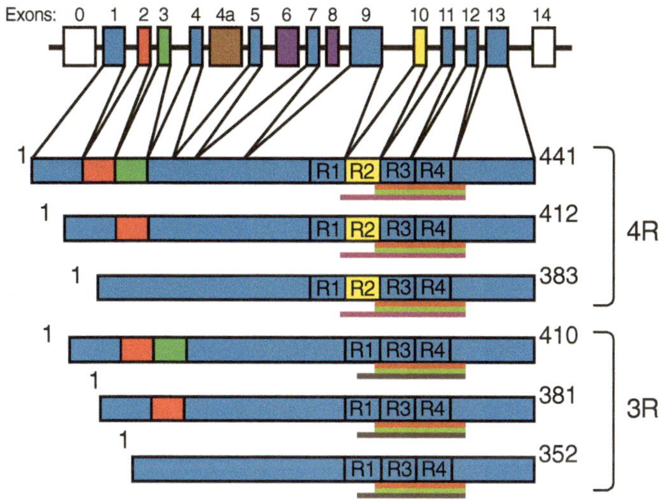

Fig. 1 Human brain tau isoforms. *MAPT* and the six tau isoforms expressed in adult human brain. *MAPT* consists of 14 exons (E). Alternative mRNA splicing of E2 (red), E3 (green) and E10 (yellow) gives rise to six tau isoforms (352–441 amino acids). The constitutively spliced exons (E1, E4, E5, E7, E9, E11, E12 and E13) are shown in dark blue. E6 and E8 (violet) are not transcribed in human brain. E4a (orange) is only expressed in the peripheral nervous system. The repeats (R1–R4) are shown, with three isoforms having four repeats (4R) and three isoforms with three repeats (3R). The core sequences of tau filaments from Alzheimer's disease (G273/G304-E380) are underlined (in orange); the core sequences of tau filaments from chronic traumatic encephalopathy (K274/S305-R379) are underlined (in green); the core sequences of tau filaments from Pick's disease (K254-F378 of 3R tau) are underlined (in gray); the core sequences of tau filaments from corticobasal degeneration are underlined (K274-E380 of 4R tau) are underlined (in cyan).

Despite the absence of a typical low-complexity domain, full-length tau can undergo liquid-liquid phase separation through electrostatic and hydrophobic interactions,[9–11] in conjunction with amyloid aggregation, at least *in vitro*. Although liquid-liquid phase separation and amyloid aggregation of tau are independent processes, they may be able to influence each other.

Six tau isoforms that range from 352 to 441 amino acids in length are expressed from a single *MAPT* gene in adult human brain[12] (Fig. 1). They differ by the presence or absence of inserts of 29 and 58 amino acids (encoded by exons 2 and 3, with exon 3 being only transcribed with exon 2) in the N-terminal half, and the inclusion, or not, of the 31 amino acid repeat, encoded by exon 10, in the C-terminal half. Inclusion of exon 10 results in the production of three isoforms with four repeats (4R) and its exclusion in a further three isoforms with three repeats (3R).

The repeats comprise residues 244–368, in the numbering of the 441 amino acid isoform. In adult human brain, similar levels of 3R and 4R tau are expressed[13]; finding that a correct isoform ratio is essential for preventing neurodegeneration and dementia came as a surprise. 2N isoforms are under-represented in comparison with isoforms that include exon 2 or exclude both exons 2 and 3; 2N, 1N and 0N isoforms make up 9%, 54% and 37%, respectively, of total brain tau. Big tau, which carries an additional large exon in the N-terminal half, is only expressed in the peripheral nervous system.

Isoform expression is not conserved between species. Thus, in adult mouse brain, only 4R tau isoforms are present, whereas adult chicken brain expresses 3R, 4R and 5R tau isoforms.[14] However, the presence of a single hyperphosphorylated 3R tau isoform that lacks N-terminal inserts is characteristic of developing vertebrates. In mice, the switch from 3R to 4R tau occurs between postnatal days 9 and 18, with tau phosphorylation decreasing over time. Isoform switching and phosphorylation are regulated differently.[15] Adult 4R tau isoforms are better at promoting microtubule assembly and at binding to microtubules than the 3R tau isoform that is expressed during development.

3. Tau assemblies

Between 1985 and 1991, the paired helical and straight filaments (PHFs and SFs) of AD were shown to be made of tau protein.[16–20] Subsequently, tau was also shown to be the major component of the filamentous inclusions of Pick's disease, progressive supranuclear palsy and corticobasal degeneration.[1] Negative-stain immuno-electron microscopy (immuno-EM) showed that antibodies specific for the N- and C-termini of tau decorate filaments. This was not the case of antibodies directed against R3 and R4, because their epitopes are occluded in the filaments.[17,19,20] Together with biochemical studies, this work established that tau filaments consist of a core region made of the repeats and a fuzzy coat.[21] They have the biophysical characteristics of amyloid.[22]

Tau filaments are post-translationally modified, with abnormal phosphorylation being the most widely studied. But other modifications are also present, including acetylation, methylation, glycation, isomerization, O-GlcNAcylation, nitration, sumoylation, ubiquitination and truncation.[1] It remains to be seen if abnormal phosphorylation or any other modification is necessary and/or sufficient for the assembly of tau into filaments. Alternatively, a change in conformation as part of filament assembly may lead to tau hyperphosphorylation and other modifications. Acetylation of lysine residues in tau has come to the fore in recent years. It reduces charge, which may play a role in filament assembly. Site-specific acetylation of K280 has been shown to enhance tau aggregation, while reducing microtubule assembly.[23] Twenty-one lysines are found in the repeat region of human tau (residues 244–368).

Because tau is highly soluble, it is not surprising that unmodified, full-length recombinant protein requires cofactors, such as sulfated glycosaminoglycans, free fatty acids and nucleic acids, to assemble into filaments.[24–27] Different cofactors and/or post-translational modifications may cause the assembly of tau in human brain.[28,29]

In AD, CTE, tangle-only dementia and several other tau proteinopathies, all six tau isoforms (3R plus 4R) are present in disease filaments. Pick bodies are made of only 3R tau. In CBD, PSP, AGD, GGT and several other diseases, 4R tau isoforms make up the filaments. Morphologies and structures of filaments vary, even when they are made of the same tau isoforms. Biochemically, the banding patterns of C-terminal fragments of tau can vary between 4R tau proteinopathies.[30]

4. Genetics of *MAPT*

The relevance of tau dysfunction for neurodegeneration became clear, when dominantly inherited mutations in *MAPT* were shown to cause a form of frontotemporal dementia that can be associated with parkinsonism (FTDP-17T, also known as familial FTLD-tau).[31–33] In FTDP-17T, abundant filamentous tau inclusions are present in either nerve cells, or in both nerve cells and glial cells. Aβ deposits, a defining feature of AD, are not present, nor are other extracellular amyloids. This work established that a pathological pathway leading from monomeric to assembled tau is sufficient to cause neurodegeneration and dementia.

Sixty-five mutations in *MAPT* have been identified in FTDP-17T (Fig. 2). Filamentous inclusions are composed of either 3R, 4R or 3R plus 4R tau.[34] Mutations are concentrated in exons 9–12 (encoding R1–R4) and the introns flanking exon 10, with a smaller number of mutations in exon 13. Two mutations (R5H and R5L) are found in exon 1. Mutations can be divided into those with a primary effect at the protein level and those affecting the alternative splicing of tau pre-mRNA. *MAPT* on chromosome 17q21.31 is characterized by two haplotypes as the result of a 900 kb inversion (H1) or noninversion (H2) polymorphism.[35] Inheritance of the H1 haplotype is a risk factor for PSP and CBD, but not for PiD.[36–39]

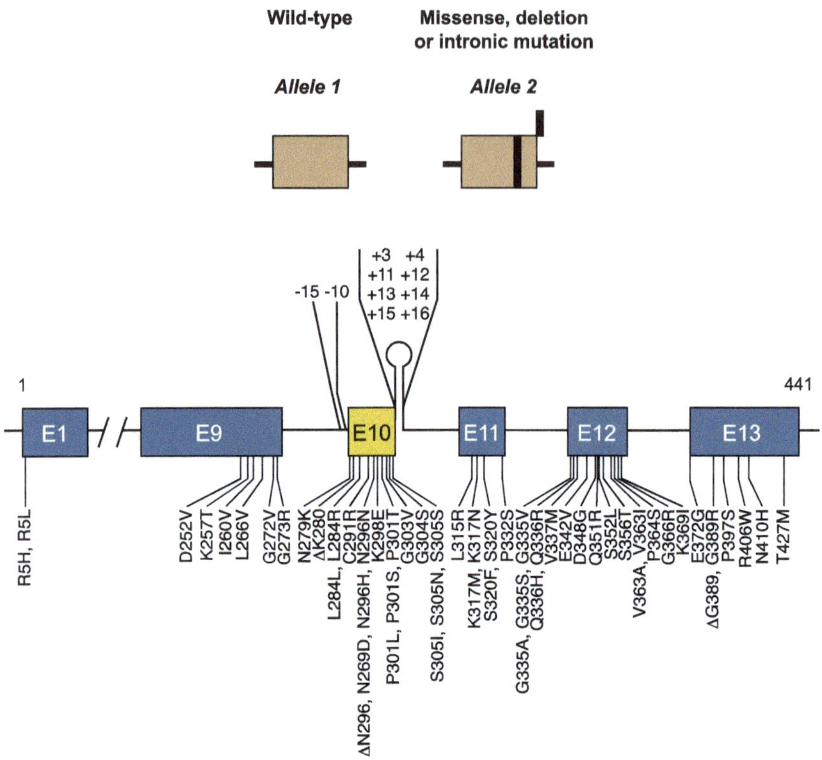

Fig. 2 Mutations in *MAPT* in FTDP-17T. Missense, deletion and intronic mutations are dominantly inherited. Fifty-five coding region and 10 intronic mutations are shown. They cluster in the sequences encoding R1–R4 (distal part of E9, E10, E11 and E12) and the intronic sequences flanking E10.

Disease-causing mutations in *MAPT* have made it possible to produce transgenic rodent lines that form tau filaments in nerve cells and show neurodegeneration.[40–42] Depending on the promoters used, tau filaments can also form in astrocytes and oligodendrocytes.[43,44] Aggregation of tau correlates with neurodegeneration.[45] Reducing aggregation and increasing removal of aggregates are therefore therapeutic objectives.

5. Neuropathological phenotypes of FTDP-17T

All cases of FTDP-17T are characterized by the presence of filamentous tau inclusions in nerve cells or in both nerve cells and glial cells (Fig. 3).[1,34] Cases with glial cell inclusions only have not been described. Tau inclusions are most abundant in cerebral cortex and hippocampal formation.

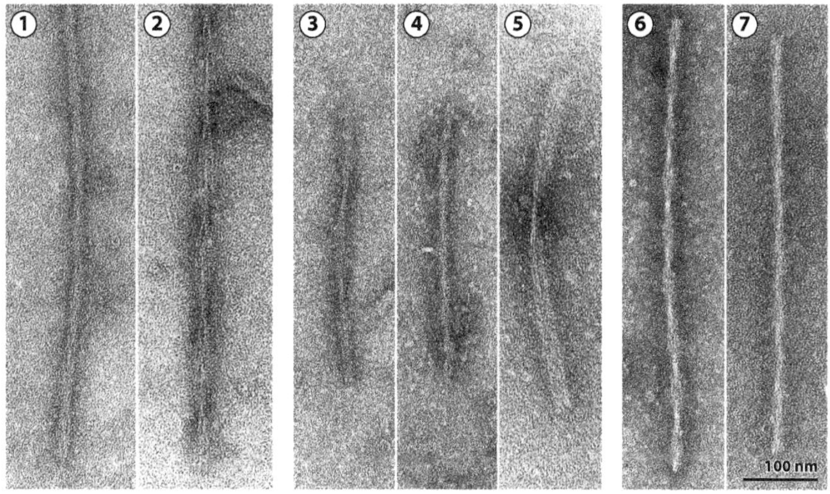

Fig. 3 Negative stain electron microscopy of tau filaments from cases of FTDP-17T. (1, 2) Tau filaments from a case with abundant Pick body-like inclusions and a G389R mutation (E13). (1) Straight filaments form the majority species and (2) strongly stranded filaments are in the minority. (3–5) Tau filaments from cases with neuronal and glial inclusions and a P301L (E10) mutation or a mutation in intron 10. (3) Narrow twisted ribbons and (4) occasional rope-like filaments. (5) Wide twisted ribbons. (6, 7) Paired helical and straight tau filaments as in Alzheimer's disease are present in cases with mutations V337M (E12) and R406W (E13).

Inclusions similar to Pick bodies are often observed in the brains of individuals with mutations in exons 9, 11, 12 and 13 of *MAPT*. Similar to sporadic PiD, inclusions associated with mutations G272V in exon 9 and ΔK280 in exon 10 are made of 3R tau and are not phosphorylated at S262.[46,47] For other mutations, such as G389R in exon 13, variable amounts of 4R tau and some phosphorylation of S262 are seen in Pick-like bodies.[48] The latter resemble Pick-like bodies of CBD. Mutation N410H in exon 13 phenocopies the tau pathology of CBD.[49]

In the work mentioned above, tau deposits are found predominantly in neurons, whereas mutations in exons 1 and 10, as well as in the introns following exons 9 and 10, are associated with abundant neuronal and glial tau inclusions.[34] Glial pathology is in the form of coiled bodies in oligodendroglia, as well as tufted astrocytes and astrocytic plaques reminiscent of PSP and CBD. Mutations in exon 10 cause the formation of inclusions made of 4R tau; most of these mutations affect exon 10 pre-mRNA splicing, altering the ratio of 3R/4R tau. *MAPT* mutations P301L, P301S and P301T, the primary effects of which are at the protein level, are exceptions.

They continue to be important for the generation of experimental models of tauopathy and illustrate the clinical and pathological heterogeneity associated with *MAPT* mutations. Although most individuals with mutations P301L and P301S develop behavioral-variant FTD, cases of primary progressive aphasia have been described.[50] A P301S carrier presented with corticobasal syndrome.[51] A P301L patient had GGT, as had individuals with mutation P301T.[50,52] GGT has emerged as a common disease associated with mutations in *MAPT*. Mutations in codon 301 affect only 20%–25% of tau molecules, with 75%–80% being wild-type, arguing against a simple loss-of-function mechanism as an important disease determinant.[53]

Intronic mutations and most mutations in exon 10 affect the ratio of 3R/4R tau, which is normally around 1:1, without changing the amount of total tau. For most mutations, this results in the relative overproduction of wild-type 4R tau and its assembly into filamentous inclusions. Tau filaments appear as twisted ribbons or half ribbons. Although these mutations often give rise to behavioral-variant FTD, cases of atypical PSP have also been described.[54] For other mutations, such as V337M (exon 12)[55] and R406W (exon 13),[56] tau inclusions resemble those of AD and filaments are made of all six brain tau isoforms.

6. Prion-like propagation of tau assemblies

Even though there is no evidence for inter-individual transmission of tau proteinopathies, several lines of investigation in cell culture and animal models have shown that assembled tau can behave like a prion.[57,58] It may propagate between cells, but it is difficult to obtain evidence for the transfer of assembled tau. Although the staging of tau pathology[59] is consistent with such a mechanism, it cannot prove its existence. Perhaps the best evidence for prion-like spreading in human brain comes from a study showing the absence of assembled tau in denervated frontal cortex from AD.[60]

Intracerebral injection of human tau inclusion-bearing brain extracts from mice transgenic for human P301S tau (0N4R P301S tau) into mice transgenic for wild-type tau (2N4R tau) induced the formation of inclusions made of wild-type human tau, followed by their spreading to distant brain regions.[61] Subsequently, several groups studied the spreading of pathological tau along the entorhinal cortex/hippocampal pathway. They used mouse models that apparently expressed human P301L tau only in the entorhinal cortex.[62–64] Several months after the appearance of the first tau inclusions, hippocampal neurons developed filamentous tau pathology.

Using this system, together with a mouse line transgenic for mutant amyloid precursor protein and presenilin-1, it has been shown that Aβ deposition in cerebral cortex can increase the spreading of aggregated tau.[65] When presymptomatic P301S tau mice were intracerebrally injected with brain extracts from symptomatic animals,[66] tau inclusions formed at the injection sites after 2 weeks. Contralateral and caudo-rostral propagations were evident in nuclei with strong afferent and efferent connections to the injection sites, showing that the spread of tau pathology was dependent on connectivity, not proximity. Evidence so far suggests that the propagation of misfolded tau through the brain is an active process and not simply the result of seeds being released following the disintegration of inclusion-bearing cells.[67]

Virally-mediated expression of tau has also been used to investigate the propagation of pathology. Tau pathology was observed in the dentate gyrus following adeno-associated virus-mediated expression of P301L tau in layer II of the entorhinal cortex.[68–70] Expression of mouse tau was not necessary for the formation of tau pathology. The intraperitoneal injection of brain extracts from symptomatic P301S tau mice into presymptomatic animals promoted the formation of cerebral tau inclusions.[71] Aggregated tau can thus promote inclusion formation in the central nervous system of transgenic mice following peripheral administration.

Intercellular transfer of tau assemblies and the adoption of self-propagating conformations have been demonstrated in cell culture.[72–76] Uptake of aggregated tau appears to rely on macropinocytosis and heparan sulfate proteoglycans.[73,77] Both monomeric and aggregated tau are taken up by cells, but only assembled tau is able to seed aggregation.[78] It has been shown that monomeric tau is taken up through clathrin-dependent endocytosis, whereas recombinant tau assemblies are taken up by clathrin-independent endocytosis.[79] These findings are seemingly at odds with a study reporting that seed-competent tau uptake and spread are regulated by LRP1 (low-density lipoprotein receptor-related protein 1), a receptor coupled to clathrin-dependent endocytosis.[80] It remains to be seen what mechanisms underlie the uptake of seeds from tau proteinopathies.

Using a cell assay, we found that the seeding potency of recombinant P301S tau aggregated with heparin was lower than that of aggregated tau from the brains of mice transgenic for human P301S tau.[78] Similar discrepancies between recombinant and brain aggregates have been described for prions, Aβ, α-synuclein and serum amyloid A.[81–86] Recombinant tau aggregates were more resistant to disaggregation by guanidine hydrochloride and digestion by proteinase K than those from transgenic mouse brain, consistent with the view that more stable aggregates possess lower seeding activity. Recombinant aggregated tau was also found to be more stable than tau filaments extracted from Alzheimer's disease brain.[87] Distinct conformations accounted for differences in seeding potency. Thus, tau filaments formed from recombinant P301S tau following seeding with aggregated tau from transgenic mouse brain showed resistance to guanidine hydrochloride similar to that of tau seeds from the brains of mice transgenic for human P301S tau.[78] The seeding potency of tau filaments was like that of brain-derived aggregated P301S tau, consistent with the prion concept of tau on non-denaturing gels.

To identify the tau species that are most seed-competent, we used sucrose gradient fractionation of brains from symptomatic mice transgenic for human P301S tau and assessed their seeding ability in human embryonic kidney cells inducibly expressing 1N4R P301S human tau.[88] The 40% fraction, with tau filaments of 179 ± 12 nm length, was most seed-competent. Fractions with longer filaments showed less seeding activity. Other high-molecular weight tau species (>10 mers) were also seed-competent, but low molecular weight and monomeric tau were unable to seed assembly. These findings demonstrate that short filaments are the major seed-competent tau species in the P301S tau transgenic mouse model. It remains to be seen if the same is true of human tau proteinopathies. Fragmentation of tau assemblies may thus increase prion-like spreading, consistent with reported findings.[89–91] Cells may interact with particles, rather than monomeric proteins. Because oligomeric species are believed to be made of fewer than a hundred and filaments of thousands of monomers, at equal monomer concentrations, many more oligomers then filaments are used.

Less is understood about the mechanisms underlying the release of assembled tau. It has been suggested that exosome-mediated secretion of tau may play a role in disease.[92,93] By contrast, another study concluded that tau is released through a pathway that is distinct from exosome-mediated secretion.[94] This pathway requires heat shock cognate 70, its co-chaperone DnaJ and the SNARE protein SNAP23. These findings suggest that chaperones may play an important role in the removal of misfolded proteins from cytoplasm. A separate study has reported that free tau interacts with components of the inner leaf of the plasma membrane, resulting in sulfated glycosaminoglycan-dependent translocation across the plasma membrane.[95] It remains to be seen by which mechanisms aggregation-competent tau is normally released. Most studies used transfected non-neuronal cells, some of which were of human origin, or primary nerve cells from rodents. However, human neurons derived from induced pluripotent stem cells can also take up seeds and show seed-induced tau aggregation.[96,97]

7. Conformers of assembled tau

Distinct conformers of assembled tau appear to exist, reminiscent of prion strains.[98,99] This may explain the variety of human tau proteinopathies. Inclusions formed after intracerebral inoculation of brain homogenates from cases of AD, tangle-only dementia, PiD, AGD, PSP and CBD into the mouse line transgenic for 2N4R tau.[98] Tau assemblies reminiscent of human disorders were observed following injection of brain homogenates from patients with the 4R tau proteinopathies AGD, CBD and PSP. A smaller number of similar inclusions also formed after the intracerebral injection of brain homogenates from human tau proteinopathies into wild-type mice.[98]

Conformationally distinct tau assemblies made of four tau repeats formed in HEK cells following exposure to seeds from human tau proteinopathies.[99] Inoculation of these seeded assemblies into the hippocampus of young P301S tau transgenic mice induced the formation of assemblies that were stable through serial transmission. When HEK cells expressing four tau repeats were seeded with homogenates from these brains, inclusions formed that were identical to those present initially.

Taking everything together, these findings suggest that distinct conformers of assembled tau underlie different human tau proteinopathies. Structural work using cryo-EM supports this view.

8. Structures of tau filaments from Alzheimer's disease

AD, the most common dementia, is an Aβ and tau proteinopathy. Tau inclusions are made of all six brain isoforms[100] and are largely neuronal. By cryo-EM, the Alzheimer fold comprises 75 amino acids; it starts at G273 for 3R and G304V for 4R tau isoforms, and ends at E380 (Fig. 4A, 5).[29,101] These sequences correspond to the C-termini of R1 and R2, the whole of R3 and R4, and 12 amino acids after R4. We determined the structures of PHFs to 3.2 Å and of SFs to 3.3 Å. Their cores are composed of eight β-strands (β1-β8) in a two-layered C-shaped arrangement. A heterotypic interface forms between β1–2 and β8. Beta-1 consists of ^{306}VQIVYK^{311}, which forms a packing interface with amino acids 373–378 from β8. Strands β2 and β8 pack against each other through a polar-zipper motif.

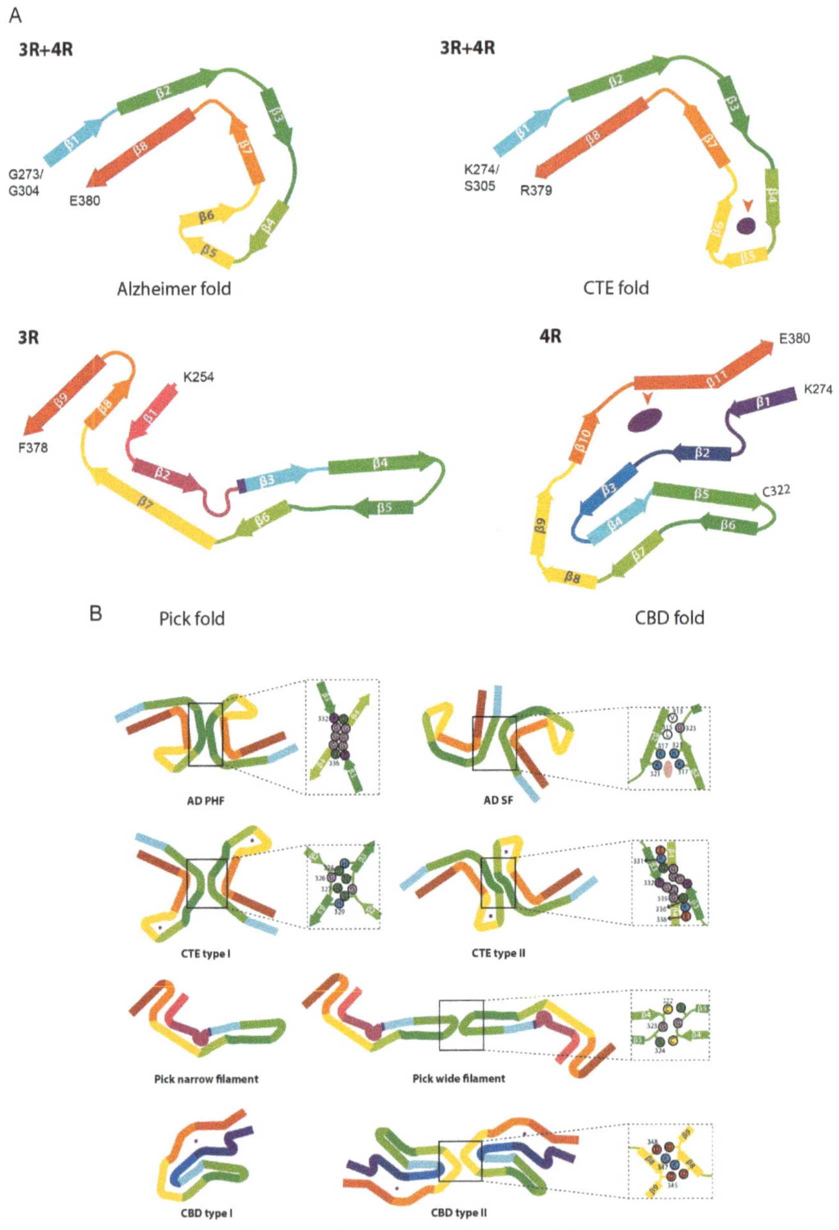

Fig. 4 Structures of tau filaments from human brain and inter-protofilament packing. (A) Tau protofilaments shown through the XY-plane, perpendicular to the helical axis. Beta-strands correspond to colored, arrowed bars. Red arrowheads point to internal densities. Fuzzy densities on the outside of protofilaments are not shown. (B) Two filament types are characteristic of each disease. Zoomed-in schematics show amino acids involved in protofilament interfaces. Pink oval represents the additional density in the straight filament of Alzheimer's disease (AD SF).

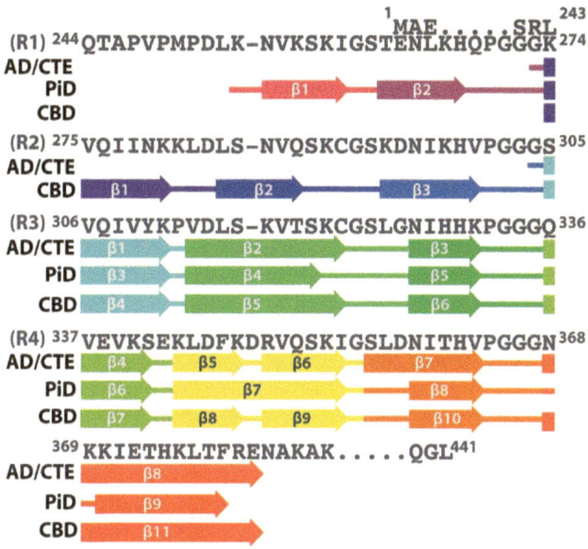

Fig. 5 Comparison of the secondary structures of tau filaments from Alzheimer's disease (AD), chronic traumatic encephalopathy (CTE), Pick's disease (PiD) and corticobasal degeneration (CBD). Beta-strands are shown as colored, arrowed bars.

A second anti-parallel β-sheet forms between β3 and β7. The two sides meet through a β-helix that consists of β4-β6 from R4.

PHFs and SFs are made of the same protofilament pairs. The ultrastructural polymorphism that distinguishes them results from differences in lateral contacts (Fig. 4B). In PHFs, the interface is stabilized by hydrogen bonding across the extended β-spiral formed by residues ^{332}PGGGQ^{336}, as well as by interactions between K331 from one protofilament, and Q336 and E338 from the other. In SFs, protofilaments pack asymmetrically and appear to be stabilized by an additional, possibly anionic, density between the side chains of K317 and K321 from both protofilaments.

The first structures were from frontal cortex of a patient with sporadic AD.[29] We have since determined the structures of tau filaments from frontal cortices of three additional patients with AD [two sporadic and one familial (V717F mutation in the amyloid precursor protein gene)].[101] Tau filament structures were identical in all four cases. We also used immuno-EM[101] to extend this analysis to frontal cortices from another 15 patients with sporadic AD (14 sporadic and one familial), occipital cortices from two patients with posterior cortical atrophy, a subtype of AD,[102] and additional brain regions (temporal cortex, occipital cortex, parietal cortex, cingulate cortex, thalamus, substantia innominata and putamen) from the first case used for cryo-EM.

Immuno-EM findings were consistent with cryo-EM results, indicating the absence of significant variation in tau filament structures between patients with AD.

9. Structures of tau filaments from chronic traumatic encephalopathy

CTE is associated with repetitive head impacts and exposure to blast waves. Tau inclusions are neuronal and glial. Nerve cell inclusions, which are in the majority, are made of all six brain tau isoforms.[103] By immunohistochemistry, astrocytic inclusions have been reported to consist of 4R tau.[104] Prior to our work,[105] nothing was known about tau filaments from CTE. By negative-stain EM, we observed a helical filament type (90%) in two ex-boxers and a former American football player, who had died with a neuropathologically confirmed diagnosis of stage IV CTE. The remaining filaments resembled Alzheimer PHFs. Filament structures were determined to a resolution of 2.3 Å. They comprise residues K274-R379 of 3R tau and S305-R379 of 4R tau, which form the ordered core of two identical C-shaped protofilaments (Figs. 4A, 5). The secondary structures of Alzheimer and CTE folds comprise eight β-strands. The tip of the C is formed by β4-β6, where the packing coincides with an opening in the CTE fold. The resulting cavity is filled with a density that is not connected to tau, consistent with the presence of additional molecules of unknown identity. The nature of the cavity suggests that these molecules are hydrophobic and non-proteinaceous.

In CTE type I filaments (90% by negative staining), two identical protofilaments pack in a staggered manner through an anti-parallel steric zipper formed by residues ^{324}SLGNIH^{329} (Fig. 4B). In CTE type II filaments (7% by negative staining, 3% were PHFs), the protofilament interface is also staggered and comprises the same residues as in AD PHFs, but a kinked conformation reduces the number of hydrogen bonds across the interface. These findings provide a unifying criterion for stage IV CTE and confirm that dementia pugilistica and CTE are one disease. They also establish that AD and CTE are distinct tau proteinopathies.

10. Structures of tau filaments from Pick's disease

PiD is a type of frontotemporal dementia with mostly neuronal inclusions that are made of 3R tau.[106] We observed narrow (>90%) and wide

(<10%) Pick filaments (NPFs and WPFs) by negative staining. With cryo-EM, we determined a 3.2 Å resolution map of the core of NPFs from frontotemporal cortex of a case of sporadic PiD (Figs. 4A, 5).[107] A single protofilament extends from K254-F378 of 3R tau, comprising 94 amino acids. It consists of the C-terminal 21 amino acids of R1, the whole of R3 and R4, and ten amino acids after R4. Nine β-strands adopt a J shape and are arranged into four cross-β packing stacks, which are connected by turns and arcs. Beta1, β8 and β9 form a three-layered motif, with the rest of the J containing two layers. In WPFs, two protofilaments pack symmetrically against each other through Van der Waals interactions at the tip of the J (Fig. 4B).

To investigate the generality of the Pick fold, we used immuno-EM of tau filaments from frontotemporal cortex of eight additional patients with sporadic PiD.[107] Most filaments were NPFs, with a minority of WPFs; they were not decorated by the repeat-specific antibodies. This showed that R1, R3 and R4 epitopes were inaccessible, indicating that they form part of the filament core. In keeping with the absence of tau phosphorylation at S262 and/or S356 in PiD,[108] the tight turn at G261 of the NPF prevents phosphorylation of S262.

11. Structures of tau filaments from corticobasal degeneration

CBD is characterized by motor and cognitive disturbances, with 4R tau inclusions in nerve cells and glial cells.[109] Astrocytic plaques are pathognomonic.[110] We observed narrow (type I) and wide (type II) tau filaments by negative staining, with ratios of 1:1 to 1:3. The structures of type I and type II filaments were determined to a resolution of 3.2 and 3.0 Å, respectively.[111] Type I filaments are composed of a protofilament that adopts a compact four-layered fold (Figs. 4A, 5). It contains a density that is surrounded by tau protein. Unlike CTE filaments, this density is located in a positively charged environment. Type II filaments consist of pairs of type I protofilaments, related by C2 symmetry. The protofilament interface is formed by anti-parallel stacking of ^{343}KLDFKDR^{349} (Fig. 4B). The CBD fold extends from K274-E380 of 4R tau, comprising 107 amino acids (Fig. 5). The core of type I filaments consists of the last residue of R1, the whole of R2-R4 and 12 amino acids after R4. They comprise 11 β-strands. The central four layers are formed by β7, β4, β3 and β10. On one side, the residues that form β7-β9 in the CBD fold are the same as β4-β6 in Alzheimer and CTE folds. On the other side, the hairpin structure formed by β5 and β6 is similar to the conformation seen between β4 and β5 of the Pick fold. In addition to the internal density, fuzzy densities are present on the outside of filaments, often in close proximity of the side chains of lysine residues and solvent-exposed hydrophobic amino acid patches. It has been suggested that some of these densities may correspond to ubiquitin and that differential ubiquitination of lysine residues could explain some of the differences between the tau folds of AD and CBD.[112]

12. Conclusion

In tau proteinopathies, monomeric, highly soluble intracellular tau becomes less soluble and assembles into filaments. During the past decade, multiple lines of evidence have shown that assembled tau behaves like a prion.[1] Tau seeds can template the aggregation of monomeric protein. Over the past 4 years, cryo-EM has provided evidence for the existence of multiple conformers of assembled tau, akin to prion strains. It appears likely that differences between tau isoforms, post-translational modifications and associated, non-proteinaceous molecules, account for this diversity. It is now possible to use cryo-EM to compare the structures of disease seeds and seeded assemblies. This comparison is essential for understanding the seeded aggregation of tau and for putting the prion concept on a more solid structural basis.

References

1. Goedert M, Eisenberg DS, Crowther RA. Propagation of tau aggregates and neurodegeneration. *Annu Rev Neurosci*. 2017;40:189–210.
2. Lee G, Newman ST, Gard DL, Band H, Panchamoorthy G. Tau interacts with src-family non-receptor tyrosine kinases. *J Cell Sci*. 1998;111:3167–3177.
3. Al-Bassam J, Ozer RS, Safer D, Halpain DS, Milligan RA. MAP2 and tau bind longitudinally along the outer ridges of microtubule protofilaments. *J Cell Biol*. 2002; 157:1187–1196.
4. Kellogg EH, Hejab NMA, Poepsel S, Downing KH, DiMaio F, Nogales E. Near-atomic model of microtubule-tau interactions. *Science*. 2018;360:1242–1246.
5. Janning D, Igaev M, Sündermann F, et al. Single-molecule tracking of tau reveals kiss-and-hop interaction with microtubules in living neurons. *Mol Biol Cell*. 2014;25: 3541–3551.
6. Niewidok B, Igaev M, Sündermann F, Janning BF, Bakota L, Brandt R. Presence of a carboxy-terminal pseudo repeat and disease-like pseudohyperphosphorylation critically influence tau's interaction with microtubules in axon-like processes. *Mol Biol Cell*. 2016;27:3537–3549.
7. Black MM, Slaughter T, Moshiach S, Obrocka M, Fischer I. Tau is enriched on dynamic microtubules in the distal region of growing axons. *J Neurosci*. 1996;16:3601–3619.

8. Qiang L, Sun X, Austin TO, et al. Tau does not stabilize axonal microtubules but rather enables them to have long labile domains. *Curr Biol.* 2018;28:2181–2189.
9. Boyka S, Qi X, Chen TH, Surewicz K, Surewicz WT. Liquid-liquid phase separation of tau protein: the crucial role of electrostatic interactions. *J Biol Chem.* 2019; 294:11054–11059.
10. Lin Y, Fichou Y, Zeng Z, Hu NY, Han S. Electrostatically driven complex coacervation and amyloid aggregation of tau are independent processes with overlapping conditions. *ACS Chem Neurosci.* 2020;11:615–627.
11. Wegmann S, Eftekharzadeh B, Tepper K, et al. Tau protein liquid-liquid phase separation can initiate tau aggregation. *EMBO J.* 2018;37, e98049.
12. Goedert M, Spillantini MG, Jakes R, Rutherford D, Crowther RA. Multiple isoforms of human microtubule-associated protein tau: sequences and localization in neurofibrillary tangles of Alzheimer's disease. *Neuron.* 1989;3:519–526.
13. Goedert M, Jakes R. Expression of separate isoforms of human tau protein: correlation with the tau pattern in brain and effects on tubulin polymerization. *EMBO J.* 1990;9:4225–4230.
14. Yoshida H, Goedert M. Molecular cloning and functional characterization of chicken brain tau: isoforms with up to five tandem repeats. *Biochemistry.* 2002;41:15203–15211.
15. Tuerde D, Kimura T, Miyasaka T, et al. Isoform-independent and -dependent phosphorylation of microtubule-associated protein tau in mouse brain during postnatal development. *J Biol Chem.* 2018;293:1781–1793.
16. Brion JP, Passareiro H, Nunez J, Flament-Durand J. Mise en évidence immunologique de la protéine tau au niveau des lésions de dégénérescence neurofibrillaire de la maladie d'Alzheimer. *Arch Biol.* 1985;95:229–235.
17. Goedert M, Wischik CM, Crowther RA, Walker JE, Klug A. Cloning and sequencing of the cDNA encoding a core protein of the paired helical filament of Alzheimer disease: identification as the microtubule-associated protein tau. *Proc Natl Acad Sci USA.* 1988;85:4051–4055.
18. Lee VM-Y, Balin BJ, Otvos L, Trojanowski JQ. A68: a major subunit of paired helical filaments and derivatized forms of normal tau. *Science.* 1991;251:675–678.
19. Wischik CM, Novak M, Thøgersen HC, et al. Isolation of a fragment of tau derived from the core of the paired helical filament of Alzheimer disease. *Proc Natl Acad Sci USA.* 1988;85:4506–4510.
20. Wischik CM, Novak M, Edwards PC, Klug A, Tichelaar W, Klug A. Structural characterization of the core of the paired helical filament of Alzheimer disease. *Proc Natl Acad Sci USA.* 1988;85:4884–4888.
21. Crowther RA. Straight and paired helical filaments in Alzheimer disease have a common structural unit. *Proc Natl Acad Sci USA.* 1991;85:4884–4888.
22. Berriman J, Serpell LC, Oberg KA, Fink AL, Goedert M, Crowther RA. Tau filaments from human brain and from in vitro assembly of recombinant protein show cross-beta structure. *Proc Natl Acad Sci USA.* 2003;100:9034–9038.
23. Haj-Yahya M, Lashuel HA. Protein semisynthesis provides access to tau disease-associated post-translational modifications (PTMs) and paves the way to deciphering the tau PTM code in health and diseased states. *J Am Chem Soc.* 2018;140:6611–6621.
24. Goedert M, Jakes R, Spillantini MG, Hasegawa M, Smith MJ, Crowther RA. Assembly of microtubule-associated protein tau into Alzheimer-like filaments induced by sulphated glycosaminoglycans. *Nature.* 1996;383:550–553.
25. Kampers T, Friedhoff P, Biernat J, Mandelkow EM, Mandelkow E. RNA stimulates aggregation of microtubule-associated protein tau into Alzheimer-like paired helical filaments. *FEBS Lett.* 1996;399:344–349.
26. Pérez M, Valpuesta JM, Medina M, Montejo de Garcini E, Avila J. Polymerization of tau into filaments in the presence of heparin: the minimal sequence required for tau-tau interaction. *J Neurochem.* 1996;67:1183–1190.

27. Wilson DM, Binder LI. Free fatty acids stimulate the polymerization of tau and amyloid beta peptides. *Am J Pathol.* 1997;150:2181–2195.
28. Fichou Y, Lin Y, Rauch JN, et al. Cofactors are essential constituents of stable and seeding-active tau fibrils. *Proc Natl Acad Sci USA.* 2018;115:13234–13239.
29. Fitzpatrick AWP, Falcon B, He S, et al. Cryo-EM structures of tau filaments from Alzheimer's disease. *Nature.* 2017;547:185–190.
30. Arai T, Ikeda K, Akiyama H, et al. Identification of amino-terminally cleaved tau fragments that distinguish progressive supranuclear palsy from corticobasal degeneration. *Ann Neurol.* 2004;55:72–79.
31. Hutton M, Lendon CL, Rizzu P, et al. Association of missense and 5′-splice-site mutations in *tau* with the inherited dementia FTDP-17. *Nature.* 1998;393:702–705.
32. Poorkaj P, Bird TD, Wijsman E, et al. Tau is a candidate gene for chromosome 17 frontotemporal dementia. *Ann Neurol.* 1998;43:815–825.
33. Spillantini MG, Murrell JR, Goedert M, Farlow MR, Klug A, Ghetti B. Mutation in the tau gene in familial multiple system tauopathy with presenile dementia. *Proc Natl Acad Sci USA.* 1998;95:7737–7741.
34. Ghetti B, Oblak AL, Boeve BF, Johnson KA, Dickerson DC, Goedert M. Frontotemporal dementia caused by microtubule-associated protein tau gene (*MAPT*) mutations: a chameleon for neuropathology and neuroimaging. *Neuropathol Appl Neurobiol.* 2015;41:24–46.
35. Stefansson H, Helgason A, Thorleifsson G, et al. A common inversion under selection in Europeans. *Nat Genet.* 2005;37:129–137.
36. Baker M, Litvan I, Houlden H, et al. Association of an extended haplotype in the tau gene with progressive supranuclear palsy. *Hum Mol Genet.* 1999;8:711–715.
37. Conrad C, Andreadis A, Trojanowski JQ, et al. Genetic evidence for the involvement of tau in progressive supranuclear palsy. *Ann Neurol.* 1997;41:277–281.
38. Di Maria E, Tabaton M, Vigo T, et al. Corticobasal degeneration shares a common genetic background with progressive supranuclear palsy. *Ann Neurol.* 2000;47:374–377.
39. Morris HR, Baker M, Yasojima K, et al. Analysis of tau haplotypes in Pick's disease. *Neurology.* 2002;59:443–445.
40. Allen B, Ingram E, Takao M, et al. Abundant tau filaments and nonapoptotic neurodegeneration in mice expressing human P301S tau protein. *J Neurosci.* 2002;22: 9340–9351.
41. Götz J, Bodea LG, Goedert M. Rodent models for Alzheimer disease. *Nat Rev Neurosci.* 2018;19:583–598.
42. Lewis J, McGowan E, Rockwood J, et al. Neurofibrillary tangles, amyotrophy and progressive motor disturbance in mice expressing mutant (P301L) tau protein. *Nat Genet.* 2000;25:402–405.
43. Götz J, Tolnay M, Barmettler R, Chen F, Probst A, Nitsch RM. Oligodendroglial tau filament formation in transgenic mice expressing G272V tau. *Eur J Neurosci.* 2001; 13:2131–2140.
44. Murakami T, Paitel E, Kawarabayashi T, et al. Cortical neuronal and glial pathology in $tgTau^{P301L}$ transgenic mice. *Am J Pathol.* 2006;169:1365–1375.
45. Macdonald JA, Bronner IF, Drynan L, et al. Assembly of transgenic human P301S tau is necessary for neurodegeneration in murine spinal cord. *Acta Neuropathol Commun.* 2019;7:44.
46. Spillantini MG, Crowther RA, Kamphorst W, Heutink P, Van Swieten JC. Tau pathology in two Dutch families with mutations in the microtubule-binding region of tau. *Am J Pathol.* 1998;153:1359–1363.
47. Van Swieten JC, Bronner IF, Azmani A, et al. The deltaK280 mutation in *MAPT* favours exon 10 skipping in vivo. *J Neuropathol Exp Neurol.* 2007;66:17–25.
48. Murrell JR, Spillantini MG, Zolo P, et al. Tau gene mutation G389R causes a tauopathy with abundant Pick body-like inclusions and axonal deposits. *J Neuropathol Exp Neurol.* 1999;58:1207–1226.

49. Kouri N, Carlomagno Y, Baker M, et al. Novel mutation in MAPT exon 13 (p.N410H) causes corticobasal degeneration. *Acta Neuropathol.* 2014;127:271–282.
50. Tacik P, Sanchez-Contreras M, DeTure M, et al. Clinicopathologic heterogeneity in frontotemporal dementia and Parkinsonism linked to chromosome 17 (FTDP-17) due to microtubule-associated protein tau (MAPT) p.P301L mutation, including a patient with globular glial tauopathy. *Neuropathol Appl Neurobiol.* 2017;43:200–214.
51. Bugiani O, Murrell JR, Giaccone G, et al. Frontotemporal dementia and corticobasal degeneration in a family with a P301S mutation in Tau. *J Neuropathol Exp Neurol.* 1999;58:667–677.
52. Erro ME, Zelaya MV, Mendioroz M, et al. Globular glial tauopathy caused by *MAPT* P301T mutation: clinical and neuropathological findings. *J Neurol.* 2019;266: 2396–2405.
53. Goedert M. The ordered assembly of tau is the gain of toxic function that causes human tauopathies. *Alzheimers Dement.* 2016;12:1040–1050.
54. Spina S, Farlow MR, Unverzagt FW, et al. The tauopathy associated with mutation +3 in intron 10 of *Tau*: characterization of the MSTD family. *Brain.* 2008;131:72–89.
55. Spillantini MG, Crother RA, Goedert M. Comparison of the neurofibrillary pathology in Alzheimer's disease and familial presenile dementia with tangles. *Acta Neuropathol.* 1996;92:42–48.
56. Reed LA, Grabowski TJ, Schmidt ML, et al. Autosomal dominant dementia with widespread neurofibrillary tangles. *Ann Neurol.* 1997;42:564–572.
57. Collinge J. Mammalian prions and their wider relevance in neurodegenerative diseases. *Nature.* 2016;529:217–226.
58. Prusiner SB. *Madness and Memory.* Yale University Press; 2014.
59. Braak H, Braak E. Neuropathological stageing of Alzheimer-related changes. *Acta Neuropathol.* 1991;82:239–259.
60. Duyckaerts C, Uchihara T, Seilhean D, He Y, Hauw JJ. Dissociation of Alzheimer type pathology in a disconnected piece of cortex. *Acta Neuropathol.* 1997;93:501–507.
61. Clavaguera F, Bolmont T, Crowther RA, et al. Transmission and spreading of tauopathy in transgenic mouse brain. *Nat Cell Biol.* 2009;11:909–913.
62. de Calignon A, Polydoro M, Súarez-Calvet M, et al. Propagation of tau pathology in a model of early Alzheimer's disease. *Neuron.* 2012;73:685–697.
63. Harris JA, Koyama A, Maeda S, et al. Human P301L-mutant tau expression in mouse entorhinal-hippocampal network causes tau aggregation and presynaptic pathology but no cognitive defects. *PLoS One.* 2012;7, e45881.
64. Liu L, Drouet V, Wu JW, Witter MP, Small SA, Clelland C. Trans-synaptic spread of tau pathology in vivo. *PLoS One.* 2012;7, e31302.
65. Pooler AM, Polydoro M, Maury EA, et al. Amyloid accelerates tau propagation and toxicity in a model of early Alzheimer's disease. *Acta Neuropathol Commun.* 2015;3:14.
66. Ahmed Z, Cooper J, Murray TK, et al. A novel in vivo model of tau propagation with rapid and progressive neurofibrillary tangle pathology: the pattern of spread is determined by connectivity, not proximity. *Acta Neuropathol.* 2014;127: 667–683.
67. Halihan GI, Vargas-Caballero M, West J, Deinhardt K. Tau misfolding efficiently propagates between individual intact hippocampal neurons. *J Neurosci.* 2019;39: 9623–9632.
68. Asai H, Ikezu S, Tsunoda S, et al. Depletion of microglia and inhibition of exosome synthesis halt tau propagation. *Nat Neurosci.* 2015;18:1584–1593.
69. Siman R, Lin YG, Malthankar-Phatak G, Dong Y. A rapid gene delivery-based mouse model for early-stage Alzheimer's disease-type tauopathy. *J Neuropathol Exp Neurol.* 2013;72:1062–1071.
70. Wegmann S, Maury EA, Kirk MJ, et al. Removing endogenous tau does not prevent tau propagation, yet reduces its neurotoxicity. *EMBO J.* 2015;34:3028–3041.

71. Clavaguera F, Hench J, Lavenir I, et al. Peripheral administration of tau aggregates triggers intracerebral tauopathy in transgenic mice. *Acta Neuropathol.* 2014;127:299–301.
72. Frost B, Jacks RL, Diamond MI. Propagation of tau misfolding from the outside to the inside of a cell. *J Biol Chem.* 2009;284:12845–12852.
73. Kfoury N, Holmes BB, Jiang H, Holtzman DM, Diamond MI. Transcellular propagation of tau aggregation by fibrillar species. *J Biol Chem.* 2012;287:19440–19451.
74. Nonaka T, Watanabe ST, Iwatsubo T, Hasegawa M. Seeded aggregation and toxicity of α-synuclein and tau. *J Biol Chem.* 2010;285:34885–34898.
75. Santa-Maria I, Varghese M, Ksiezak-Reding H, Dzhun A, Wang J, Pasinetti GM. Paired helical filaments from Alzheimer disease brain induce intracellular accumulation of tau protein in aggresomes. *J Biol Chem.* 2012;287:20522–20533.
76. Wu JW, Herman M, Liu L, et al. Small misfolded tau species are internalized via bulk endocytosis and anterogradely and retrogradely transported in neurons. *J Biol Chem.* 2013;288:1856–1870.
77. Holmes BB, DeVos SL, Kfoury N, et al. Heparan sulphate proteoglycans mediate internalization and propagation of specific proteopathic seeds. *Proc Natl Acad Sci U S A.* 2013;110:E3138–E3147.
78. Falcon B, Cavallini A, Angers R, et al. Conformation determines the seeding potencies of native and recombinant tau aggregates. *J Biol Chem.* 2015;290:1049–1065.
79. Falcon B, Noad J, McMahon H, Randow F, Goedert M. Galectin 8-mediated selective autophagy protects against seeded tau aggregation. *J Biol Chem.* 2018;293:2438–2451.
80. Rauch JN, Luna G, Guzman E, et al. LRP1 is a master regulator of tau uptake and spread. *Nature.* 2020;580:381–385.
81. Recasens A, Dehay B, Bové J, et al. Lewy body extracts from Parkinson disease brains trigger α-synuclein pathology and neurodegeneration in mice and monkeys. *Ann Neurol.* 2014;75:351–362.
82. Luk KC, Kehm VM, Zhang B, O'Brien P, Trojanowski JQ. Intracerebral inoculation of pathological α-synuclein initiates a raoidly progressive neurodegenerative α-synucleinopathy in mice. *J Exp Med.* 2012;209:975–986.
83. Meyer-Luehmann M, Coomaraswamy J, Bolmont T, et al. Exogenous induction of cerebral beta-amyloidogenesis is governed by agent and host. *Science.* 2006;313:1781–1784.
84. Prusiner SB. Biology and genetics of prions causing neurodegeneration. *Annu Rev Genet.* 2013;47:601–623.
85. Stöhr J, Watts JC, Mensinger ZL, et al. Purified and synthetic Alzheimer's amyloid beta (Aβ) prions. *Proc Natl Acad Sci USA.* 2012;109:11025–11030.
86. Zhang B, Une Y, Fu X, et al. Fecal transmission of AA amyloidosis in the cheetah contributes to high incidence of disease. *Proc Natl Acad Sci USA.* 2008;105:7263–7268.
87. Morozova OA, March ZM, Robinson AS, Colby DW. Conformational features of tau fibrils from Alzheimer's disease brain are faithfully propagated by unmodified recombinant protein. *Biochemistry.* 2013;52:6960–6967.
88. Jackson SJ, Kerridge C, Cooper J, et al. Short fibrils constitute the major species of seed-competent tau in the brains of mice transgenic for human mutant P301S tau. *J Neurosci.* 2016;36:762–772.
89. Cliffe R, Sang JC, Kundel F, Finley D, Klenerman D, Ye Y. Filamentous aggregates are fragmented by the proteasome holoenzyme. *Cell Rep.* 2019;26:2140–2149.
90. Kundel F, Hong L, Falcon B, et al. Measurement of tau filament fragmentation provides insights into prion-like spreading. *ACS Chem Neurosci.* 2018;9:1276–1282.
91. Nachman E, Wentink AS, Madiona K, et al. Disassembly of tau fibrils by the human Hsp70 disaggregation machinery generates small seeding-competent species. *J Biol Chem.* 2020;29:9676–9690. https://doi.org/10.1074/jbc.RA120.013478.
92. Polanco JC, Scicluna J, Hill AF, Götz J. Extracellular vesicles isolated from the brains of rTg4510 mice seed tau protein aggregation in a threshold-dependent manner. *J Biol Chem.* 2016;291:12445–12466.

93. Saman S, Kim W, Raya M, et al. Exosome-associated tau is secreted in tauopathy models and is selectively phosphorylated in cerebrospinal fluid in early Alzheimer disease. *J Biol Chem*. 2012;287:3842–3849.
94. Fontaine SH, Zheng D, Sabbagh JJ, et al. DnaJ/Hsc70 chaperone complexes control the extracellular release of neurodegenerative-associated proteins. *EMBO J*. 2016;35: 1537–1549.
95. Katsinelos T, Zeitler M, Dimou E, et al. Unconventional secretion mediates the trans-cellular spreading of tau. *Cell Rep*. 2018;23:2039–2055.
96. Usenovic M, Niroomand S, Drolet RE, et al. Internalized tau oligomers cause neurodegeneration by inducing accumulation of pathogenic tau in human neurons derived from induced pluripotent stem cells. *J Neurosci*. 2015;35:14234–14250.
97. Verheyen A, Diels A, Dijkmans J, et al. Using human iPSC-derived neurons to model tau aggregation. *PLoS One*. 2015;10:30146127.
98. Clavaguera F, Akatsu H, Fraser G, et al. Brain homogenates from human tauopathies induce tau inclusions in mouse brain. *Proc Natl Acad Sci USA*. 2013;110:9535–9541.
99. Sanders DW, Kaufman SK, DeVos SL, et al. Distinct tau protein strains propagate in cells and mice and define different tauopathies. *Neuron*. 2014;82:1271–1288.
100. Goedert M, Spillantini MG, Cairns NJ, Crowther RA. Tau proteins of Alzheimer paired helical filaments: abnormal phosphorylation of all six brain isoforms. *Neuron*. 1992;8:159–168.
101. Falcon B, Zhang W, Schweighauser M, et al. Tau filaments from multiple cases of sporadic and inherited Alzheimer's disease adopt a common fold. *Acta Neuropathol*. 2018;136:699–708.
102. Crutch SJ, Lehmann M, Schott JM, Rabinovici GD, Rossor MN, Fox NC. Posterior cortical atrophy. *Lancet Neurol*. 2012;11:170–178.
103. Schmidt ML, Zhukareva V, Newell KL, Lee VM-Y, Trojanowski JQ. Tau isoform profile and phosphorylation state in in dementia pugilistica recapitulate Alzheimer's disease. *Acta Neuropathol*. 2001;101:518–524.
104. Cherry JD, Kim SH, Stein TD, et al. Evolution of neuronal and glial tau isoforms in chronic traumatic encephalopathy. *Brain Pathol*. 2020. https://doi.org/10.1111/bpa.12867.
105. Falcon B, Zivanov J, Zhang W, et al. Novel tau filament fold in chronic traumatic encephalopathy encloses hydrophobic molecules. *Nature*. 2019;568:420–423.
106. Delacourte A, Robitaille Y, Sergeant N, et al. Specific pathological tau protein variants characterize Pick's disease. *J Neuropathol Exp Neurol*. 1996;55:159–168.
107. Falcon B, Zhang W, Murzin AG, et al. Structures of filaments from Pick's disease reveal a novel tau protein fold. *Nature*. 2018;561:137–140.
108. Probst A, Tolnay M, Langui D, Goedert M, Spillantini MG. Pick's disease: hyperphosphorylated tau protein segregates to the somatoaxonal compartment. *Acta Neuropathol*. 1996;92:588–596.
109. Sergeant N, Wattez A, Delacourte A. Neurofibrillary degeneration in progressive supranuclear palsy and corticobasal degeneration: tau pathologies with exclusively "exon 10" isoforms. *J Neurochem*. 1999;72:1243–1249.
110. Feany MB, Dickson DW. Widespread cytoskeletal pathology characterizes corticobasal degeneration. *Am J Pathol*. 1995;146:1388–1396.
111. Zhang W, Tarutani A, Newell KL, et al. Novel tau filament fold in corticobasal degeneration. *Nature*. 2020;580:283–287.
112. Arakhamia T, Lee CE, Carlomagno Y, et al. Posttranslational modifications mediate the structural diversity of tauopathy strains. *Cell*. 2020;180:633–644.

CHAPTER ELEVEN

The prion-like nature of amyotrophic lateral sclerosis

L. McAlary[a,b], J.J. Yerbury[a,b], and N.R. Cashman[c,*]
[a]Illawarra Health and Medical Research Institute, Wollongong, NSW, Australia
[b]Molecular Horizons and School of Chemistry and Molecular Bioscience, University of Wollongong, Wollongong, NSW, Australia
[c]Djavad Mowafaghian Centre for Brain Health, University of British Columbia, Vancouver, BC, Canada
*Corresponding author: e-mail address: neil.cashman@vch.ca

Contents

Abstract

The misfolding, aggregation, and deposition of specific proteins is the key hallmark of most progressive neurodegenerative disorders such as Alzheimer's disease, Parkinson's disease, and amyotrophic lateral sclerosis (ALS). ALS is characterized by the rapid and progressive degenerations of motor neurons in the spinal cord and motor cortex, resulting in paralysis of those who suffer from it. Pathologically, there are three major aggregating proteins associated with ALS, including TAR DNA-binding protein of 43 kDa (TDP-43), superoxide dismutase-1 (SOD1), and fused in sarcoma (FUS). While there are ALS-associated mutations found in each of these proteins, the most prevalent aggregation pathology is that of wild-type TDP-43 (97% of cases), with the remaining

Progress in Molecular Biology and Translational Science, Volume 175
ISSN 1877-1173
https://doi.org/10.1016/bs.pmbts.2020.07.002

split between mutant forms of SOD1 (~2%) and FUS (~1%). Considering the progressive nature of ALS and its association with the aggregation of specific proteins, a growing notion is that the spread of pathology and symptoms can be explained by a prion-like mechanism. Prion diseases are a group of highly infectious neurodegenerative disorders caused by the misfolding, aggregation, and spread of a transmissible conformer of prion protein (PrP). Pathogenic PrP is capable of converting healthy PrP into a toxic form through template-directed misfolding. Application of this finding to other neurodegenerative disorders, and in particular ALS, has revolutionized our understanding of cause and progression of these disorders. In this chapter, we first provide a background on ALS pathology and genetic origin. We then detail and discuss the evidence supporting a prion-like propagation of protein misfolding and aggregation in ALS with a particular focus on SOD1 and TDP-43 as these are the most well-established models in the field

1. Neurodegenerative diseases are associated with protein misfolding and aggregation

Neurodegenerative diseases are characterized by the progressive loss of neuronal cells in the central nervous system (CNS) and typically occur in older individuals.[1] The major neurodegenerative disorders are Alzheimer's disease (AD), Parkinson's disease (PD), multiple system atrophy (MSA), dementia with Lewy bodies, spinocerebellar ataxia, frontotemporal lobar degeneration (FTLD), amyotrophic lateral sclerosis (ALS), and Huntington's disease.[2] In concordance with neuronal loss, symptoms can manifest as cognitive, behavioral, and/or motor deficits as the disease progresses.[1] The loss of neurons appears to be specific for each disease suggesting different types of neuronal cells are selectively vulnerable in each disease,[3] although common cellular pathways are found to be affected in most neurodegenerative diseases.[4–6]

Although there are differences in the symptoms, CNS regions, and cell types affected, the major hallmark pathology of neurodegenerative disease is the deposition of misfolded and aggregated proteins into insoluble structures in and around affected cells.[7] Indeed, absolute confirmation of an individual suffering from a specific neurodegenerative disease is typically only made post-mortem once the characteristic pathology of that disease can be confirmed.[8] Similar to the cell type specificity, each neurodegenerative disease is associated with few pathological protein markers. In AD, the primary pathological markers are extracellular plaques composed of amyloid-β[9] and neurofibrillary tangles composed of tau-protein.[10] For PD and MSA the primary pathological marker is α-synuclein.[11–13] FTLD is characterized by either tau-protein or transactive response DNA-binding of 43 kDa (TDP-43) pathological inclusions.[14] Finally, ALS is associated with either superoxide dismutase-1 (SOD1) positive inclusions,[15] TDP-43 positive inclusions,[14] or fused in sarcoma (FUS—also known as translocated in liposarcoma) positive inclusions.[16,17]

Considering that protein inclusions are the major hallmark and signifier of neurodegenerative disease, a great deal of research has been focused on understanding why and how these inclusions form.[7] Perhaps the most enlightening hint toward explaining the pathobiology of neurodegeneration comes from the understanding of the clinically and pathologically similar prion diseases.

2. Prion diseases are clinically and pathologically similar to other neurodegenerative diseases

Prion diseases are a set of incurable progressive neurodegenerative disorders which includes Creutzfeldt-Jakob disease (CJD), fatal familial insomnia, kuru, bovine spongiform encephalopathy, and others.[18] CJD is the most common prion disease, and similar to other age-related neurodegenerative diseases it can occur sporadically or within genetic lineages, and it also primarily affects individuals later in life.[19,20]

Clinically, prion diseases are associated with rapid and significant decline in neurological functions and can include deficits in motor function (spasms, tremors, muscle seizing), deficits in memory and cognitive function (dementia, hallucinations, confusion), as well as seizures, although the latter symptom is less common.[21] Similar to the above mentioned neurodegenerative diseases, prion diseases are confirmed upon histopathological examination, however, a probable diagnosis can be made if the patient presents with a rapidly progressing dementia, a short disease course (less than 2 years) and two other symptoms such as myoclonia, movement deficits, impaired vision, and extra- or pyramidal signs.[21]

Pathologically, prion diseases are associated with the accumulation of proteins into inclusions within the CNS, significant degeneration of neuronal tissues, and a characteristic spongiform appearance of affected tissues.[22] Considering these common signs, there can be a great deal of clinical and pathological heterogeneity in sufferers of prion diseases. For example, in CJD, genetic variants of prion protein (PrP) can result in dramatically different presentation and disease courses in patients.[23–25] Regardless of heterogeneity, the key pathological marker of prion disease is the detection of pathologically misfolded PrP in the brain at autopsy. It is suggested that although PrP is associated with all prion diseases, the type or subtype of disease that develops is dependent on the "strain" of misfolded prion protein that is pathologically propagating in that case.[26] Indeed, the key finding underpinning our understanding of prion disease is that a protein may encipher its pathological properties onto a benign protein and generate an ever increasing amount of toxic misfolded protein.[27] Application of this theory to neurodegeneration in general has transformed our understanding and led to the development of new therapeutic targets and strategies aimed at alleviating the toll these diseases take on human well-being.

3. Prions and prion-like proteins—An important distinction

The hypothesis that a protein could be self-replicating and infectious in the absence of nucleic acids was first proposed in the 1960s,[28,29] however, it was not until 1982 that the culprit protein was isolated by Prusiner.[27] Prusiner named the newly isolated protein as a "prion" for *pro*tein *in*fectious particle and the protein itself became known as prion protein (PrP) after the protein was discovered to be normally physiologically expressed. Subsequent research that established that the normal alpha helical form of PrP (named PrP^{C} where *C* is for "cellular") could be converted to the beta-sheet rich infectious disease-associated form through template-directed misfolding interactions with exogenous misfolded PrP (named PrP^{Sc} where *Sc* is for "scrapie").[30]

In regards to human age-associated neurodegenerative disorders, the term prion is not typically applied due to a lack of evidence of organism-to-organism infectivity, which prompts usage of the term "prion-like." Although there is growing debate surrounding this notion as some evidence suggests that there is possible infectivity from the pathological forms of the proteins associated with these diseases.[31,32] At the very least, there is strong evidence suggesting that the human brain-derived pathologically aggregated proteins in AD, PD, MSA, and ALS/FTLD are capable of seeding protein misfolding in recipient model animals and cells.[33–38]

The term "prion-like" can also be applied to a set of proteins whose function(s) can be dependent on self-polymerization and are considered as a form of non-Mendelian inheritance.[39] This idea was proposed from the basis of research in yeast (primarily *Saccharomyces cerevisiae*) in which a trait termed [PSI +] was observed to be inherited in a manner unrelated to mutation.[40] Following work revealed that the inheritance pattern was dependent on the conversion of a protein called Sup35 to an amyloidogenic state which could propagate between cells similar to PrP^{Sc}.[41] Some proteins are termed prion-like on the basis that they share similar amino acid sequence properties with Sup35 protein (enriched for asparagine, glutamine, glycine, and tyrosine), and this can lead to some confusion as these proteins are increasingly being associated with neurodegenerative disease and are suggested to also be prion-like in a similar manner to PrP^{Sc}. This is evidenced by proteins such as TDP-43, which is associated with the majority of cases of ALS and about half of FTLD cases,[42] having a prion-like domain (PrLD)[43] but also being capable of pathological propagation in cells and animals.[33,44] PrLDs are domains with a low-complexity of amino acids that results in them typically being intrinsically disordered.[43]

Considering that "prion-like" as a term is often used loosely to describe both functional and pathologic conformational protein propagation, we will distinguish the PrP^{SC}-like properties and actions from the Sup35-like properties and actions by using the terms prion-like and PrLD respectively. This distinction is necessary when discussing ALS/FTLD, diseases which have strong evidence supporting pathological prion-like propagation of protein misfolding, and also genetic alterations with proteins containing PrLDs.

4. Amyotrophic lateral sclerosis—Spread and progression of clinical symptoms

Amyotrophic lateral sclerosis (ALS), also known as motor neuron disease or Lou Gehrig's disease, is an ultimately fatal neuromuscular disorder in which motor neurons preferentially and progressively degenerate, leading to paralysis. Diagnosis of a patient with ALS is devastating to emotional well-being due to the nature of the disease leaving the patient fully aware of the decline in their autonomy, which in combination with the lack of any effective treatment often results in an aggressive disease course that imparts significant hardship on patients.

Clinically, the first manifestation of ALS symptoms can be seen in either upper or lower motor systems,[45,46] although most patients present with the key symptoms of lower motor neuron degeneration which includes weakness in limbs, cramping, and muscle fasciculation.[47] In some cases symptoms for degeneration of upper motor neurons are observed initially, which can include limb weakness and spasticity.[48,49] Following the initial presentation of symptoms, weakness will progressively increase and spread to other limbs until ultimately the patient can no longer swallow or respirate due to degeneration of the motor neurons that innervate the muscles that control respiration and the swallow reflex, where this process typically takes 1 − 2 years after diagnosis.[46]

A curious observation of the progression of the clinical symptoms of ALS is that they appear to spread in a manner consistent with the interconnection of the motor system.[50,51] Indeed, the spatiotemporal spread of ALS symptoms is found to be primarily contiguous, however, there are some cases where the spread has been suggested to be discontiguous.[52,53] The site of onset and subsequent speed of symptom spread can even predict survival.[54] Importantly, symptoms at the site of initial presentation gradually worsen as the disease spreads, indicating that whatever underlies disease onset and pathology is capable of spreading. This observation, in conjunction with an increasing understanding of the biophysical properties of the key proteins implicated in ALS pathogenesis supports the notion that a pathological prion-like mechanism may play a role in ALS.

5. Amyotrophic lateral sclerosis—Pathological manifestation

ALS shares many pathological features with other neurodegenerative diseases, such as the degeneration of neuronal cells within the CNS and the deposition of proteins into inclusion bodies, however, the manifestation of pathology can be both morphologically and biochemically heterogeneous. The ALS-associated inclusion types includes spheroids, hyaline inclusions, skein-like inclusions, basophilic inclusions, and bunina bodies (reviewed in Ref. 55). This difference in inclusion morphology suggests that each type of inclusion may be constituted of a different sub-proteome or be created by a specific biochemical pathway, however, this remains to be elucidated.

Although the proteins found to be colocalized within ALS inclusions are found to be diverse,[56] the proteins that are the hallmark of ALS are TDP-43,[14] SOD1,[15] and FUS.[16,17] TDP-43 positive inclusions are found in ~97% of all ALS cases, SOD1 is found in ~2% of cases, and FUS is found in ~1% of cases.[42] TDP-43 and FUS inclusions are also present in patients who have suffered from frontotemporal lobar degeneration (FTLD).[14,42,57] FTLD is thought to exist on a disease spectrum with ALS owing to mutations in several genes, and similar TDP-43 and FUS pathology, being associated with both diseases.[42]

TDP-43 in inclusions is also found to be hyperphosphorylated in the majority of ALS cases,[58] and some evidence suggests that in cases of C9ORF72-associated ALS that the deposition of TDP-43 is promoted by dipeptide repeat (DPR) polypeptides translated from the hexanucleotide repeat G_4C_2 thought to be responsible for C9ORF72-ALS.[59–62] DPRs are capable of forming intracellular inclusions as well that often have a "star-like" morphology, where these can form in the hippocampus and cerebellum.[63]

Other major proteins found in both sALS and fALS inclusions include the signaling protein ubiquitin[64,65] and the ubiquitin binding protein p62[66] which is also known as sequestosome 1 (SQSTM1) and contains ALS-associated mutations.[67,68] Importantly, it appears as though the inclusion type (skein, lewy body-like, basophilic, bunina bodies) dictates the proteins contained within. For example, p62 and ubiquitin are not found to be colocalized to bunina bodies.[66]

Inclusions in ALS are predominantly found in the cytoplasm of dead or dying motor neurons and in the cytoplasm of surrounding glial cells,[55] however, other cell types and regions within the CNS can also be affected by the formation of pathological inclusions although there seems to be some CNS region specificity.[69]

In regards to the regional burden of ALS pathology, application of Braak staging[70] to ALS has suggested that the burden of hyperphosphorylated TDP-43 (pTDP-43) in the CNS increases and spreads in a spatiotemporal manner consistent with axonal spread.[69] Brettschneider and colleagues have suggested that pTDP-43 pathology can be segregated into four stages where the 1st stage is characterized by neurons containing pTDP-43 inclusions being detected in the agranular motor neocortex, the spinal cord ventral horn, and in cranial nerves responsible for tongue movement (XII), head/shoulder movement (XI), vagus (X), facial expression and mastication (VII), and the trigeminal nerve (V). The second stage then includes the reticular formation and precerebellar nuclei, the third stage includes the prefrontal neocortex and basal ganglia, and, finally, the fourth stage is characterized by lesions in the temporal lobe and hippocampal formation.[69]

That ALS is characterized by the deposition of protein inclusions throughout the CNS, and that these inclusions (in particular pTDP-43) appear to spread through axonally connected CNS regions supports the notion that a pathological prion-like mechanism may operate in ALS, although it remains to be fully elucidated as to whether the inclusions are a byproduct of some other disease causing agent or if they are the toxic species themselves.

6. Amyotrophic lateral sclerosis—Diversity of associated genes

Similar to the diverse morphological and biochemical properties of the pathological inclusions in ALS, there is a large and growing, diverse set of genes found to be ALS-associated.[71] Stratification of these ALS-associated genes yields three major processes that may play a role in disease pathogenesis including RNA metabolism, protein quality control, and cytoskeletal dynamics/transport (Table 1). These processes do not occur independently from one another, but rather represent a global cellular maintenance of protein homeostasis (proteostasis) via ensuring that proteins are expressed at the right concentration, present at the right location, and properly folded.

To date, the majority of the proteins involved in RNA metabolism that are associated with ALS contain PrLDs (Table 1), where ALS-associated mutations appear to be concentrated in these domains.[92] That mutations in these PrLDs are associated with ALS pathogenesis suggests that alterations to the PrLD physicochemical properties, and perhaps function, may play a role in disease. Our understanding of the function of PrLDs has significantly grown in recent years, where research has indicated that these domains control the dynamic functional polymerization of proteins into higher order amyloid-like structures[93] and can also govern the ability to form membraneless organelles via liquid-liquid phase transition (reviewed in Ref. 94). Correct formation and dissolution of some biomolecular condensates appears to be integral to not only maintaining proper RNA metabolism, but for the prevention of the formation of pathological aggregates.[82,95–102]

Genes responsible for protein quality control are also found to harbor ALS-associated mutations (Table 1), suggesting that a failure or impairment of the processes that prevent protein aggregation play a central role in ALS. Indeed, a growing notion in disorders associated with protein aggregation is that specific proteins are supersaturated, which means they are expressed at levels greater than their expected solubility.[103] This is also suggested to be the case in ALS, where the primary aggregating proteins SOD1, TDP-43, and FUS are supersaturated in motor neurons.[56] If the cellular pathways that prevent protein aggregation from occurring are impaired via mutation, then the subsequent aggregation of the most supersaturated proteins is expected to occur.

Table 1 Genes associated with amyotrophic lateral sclerosis and their functions.

Gene	Protein	Functions	References
SOD1	Cu/Zn superoxide dismutase (SOD1)	• Free radical scavenger	72
TARDBP	TAR DNA-binding protein (TDP-43)	• RNA-binding protein with PrLD • Repression of cryptic exons	73,74
FUS	Fused in sarcoma (FUS)	• RNA-binding protein with PrLD • Transcriptional regulator	16,17
C9ORF72	C9ORF72	• Possible GDP/GTP exchange factor for Rab-GTPases • Potential roles in autophagy and vesicle transport	75,76
OPTN	Optineurin	• Regulation of Golgi complex • Membrane trafficking and exocytosis	77
DCTN1	Dynactin	• Binds dynein motor complexes • Mitotic pole focusing	78
VCP	Valosin-containing protein (VCP)	• AAA ATPase family member • Protein degradation via the ubiquitin proteasome	79
HNRNPA1	Heterogeneous nuclear ribonucleoprotein A1 (HNRNPA1)	• RNA-binding protein with PrLD • Packaging of pre-mRNA into hnRNP particles	80,81
TIA1	TIA1 cytotoxic granule associated RNA-binding protein (TIA1)	• RNA-binding protein with PrLD • Stress granule marker • Regulates alternative splicing	82
MATR3	Matrin 3	• RNA-binding protein • Possible role in internal fibrogranular network	83

Continued

Table 1 Genes associated with amyotrophic lateral sclerosis and their functions.—cont'd

Gene	Protein	Functions	References
TAF15	TATA-box binding protein associated factor 15 (TAF15)	• DNA/RNA-binding protein with PrLD • Transcription initiation	84
ANXA11		• Phospholipid-binding protein with PrLD • Tethers RNA-granules to lysosomes	85
DNAJC7	Heat shock protein 40 (HSP40)	• ATP-dependent binding of unfolded proteins • Regulate complex formation with HSP70	86
SQSTM1	Sequestosome-1 (also known as p62)	• Autophagy receptor that facilitates transfer of ubiquitinated cargo to the autophagosome	67,68
UBQLN2	Ubiquilin 2 (UBQLN2)	• Links the ubiquitination machinery to the proteasome • Possible role in removing ubiquitinated proteins from membraneless organelles	87
ANG	Angiogenin (ANG)	• Stimulates angiogenesis • Ribonuclease function	88
PFN1	Profilin 1 (PFN1)	• Actin monomer-binding protein • Possible signaling protein	89
TUBA4A	A-tubulin 4A (TUBA4A)	• Major component of microtubules	90
TBK1	Tank-binding kinase 1 (TBK1)	• Serine/threonine protein kinase • Central node for several signaling pathways	91

7. Prion protein and ALS-associated proteins are capable of forming amyloidogenic assemblies

Amyloid fibrils composed of disease-associated proteins are intimately linked with neurodegenerative disease.[7,104,105] Amyloid fibrils are characterized by their cross-β spine structure that typically spans 7–13 nm and their ability to bind certain dyes such as thioflavin-T and Congo red, although there are cases where thioflavin-T binding is negative but fibrils can be observed via electron microscopy.[106] A great deal of effort has been applied to understand the processes that underpin the formation of these highly stable multimeric assemblies and key insights into potential pathological mechanisms have been made from the study of purified proteins *in vitro*.[106–109] Essentially, a protein will misfold and form disordered and dynamic oligomeric species that will eventually form the more ordered and rigid fibrillar species and it is at this point the reaction transitions from one of oligomer formation, to one of template-directed fibril growth.[107] The longer fibrils become, the more likely they are to fragment which provides more ends from which the template-directed growth can occur, thus increasing the rate at which amyloidogenic proteins will fibrillize.[110] This kinetic process is thought to occur within cells where the resultant amyloid fibrils are capable of transferring to nearby cells to seed aggregation, suggesting that the amyloid state may explain the progressive nature of most neurodegenerative diseases.

The process of amyloid formation is highly similar to the process of prion conversion. The key biophysical characteristic of prion diseases is the conversion of the normally alpha helical PrP^{C} to the pathological beta-sheet enriched PrP^{Sc}.[30] PrP^{Sc} is capable of catalyzing the conversion of PrP^{C} to PrP^{Sc} in a template-directed manner much similar to amyloid, and PrP can even form amyloid-like assemblies that bind thioflavin-T,[111,112] suggesting that perhaps that the amyloid state of PrP is responsible for prion diseases.

In the case of ALS, the major aggregating proteins (SOD1, TDP-43, and FUS) are all capable of forming amyloid-like fibrils *in vitro*,[93,113,114] suggesting a capacity for these proteins to act in a pathological prion-like manner in disease. Histopathological examination of fALS and sALS patient tissue has not fully elucidated the structure that SOD1, TDP-43, or FUS adopt when they are observed in a pathological state. TDP-43 has been reported to adopt a filamentous amyloid-like structure that can bind thioflavin-S in some cases.[65,115,116] To our knowledge, there is no evidence for FUS forming amyloid fibrils in ALS patient tissue, however, evidence from FUS-associated FTLD has shown that FUS-reactive antibodies label filamentous structures within inclusions,[117] although further confirmation is required as well as a more exhaustive examination in ALS patient tissue. Similar to FUS, the underlying structure of SOD1 inclusions in patients is not fully understood, however, they have been described as amorphous and granular[118] which in conjunction with negative thioflavin-S and Congo red binding suggests that they are not classical amyloid.[119] The lack of conclusive data on whether SOD1 forms amyloid in ALS would suggest perhaps that it could not act in a pathological prion-like manner, however, SOD1 has the greatest collection of evidence to support a pathological prion-like mechanism in ALS.

8. Cu/Zn superoxide dismutase has pathological prion-like properties *in vitro* and *in vivo*

SOD1 was the first gene found to carry ALS-associated mutations[72] and as such is the most well studied where now there are over 160 known mutations associated with ALS.[120] The primary cellular role of SOD1 is to catalyze the dismutation of superoxide anion to less harmful hydrogen peroxide or molecular oxygen.[121] The enzymatic activity of SOD1 is dependent on its ability to fold into its native state, a process which requires zinc-binding, copper-binding, the formation of an intramolecular disulfide, and homo-dimerization (reviewed in Ref. 122). Mutations are found throughout the SOD1 amino acid sequence and are thought to decrease the folding stability and perturb metal binding, disulfide formation, and/or dimerization,[123–125] leading to misfolding and subsequent aggregation.[126,127] Importantly, SOD1 misfolding is thought to lead to a gain of toxic function rather than a loss of function.[128–130]

Studies on purified SOD1 have revealed much about the events necessary to initiate and/or modify its aggregation. This has included studies investigating the role that metal binding and disulfide reduction may play in SOD1 aggregation,[113,126,131,132] the role that single residues might have on SOD1 aggregation,[133–136] the role of post-translational modifications in aggregation,[137–139] and the role of specific sequence segments on aggregation.[140,141] Collectively, these data strongly suggest that SOD1 aggregation is promoted by structural destabilization and/or aberrant physiological post-translational modification (acetylation, oxidation, deamidation), which provides a plausible explanation for the ability of such a stable protein to misfold and aggregate. In regards to the prion-like nature of SOD1, a key finding resulting from the usage of purified SOD1 was that spinal cord homogenate from mice carrying the transgene for human SOD1-G93A was capable of seeding the aggregation of recombinant SOD1 *in vitro* under denaturing conditions,[142] suggesting that *in vivo* aggregated SOD1 could propagate via a pathological prion-like template-directed interaction with similarly destabilized forms of SOD1 *in vivo*.

Although highly useful in understanding SOD1 aggregation in a controlled environment, *in vitro* assays lack the molecular complexity and physiological processes that cells provide. Therefore, in cell experimentation of SOD1 misfolding and aggregation has provided more physiologically relevant system with which to assess SOD1 aggregation and prion-like propagation. A commonly used model used to study SOD1 in cells is to transiently overexpress a fluorescently tagged construct of SOD1,[126,136,143–147] which has been reported to not significantly affect SOD1 folding and function.[144] Similarly to purified recombinant SOD1, this fluorescent fusion system has been exploited to rapidly examine the effect of mutation and modification to SOD1 aggregation.[126,136,143,145] In cells, overexpressed recombinant wild-type SOD1 (SOD1-WT) rarely forms inclusions whereas some ALS-associated mutants of SOD1 readily form inclusions.[126,136,143,145]

Münch and colleagues[148] exploited such a fluorescent fusion system to first establish the capacity of mutant SOD1 to propagate in a pathologically prion-like manner in immortalized mouse neuronal-like cells (Neuro-2a). Purified recombinant SOD1-H46R was induced to aggregate, fluorescently labeled, and subsequently used to inoculate cells stably expressing SOD1-WT-GFP or SOD1-H46R-GFP. Addition of aggregates to cells expressing SOD1-H46R-GFP resulted in the colocalization of GFP signal with the internalized recombinant SOD1 aggregates, however, this was not seen in cells expressing SOD1-WT-GFP. Furthermore, the aggregation phenotype in the SOD1-H46R-GFP cells was heritable and proceeded to be observed across cell passages beyond detection of the initial aggregate seeds, suggesting that propagation was now occurring through the endogenously expressed SOD1-H46R-GFP fusion protein. They further established mechanisms by which the exogenous SOD1 aggregates were being internalized finding that the key process was via macropinocytosis.[148–151] That colocalization of SOD1-WT-GFP with internalized SOD1-H46R aggregates was not observed in this work, suggested that SOD1-WT may not be a competent substrate for prion-like conversion in ALS. Indeed, SOD1-WT expression in cells rarely results in observable inclusion formation,[126,136,143,145] however, the sensitivity of fluorescent tags may be too low to detect SOD1-WT misfolding or aggregation.

Although there is a prominent focus on mutant SOD1 in ALS research, there exists strong evidence from immunohistological experiments that suggests that SOD1-WT becomes misfolded and potentially aggregated in cases of sporadic ALS.[152–157] In addition, animal studies have shown that cross-breeding of mice expressing human SOD1-WT with mouse lines expressing mutant SOD1 can exacerbate disease,[15,158,159] suggesting that SOD1-WT has cytotoxic potential. Interestingly, the toxicity of astrocytes, or their conditioned media, derived from human sALS patients, to cultured motor neurons has been found to be alleviated by viral-mediated delivery of a SOD1 targeting siRNA to the sALS astrocytes.[160] This suggests that astrocytes can excrete a toxic factor that is dependent on the presence of SOD1, or that SOD1 itself could be the toxic agent being secreted. What might trigger initial SOD1 misfolding is not fully clear yet, although evidence has suggested that the mislocalization of TDP-43 or FUS can promote SOD1 misfolding,[154] or that perhaps an environmental factor such as non-protein amino acids could promote misfolding.[161]

Grad and colleagues[162] previously explored the ability of SOD1-WT to misfold and propagate in a prion-like manner. Transient expression of immortalized mesenchymal and neuronal cell lines with untagged fALS-associated mutants SOD1-G85R and SOD1-G127X expressing constructs was found to result in the prominent misfolding of the endogenous SOD1 in human cell lines. SOD1 misfolding was detected using misfolding specific antibodies that are immunoreactive for epitopes that are buried in native state SOD1.[163,164] Extensive misfolding of SOD1 was observable across multiple passages of the inoculated cells, showing that it was a heritable trait.

Curiously, Grad and colleagues[162] found that murine-derived cell lines transfected to overexpress human SOD1 mutant protein did not show a significant misfolded SOD1 signal even though the detection epitopes are highly similar between humans and mice, a result similarly observed in mutant SOD1 mouse models of ALS.[15,165,166] This species barrier was mapped specifically to the tryptophan at position 32 in the human SOD1 protein sequence (it is serine in mice) where it was determined that a W32S substitution in the overexpressed human mutant SOD1 ablated the misfolding of endogenous SOD1-WT in human cells. Tryptophan-32 was previously implicated as an aggregation and toxicity modifying residue[167] and has more recently become the target of small molecule intervention,[136,168] as well as being implicated in controlling strain formation of prion-like SOD1.[169] Further cell culture work established that spinal cord homogenates from fALS patients with SOD1 mutations could seed the aggregation of transiently expressed SOD-G85R-GFP in immortalized cells[147] and this could be inhibited by small molecule blockade of W32 in SOD1. The finding of W32 as a key residue in the ability of SOD1, both as seed and substrate, to propagate in a prion-like manner is remarkably similar to the species barriers seen for other pathologically propagating prions[170] and suggests that W32 plays a significant role in the formation of a toxic and transmissible SOD1 aggregate conformer or strain.

Strains of SOD1 aggregates have been best studied *in vivo* using mouse models which have been inoculated with either recombinant aggregated SOD1 or homogenates from ALS patients or mice.[150,171–174] A model that has been highly effective for dissecting pathological prion-like propagation of SOD1 *in vivo*, that mirrored the cellular model used by Münch and colleagues,[148] is the sub-disease level expression of human SOD1-G85R tagged on the C-terminus with yellow fluorescent protein (YFP). Using this model Ayers and colleagues[171] have shown that injection of spinal cord homogenate from symptomatic SOD1 ALS mice into specific CNS regions (sciatic nerve, cervical spinal cord, and cerebral ventricles) promotes an ALS-like phenotype in injected mice. This phenotype was accompanied by substantial aggregation of the SOD1-G85R-YFP fusion protein in the spinal cord and motor cortex, and could also be passaged across generations of mice. This model was further used to establish that recombinant aggregated SOD1-WT could be injected to give rise to heterogeneous ALS-like clinical phenotypes with distinct morphologies of SOD1-G85R-YFP inclusion in the CNS,[172] suggesting that certain strains with different pathological prion-like properties were propagating in the animals through interconnected anatomical pathways.[173]

SOD1 strains from ALS model mice have been analyzed using a binary epitope mapping approach where it has been suggested that there are two main strains of SOD1 aggregate, termed strain A and strain B, that can propagate in the models used.[34,166,174,175] Strain A was characterized by residues ~4–39 and ~80–127, corresponding to beta-strands I–III and V–VII respectively, being inaccessible to antibody detection, whereas strain B was characterized by residues ~4–57 being antibody inaccessible.[166] The N-terminus of SOD1 has previously been implicated to form fibrillar cores in the aggregation of recombinant SOD1-WT and ALS-associated mutants,[176] similar to both strain A and strain B burying a significant portion of the N-terminus. Strain A was predominantly found to occur in mice expressing SOD1-WT, SOD1-G85R, or SOD1-G93A, whereas strain A and strain B were found to both occur in mice expressing SOD1-D90A, suggesting an effect of polymorphism on the kinetics of SOD1 aggregation controlling strain selection. Furthermore, assessment of the structural stability of the strains showed that strain A was more stable and was associated with a slower and less aggressive clinical phenotype when compared to strain B.[166] Akin to the work by Ayers and colleagues,[171,172] injection of homogenates from mice carrying strain A or B aggregates showed that these strains were capable of transmitting disease and also their own conformation into recipient mice.[174] Finally, injection of homogenates from ALS patients carrying the G127X mutation into recipient mice was found to induce ALS-like symptoms with aggregates having strain A-like biochemical properties, suggesting perhaps that SOD1 mouse models may generate similar strains to humans suffering from SOD1-associated ALS,[34] however, samples from patients suffering from other SOD1 mutations would have to be examined to determine this. Further understanding of the key amino acid sequence regions within SOD1 that primarily partake in the formation of aggregates *in vivo* would allow for more targeted biophysical studies to determine potential structures of these strains.

Collectively, the evidence that supports the pathological prion-like propagation of mutant SOD1 in SOD1-associated fALS is substantial, and there is growing evidence implicating SOD1-WT in sALS cases that warrants further investigation. Although SOD1 inclusions are thought to be associated with only ~2% of ALS cases, the combined biophysical, biochemical, cell biology, and *in vivo* evidence that proposes a role for SOD1 in sporadic ALS suggests that perhaps SOD1-WT should remain on the radar for ALS researchers in their search for a therapeutic target. Some strategies that may alleviate pathological prion-like spread and perhaps even onset are already being pursued, such as gene silencing,[177,178] small molecule-based stabilization of SOD1,[179,180] and upregulation of cellular chaperones[181] among others.[182] By reducing the level of misfolded SOD1 through either silencing or stabilization, the amount of substrate available for conversion to the pathological prion-like form would be minimal and could delay or prevent disease.

9. TAR DNA-binding protein of 43 kDa has pathological prion-like properties *in vitro* and *in vivo*

The finding that TDP-43 was the primary ubiquitinated protein observed in the majority of ALS patients shifted the research focus from SOD1 onto this DNA/RNA-binding protein. As a DNA/RNA-binding protein, TDP-43 has many roles within the cell related to RNA regulation and processing[183–187] and some of these are essential for development as knockout of TDP-43 is embryonic lethal.[188–190] TDP-43 is primarily localized to the nucleus, however, it can shuttle between the cytoplasm and nucleus to fulfill its biological roles. Structurally, TDP-43 is composed of four domains; an N-terminal domain (NTD), tandem RNA-recognition motifs (RRMs), and a C-terminal low-complexity domain (LCD). The N-terminal domain contains the nuclear localization sequence and is thought to facilitate both TDP-43 homodimer formation[191,192] and functional polymerization.[193,194] The RRMs bind, and are stabilized by, TG/UG rich nucleotide sequences.[195–198] Finally, the LCD is an intrinsically disordered functional PrLD that contains a helix-turn-helix motif that facilitates interactions that allow TDP-43 to take part in the formation of biomolecular condensates[199,200] and interact with other LCD containing RBPs.

ALS and FTLD mutations currently identified within the TDP-43 sequence primarily cluster in the LCD, whereas the NTD currently has a single known ALS-associated mutation in the NLS and RRM1 contains two ALS-associated mutations. Furthermore, a key pathological marker of both ALS and FTLD is the detection of fragmented TDP-43 in both brain and spinal cord homogenates of patients,[14] where the LCD is the major constituent of these fragments by mass, however, the cause of fragmentation and the role it might play in disease etiology is still unclear.[201] The existence of fragmented TDP-43 as a pathological marker or potential causative agent, and the abundance of disease-associated mutations in the LCD, has led to research focusing on the potential mechanisms by which this domain might induce TDP-43-associated toxicity. Indeed, the TDP-43 LCD has been predicted and experimentally shown to contain several sequence segments that are highly amyloidogenic.[114,202–206] Furthermore, the 35 and 25 kDa fragments of TDP-43 are found to be highly aggregation prone and toxic when overexpressed in cultured cells.[44,207–209] This evidence suggests that the LCD plays a major role in the initiation and pathological propagation of TDP-43 in ALS and FTLD.

Furukawa and colleagues[210] were the first to show the pathological prion-like properties of TDP-43 in cultured cells by inoculating them with fibrils generated from recombinant TDP-43. Inoculation of cultured cells transiently overexpressing HA-tagged TDP-43 with exogenous TDP-43 fibrils resulted in the formation of HA-tagged TDP-43 positive cytoplasmic inclusions that were sarkosyl insoluble and polyubiquitinated, which are features of TDP-43 inclusion in patients. This recapitulation of human pathology in cultured cells via addition of exogenous aggregates was strongly suggestive of pathological prion-like properties of TDP-43, however, the use of recombinant protein may bias aggregate morphologies to ones not observed *in vivo*[166,175] and the transient overexpression of TDP-43 potentially results in artifactual aggregation.

Nonaka and colleagues[44] overcame the issue of using recombinant protein aggregates by using the sarkosyl insoluble protein fraction from patients suffering from either sporadic ALS or FTLD as a seed. Inoculation of cells transiently overexpressing HA-tagged TDP-43 resulted in the development of intracellular inclusions that were immunoreactive for ubiquitin and phospho-TDP-43. After recapitulating the human pathology in cultured cells using human-derived inoculum, Nonaka and colleagues[44] established that the inclusions formed in cell culture could be used to propagate inclusion formation in naive cells, providing strong evidence that pathologically aggregated TDP-43 can propagate in the prion-like manner. They further determined that exosomal transport may be one way through which pathological prion-like TDP-43 potentially transmits from cell-to-cell.[44] This work provided key evidence to support that a prion-like transmission of TDP-43 aggregation is a feature of both ALS and FTLD and features of this pathological prion-like of TDP-43 between cells have been further studied using various models and methods.[33,211–214]

As stated previously, the TDP-43 LCD and parts of RRM2 are capable of forming amyloid,[215] and these may give rise to different amyloidogenic assemblies and strains of aggregate. Shimonaka and colleagues[216] generated fibrils from synthetic peptides of regions of RRM2 and the LCD of TDP-43 and inoculated cells with them to induce the aggregation of both overexpressed TDP-43 or TDP-43^{dNLS} (containing a mutated nuclear localization sequence). They observed that TDP-43 fragments corresponding to either amino acids 274–313 and 314–354 were capable of significant induction of TDP-43 aggregation in recipient cells, but that fragments corresponding to 234–273 and 353–392 were not.[216] Furthermore, they determined that transiently expressed TDP-43 lacking the sequence corresponding to the fibrillogenic segments could not be seeded by the corresponding segment, indicating a template-directed interaction between seed and substrate at specific sequence segments.[216] This evidence strongly suggested that different strains of pathological TDP-43 could exist and propagate in a manner dependent on their underlying molecular architecture, an important finding considering the involvement of TDP-43 in both FTLD and ALS, as well as other neurodegenerative diseases.

More recently evidence suggesting that different pathological strains of TDP-43 occur in patients suffering from FTLD has been presented. Laferrière and colleagues[211] established that insoluble TDP-43 extracted from patients, who were diagnosed to be suffering from different TDP-43-associated FTLD subtypes, was differentially toxic to cultured cells dependent on the disease subtype. They further examined the biophysical properties of the insoluble TDP-43 extracted from the patients, finding that disease subtypes gave rise to TDP-43 assemblies with distinct physical properties.[211] The difference observed in the properties of the TDP-43 inoculum, and the corresponding differential toxicity, is strongly suggestive that different strains of TDP-43 are propagating at least in FTLD. Akin to standard procedures in the research of pathological PrP propagation, it will be interesting to determine if the disease and assembly characteristics are transmitted *in vivo* to recipient organisms.

There is already evidence that extracts from FTLD patients are capable of seeding the aggregation of both endogenous murine TDP-43 and transgenic TDP-43 in mice.[33] In this work Porta and colleagues[33] screened for FTLD extracts that could efficiently seed TDP-43 in cultured cells prior to injecting them into the CNS of both WT and transgenic mice. They found that the extracts capable of efficient seeding were proteinase-K digestion resistant and gave rise to a distinct banding pattern when examined via immunoblot, again suggestive that the morphology and underlying structure of the seed dictates the seeding potency. The transgenic animals used in this study expressed either human TDP-43^{dNLS}[217,218] or a C-terminal fragment of human TDP-43 (residues 208–414),[219] both under the control of an inducible promoter. They found that human TDP-43^{dNLS}, TDP-43$^{208–414}$, and endogenous mouse TDP-43 were capable of being seeded by the injection of seeding competent FTLD brain extracts, although to varying extents where full-length TDP-43^{dNLS} gave rise to the most obvious pathology, followed by TDP-43$^{208–414}$, then murine TDP-43.

Importantly, the cell culture and *in vivo* experiments mentioned above[33,211] were primarily performed using FTLD extracts. It remains to be elucidated fully if TDP-43 extracts from ALS patients would result in similar pathological prion-like properties in these models. It is interesting to consider the nature of ALS and FTLD as two extreme manifestations of the same disease spectrum[42] and that perhaps the development of either disorder is dependent on the generation of a particular TDP-43 strain.

Overall, the current evidence that suggests a pathological prion-like propagation of TDP-43 in ALS is strong, however, future studies will be needed to establish the capability of TDP-43 extracts from patients, as well as *in vitro* recombinant TDP-43, to seed the aggregation of TDP-43 in either WT or transgenic animals. These experiments would be essential to not only further establishing the prionogenicity of TDP-43 in ALS, but providing models for examining the underlying determinants of TDP-43-associated ALS and to provide models in which therapies targeted against TDP-43 aggregation and spread can be tested preclinically. Methods to prevent the aggregation and prion-like propagation of TDP-43 currently being pursued include reducing TDP-43 phosphorylation,[220] small molecules to inhibit TDP-43 aggregation,[221] and potential immunotherapies targeting cytoplasmic TDP-43.[222]

10. Fused in sarcoma has only modest evidence to suggest its pathological prion-like propagation in ALS

Fused in sarcoma (FUS) is similar to TDP-43 in that it is an RNA/DNA-binding protein with roles in RNA metabolism, transport, and DNA repair.[223,224] Structurally, FUS contains an N-terminal functional PrLD, several RGG rich regions, an RRM, a zinc finger motif, and a C-terminal nuclear localization signal, where all of these features make it a member of the TET protein family.[225] ALS and FTLD-associated mutations are clustered into the first RGG rich domain and the C-terminal NLS,[16,17] and as such suggests that dysfunction in RNA-binding and/or aberrant localization likely play a role in FUS-associated disease.

Unlike SOD1 and TDP-43, the evidence suggesting a prion-like propagation of FUS in ALS is currently limited. It is established that purified recombinant FUS is capable of forming amyloid-like fibers that potentially induce the aggregation of soluble FUS *in vitro*,[226] however, the timescale of aggregation in the performed seeding experiments showed that aggregation of soluble FUS occurred rapidly in minutes, suggesting perhaps the functional assembly of FUS,[93,96,227] rather than pathologic, was occurring in these assays. Further work investigating if FUS is capable of forming detergent and heat resistant pathological amyloid remains to be determined. These experiments would be essential in allowing for the testing of whether these structures can induce FUS aggregation in cultured cells.

Histopathological examination of FUS in ALS has been difficult due to the rarity of cases of patients carrying FUS mutations and sporadic patients presenting FUS pathology for examination. Furthermore, differences in the methodologies employed by researchers makes comparison difficult across the published studies, with some reporting that some features of FUS pathology correlates with disease severity[228] and others no correlation between FUS pathology and disease severity.[229] It will be important in future studies of FUS-associated ALS to determine if the presentation of FUS pathology does indeed correlate with disease progression, spread, and severity in the case of ALS.

11. Conclusions

It is clear that ALS has all the hallmarks of a pathological prion-like disease in which the initiation and spread protein misfolding and aggregation is a key feature of disease etiology and progression.[215] There is strong evidence supporting the pathological prion-like characteristics of SOD1 and TDP-43, however, the evidence supporting a similar role for FUS is less clear and it remains to be established if FUS is acting in a pathological prion-like manner in ALS. Nevertheless, the majority of cases of ALS are associated with TDP-43 pathology and models of its pathological prion-like propagation are only recently established. It will be interesting to apply high-throughput drug screening technologies to the cell culture models of TDP-43 seeding and propagation[33,211] as prevention of initial aggregation and seed formation may be impossible, however, limiting the spread of pathological TDP-43 may provide a plausible method of intervention.

A major challenge in both understanding the pathobiology of ALS, for both SOD1 and TDP-43-associated cases, is that the toxic species responsible for motor neuron degeneration is currently undefined. Although some have highlighted potential regions of either SOD1[166] or TDP-43[211] that may contribute to the formation of toxic species, high resolution structures of these species would be indispensable in the rational design of therapies. A future challenge in the field of ALS will be to determine the structures of aggregates composed of the key pathological proteins, as has been done for tau-protein in Alzheimer's disease[230] and α-synuclein in Parkinson's disease.[231] Knowing the molecular architecture of the fibrillar species would be paramount in understanding the template-directed interactions that allows for pathological conversion of substrate protein, and this understanding would result in new methods to examine molecules to block these interactions.

References

1. Erkkinen MG, Kim M-O, Geschwind MD. Clinical neurology and epidemiology of the major neurodegenerative diseases. *Cold Spring Harb Perspect Biol*. 2018;10(4): a033118. https://doi.org/10.1101/cshperspect.a033118.
2. Prince M, Bryce R, Albanese E, Wimo A, Ribeiro W, Ferri CP. The global prevalence of dementia: a systematic review and metaanalysis. *Alzheimers Dement*. 2013;9(1): 63–75.e2. https://doi.org/10.1016/j.jalz.2012.11.007.
3. Saxena S, Caroni P. Selective neuronal vulnerability in neurodegenerative diseases: from stressor thresholds to degeneration. *Neuron*. 2011;71(1):35–48. https://doi.org/10.1016/j.neuron.2011.06.031.
4. Wallace DC. A mitochondrial paradigm of metabolic and degenerative diseases, aging, and cancer: a dawn for evolutionary medicine. *Annu Rev Genet*. 2005;39(1):359–407. https://doi.org/10.1146/annurev.genet.39.110304.095751.
5. Morimoto RI. Proteotoxic stress and inducible chaperone networks in neurodegenerative disease and aging. *Genes Dev*. 2008;22(11):1427–1438. PMC2732416. https://doi.org/10.1101/gad.1657108.
6. Matus S, Glimcher LH, Hetz C. Protein folding stress in neurodegenerative diseases: a glimpse into the ER. *Curr Opin Cell Biol*. 2011;23(2):239–252. https://doi.org/10.1016/j.ceb.2011.01.003.
7. Chiti F, Dobson CM. Protein misfolding, amyloid formation, and human disease: a summary of progress over the last decade. *Annu Rev Biochem*. 2017;86:27–68. https://doi.org/10.1146/annurev-biochem-061516-045115.
8. Love S. Neuropathological investigation of dementia: a guide for neurologists. *J Neurol Neurosurg Psychiatry*. 2005;76(Suppl. 5):v8–14. PMC1765714. https://doi.org/10.1136/jnnp.2005.080754.
9. Glenner GG, Wong CW. Alzheimer's disease: initial report of the purification and characterization of a novel cerebrovascular amyloid protein. *Biochem Biophys Res Commun*. 1984;120(3):885–890. https://doi.org/10.1016/s0006-291x(84)80190-4.
10. Grundke-Iqbal I, Iqbal K, Tung YC, Quinlan M, Wisniewski HM, Binder LI. Abnormal phosphorylation of the microtubule-associated protein tau (tau) in Alzheimer cytoskeletal pathology. *Proc Natl Acad Sci U S A*. 1986;83(13):4913–4917. https://doi.org/10.1073/pnas.83.13.4913.
11. Spillantini MG, Schmidt ML, Lee VM, Trojanowski JQ, Jakes R, Goedert M. α-Synuclein in Lewy bodies. *Nature*. 1997;388(6645):839–840. https://doi.org/10.1038/42166.
12. Tu P-H, Galvin JE, Baba M, et al. Glial cytoplasmic inclusions in white matter oligodendrocytes of multiple system atrophy brains contain insoluble α-synuclein. *Ann Neurol*. 1998;44(3):415–422. https://doi.org/10.1002/ana.410440324.
13. Wakabayashi K, Yoshimoto M, Tsuji S, Takahashi H. α-Synuclein immunoreactivity in glial cytoplasmic inclusions in multiple system atrophy. *Neurosci Lett*. 1998; 249(2–3):180–182. https://doi.org/10.1016/s0304-3940(98)00407-8.
14. Neumann M, Sampathu DM, Kwong LK, et al. Ubiquitinated TDP-43 in frontotemporal lobar degeneration and amyotrophic lateral sclerosis. *Science*. 2006;314(5796): 130–133. https://doi.org/10.1126/science.1134108.
15. Bruijn LI, Houseweart MK, Kato S, et al. Aggregation and motor neuron toxicity of an ALS-linked SOD1 mutant independent from wild-type SOD1. *Science*. 1998; 281(5384):1851–1854. https://doi.org/10.1126/science.281.5384.1851.
16. Kwiatkowski Jr TJ, Bosco DA, Leclerc AL, et al. Mutations in the FUS/TLS gene on chromosome 16 cause familial amyotrophic lateral sclerosis. *Science*. 2009;323(5918): 1205–1208. https://doi.org/10.1126/science.1166066.
17. Vance C, Rogelj B, Hortobágyi T, et al. Mutations in FUS, an RNA processing protein, cause familial amyotrophic lateral sclerosis type 6. *Science*. 2009;323(5918): 1208–1211. PMC4516382. https://doi.org/10.1126/science.1165942.

18. Geschwind MD. Prion diseases. *Continuum*. 2015;21(6 Neuroinfectious Disease): 1612–1638. PMC4879966. https://doi.org/10.1212/CON.0000000000000251.
19. Puoti G, Bizzi A, Forloni G, Safar JG, Tagliavini F, Gambetti P. Sporadic human prion diseases: molecular insights and diagnosis. *Lancet Neurol*. 2012;11(7):618–628. https://doi.org/10.1016/S1474-4422(12)70063-7.
20. Collins SJ, Sanchez-Juan P, Masters CL, et al. Determinants of diagnostic investigation sensitivities across the clinical spectrum of sporadic Creutzfeldt-Jakob disease. *Brain*. 2006;129(Pt. 9):2278–2287. https://doi.org/10.1093/brain/awl159.
21. Imran M, Mahmood S. An overview of human prion diseases. *Virol J*. 2011;8(1):559. https://doi.org/10.1186/1743-422x-8-559.
22. Kovacs G, Budka H. Molecular pathology of human prion diseases. *Int J Mol Sci*. 2009;10(3):976–999. https://doi.org/10.3390/ijms10030976.
23. Bratosiewicz-Wasik J, Liberski PP, Golanska E, Jansen GH, Wasik TJ. Regulatory sequences of the PRNP gene influence susceptibility to sporadic Creutzfeldt-Jakob disease. *Neurosci Lett*. 2007;411(3):163–167. https://doi.org/10.1016/j.neulet.2006.08.001.
24. Mead S, Mahal SP, Beck J, et al. Sporadic—but not variant—Creutzfeldt-Jakob disease is associated with polymorphisms upstream of PRNP exon 1. *Am J Hum Genet*. 2001;69(6):1225–1235. https://doi.org/10.1086/324710.
25. Palmer MS, Dryden AJ, Hughes JT, Collinge J. Homozygous prion protein genotype predisposes to sporadic Creutzfeldt-Jakob disease. *Nature*. 1991;352(6333):340–342. https://doi.org/10.1038/352340a0.
26. Collinge J, Clarke AR. A general model of prion strains and their pathogenicity. *Science*. 2007;318(5852):930–936. https://doi.org/10.1126/science.1138718.
27. Prusiner SB. Novel proteinaceous infectious particles cause scrapie. *Science*. 1982; 216(4542):136–144. https://doi.org/10.1126/science.6801762.
28. Alper T, Cramp WA, Haig DA, Clarke MC. Does the agent of scrapie replicate without nucleic acid? *Nature*. 1967;214(5090):764–766. https://doi.org/10.1038/214764a0.
29. Griffith JS. Nature of the scrapie agent: self-replication and scrapie. *Nature*. 1967; 215(5105):1043–1044. https://doi.org/10.1038/2151043a0.
30. Pan KM, Baldwin M, Nguyen J, et al. Conversion of alpha-helices into beta-sheets features in the formation of the scrapie prion proteins. *Proc Natl Acad Sci U S A*. 1993; 90(23):10962–10966. PMC47901. https://doi.org/10.1073/pnas.90.23.10962.
31. Kim S, Kwon S-H, Kam T-I, et al. Transneuronal propagation of pathologic α-synuclein from the gut to the brain models Parkinson's disease. *Neuron*. 2019;103(4): 627–641.e7. https://doi.org/10.1016/j.neuron.2019.05.035.
32. Jaunmuktane Z, Mead S, Ellis M, et al. Evidence for human transmission of amyloid-β pathology and cerebral amyloid angiopathy. *Nature*. 2015;525(7568):247–250. https://doi.org/10.1038/nature15369.
33. Porta S, Xu Y, Restrepo CR, et al. Patient-derived frontotemporal lobar degeneration brain extracts induce formation and spreading of TDP-43 pathology in vivo. *Nat Commun*. 2018;9(1):4220. PMC6181940. https://doi.org/10.1038/s41467-018-06548-9.
34. Ekhtiari Bidhendi E, Bergh J, Zetterström P, et al. Mutant superoxide dismutase aggregates from human spinal cord transmit amyotrophic lateral sclerosis. *Acta Neuropathol*. 2018;136(6):939–953. PMC6280858. https://doi.org/10.1007/s00401-018-1915-y.
35. Kane MD, Lipinski WJ, Callahan MJ, et al. Evidence for seeding of beta-amyloid by intracerebral infusion of Alzheimer brain extracts in beta-amyloid precursor protein-transgenic mice. *J Neurosci*. 2000;20(10):3606–3611. PMC6772682.
36. Walker LC, Callahan MJ, Bian F, Durham RA, Roher AE, Lipinski WJ. Exogenous induction of cerebral beta-amyloidosis in betaAPP-transgenic mice. *Peptides*. 2002; 23(7):1241–1247. https://doi.org/10.1016/s0196-9781(02)00059-1.
37. Meyer-Luehmann M, Coomaraswamy J, Bolmont T, et al. Exogenous induction of cerebral beta-amyloidogenesis is governed by agent and host. *Science*. 2006;313(5794): 1781–1784. https://doi.org/10.1126/science.1131864.

38. Watts JC, Giles K, Oehler A, et al. Transmission of multiple system atrophy prions to transgenic mice. *Proc Natl Acad Sci U S A*. 2013;110(48):19555–19560. PMC3845125. https://doi.org/10.1073/pnas.1318268110.
39. Shorter J, Lindquist S. Prions as adaptive conduits of memory and inheritance. *Nat Rev Genet*. 2005;6(6):435–450. https://doi.org/10.1038/nrg1616.
40. Cox BS, Tuite MF, McLaughlin CS. The psi factor of yeast: a problem in inheritance. *Yeast*. 1988;4(3):159–178. https://doi.org/10.1002/yea.320040302.
41. Paushkin SV, Kushnirov VV, Smirnov VN, Ter-Avanesyan MD. Propagation of the yeast prion-like [psi+] determinant is mediated by oligomerization of the SUP35-encoded polypeptide chain release factor. *EMBO J*. 1996;15(12):3127–3134. PMC450255.
42. Ling S-C, Polymenidou M, Cleveland DW. Converging mechanisms in ALS and FTD: disrupted RNA and protein homeostasis. *Neuron*. 2013;79(3):416–438. PMC4411085. https://doi.org/10.1016/j.neuron.2013.07.033.
43. King OD, Gitler AD, Shorter J. The tip of the iceberg: RNA-binding proteins with prion-like domains in neurodegenerative disease. *Brain Res*. 2012;1462:61–80. PMC3372647. https://doi.org/10.1016/j.brainres.2012.01.016.
44. Nonaka T, Masuda-Suzukake M, Arai T, et al. Prion-like properties of pathological TDP-43 aggregates from diseased brains. *Cell Rep*. 2013;4(1):124–134. https://doi.org/10.1016/j.celrep.2013.06.007.
45. Al-Chalabi A, Hardiman O, Kiernan MC, Chiò A, Rix-Brooks B, van den Berg LH. Amyotrophic lateral sclerosis: moving towards a new classification system. *Lancet Neurol*. 2016;15(11):1182–1194. https://doi.org/10.1016/S1474-4422(16)30199-5.
46. Grad LI, Rouleau GA, Ravits J, Cashman NR. Clinical spectrum of amyotrophic lateral sclerosis (ALS). *Cold Spring Harb Perspect Med*. 2017;7(8):a024117. PMC5538408. https://doi.org/10.1101/cshperspect.a024117.
47. Statland JM, Barohn RJ, McVey AL, Katz JS, Dimachkie MM. Patterns of weakness, classification of motor neuron disease, and clinical diagnosis of sporadic amyotrophic lateral sclerosis. *Neurol Clin*. 2015;33(4):735–748. PMC4629510. https://doi.org/10.1016/j.ncl.2015.07.006.
48. Tartaglia MC, Rowe A, Findlater K, Orange JB, Grace G, Strong MJ. Differentiation between primary lateral sclerosis and amyotrophic lateral sclerosis: examination of symptoms and signs at disease onset and during follow-up. *Arch Neurol*. 2007;64(2):232–236. https://doi.org/10.1001/archneur.64.2.232.
49. Van den Berg-Vos RM, Visser J, Kalmijn S, et al. A long-term prospective study of the natural course of sporadic adult-onset lower motor neuron syndromes. *Arch Neurol*. 2009;66(6):751–757. https://doi.org/10.1001/archneurol.2009.91.
50. Ravits J, Paul P, Jorg C. Focality of upper and lower motor neuron degeneration at the clinical onset of ALS. *Neurology*. 2007;68(19):1571–1575. https://doi.org/10.1212/01.wnl.0000260965.20021.47.
51. Ravits JM, La Spada AR. ALS motor phenotype heterogeneity, focality, and spread: deconstructing motor neuron degeneration. *Neurology*. 2009;73(10):805–811. PMC2739608. https://doi.org/10.1212/WNL.0b013e3181b6bbbd.
52. Walhout R, Verstraete E, van den Heuvel MP, Veldink JH, van den Berg LH. Patterns of symptom development in patients with motor neuron disease. *Amyotroph Lateral Scler Frontotemporal Degener*. 2018;19(1–2):21–28. https://doi.org/10.1080/21678421.2017.1386688.
53. Zhenfei L, Shiru D, Xiaomeng Z, Cuifang C, Yaling L. Discontiguous or contiguous spread patterns affect the functional staging in patients with sporadic amyotrophic lateral sclerosis. *Front Neurol*. 2019;10:523. PMC6540586. https://doi.org/10.3389/fneur.2019.00523.

54. Gargiulo-Monachelli GM, Janota F, Bettini M, Shoesmith CL, Strong MJ, Sica REP. Regional spread pattern predicts survival in patients with sporadic amyotrophic lateral sclerosis. *Eur J Neurol.* 2012;19(6):834–841. https://doi.org/10.1111/j.1468-1331.2011.03616.x.
55. Saberi S, Stauffer JE, Schulte DJ, Ravits J. Neuropathology of amyotrophic lateral sclerosis and its variants. *Neurol Clin.* 2015;33(4):855–876. PMC4628785. https://doi.org/10.1016/j.ncl.2015.07.012.
56. Ciryam P, Lambert-Smith IA, Bean DM, et al. Spinal motor neuron protein supersaturation patterns are associated with inclusion body formation in ALS. *Proc Natl Acad Sci U S A.* 2017;114(20):E3935–E3943. PMC5441770. https://doi.org/10.1073/pnas.1613854114.
57. Neumann M, Rademakers R, Roeber S, Baker M, Kretzschmar HA, Mackenzie IRA. A new subtype of frontotemporal lobar degeneration with FUS pathology. *Brain.* 2009;132(11):2922–2931. https://doi.org/10.1093/brain/awp214.
58. Hasegawa M, Arai T, Nonaka T, et al. Phosphorylated TDP-43 in frontotemporal lobar degeneration and amyotrophic lateral sclerosis. *Ann Neurol.* 2008;64(1):60–70. PMC2674108. https://doi.org/10.1002/ana.21425.
59. Solomon DA, Stepto A, Au WH, et al. A feedback loop between dipeptide-repeat protein, TDP-43 and karyopherin-α mediates C9orf72-related neurodegeneration. *Brain.* 2018;141(10):2908–2924. PMC6158706. https://doi.org/10.1093/brain/awy241.
60. Nonaka T, Masuda-Suzukake M, Hosokawa M, et al. C9ORF72 dipeptide repeat poly-GA inclusions promote intracellular aggregation of phosphorylated TDP-43. *Hum Mol Genet.* 2018;27(15):2658–2670. https://doi.org/10.1093/hmg/ddy174.
61. Khosravi B, Hartmann H, May S, et al. Cytoplasmic poly-GA aggregates impair nuclear import of TDP-43 in C9orf72 ALS/FTLD. *Hum Mol Genet.* 2017;26(4):790–800. PMC5409121. https://doi.org/10.1093/hmg/ddw432.
62. Cooper-Knock J, Hewitt C, Highley JR, et al. Clinico-pathological features in amyotrophic lateral sclerosis with expansions in C9ORF72. *Brain.* 2012;135(Pt. 3):751–764. PMC3286332. https://doi.org/10.1093/brain/awr365.
63. Fratta P, Poulter M, Lashley T, et al. Homozygosity for the C9orf72 GGGGCC repeat expansion in frontotemporal dementia. *Acta Neuropathol.* 2013;126(3):401–409. PMC3753468. https://doi.org/10.1007/s00401-013-1147-0.
64. Leigh PN, Anderton BH, Dodson A, Gallo JM, Swash M, Power DM. Ubiquitin deposits in anterior horn cells in motor neurone disease. *Neurosci Lett.* 1988;93(2–3):197–203. https://doi.org/10.1016/0304-3940(88)90081-x.
65. Lowe J, Lennox G, Jefferson D, et al. A filamentous inclusion body within anterior horn neurones in motor neurone disease defined by immunocytochemical localisation of ubiquitin. *Neurosci Lett.* 1988;94(1–2):203–210. https://doi.org/10.1016/0304-3940(88)90296-0.
66. Mizuno Y, Amari M, Takatama M, Aizawa H, Mihara B, Okamoto K. Immunoreactivities of p62, an ubiqutin-binding protein, in the spinal anterior horn cells of patients with amyotrophic lateral sclerosis. *J Neurol Sci.* 2006;249(1):13–18. https://doi.org/10.1016/j.jns.2006.05.060.
67. Fecto F, Yan J, Vemula SP, et al. SQSTM1 mutations in familial and sporadic amyotrophic lateral sclerosis. *Arch Neurol.* 2011;68(11):1440–1446. https://doi.org/10.1001/archneurol.2011.250.
68. Teyssou E, Takeda T, Lebon V, et al. Mutations in SQSTM1 encoding p62 in amyotrophic lateral sclerosis: genetics and neuropathology. *Acta Neuropathol.* 2013;125(4):511–522. https://doi.org/10.1007/s00401-013-1090-0.
69. Brettschneider J, Del Tredici K, Toledo JB, et al. Stages of pTDP-43 pathology in amyotrophic lateral sclerosis. *Ann Neurol.* 2013;74(1):20–38. PMC3785076. https://doi.org/10.1002/ana.23937.

70. Braak H, Del Tredici K, Rüb U, de Vos RAI, Jansen EN, Braak E. Staging of brain pathology related to sporadic Parkinson's disease. *Neurobiol Aging*. 2003;24(2): 197–211. https://doi.org/10.1016/s0197-4580(02)00065-9.
71. Taylor JP, Brown Jr RH, Cleveland DW. Decoding ALS: from genes to mechanism. *Nature*. 2016;539(7628):197–206. PMC5585017. https://doi.org/10.1038/nature20413.
72. Rosen D. Mutations in Cu/Zn superoxide dismutase gene are associated with familial amyotrophic lateral sclerosis. *Nature*. 1993;364(6435):362. https://doi.org/10.1038/364362c0.
73. Sreedharan J, Blair IP, Tripathi VB, et al. TDP-43 mutations in familial and sporadic amyotrophic lateral sclerosis. *Science*. 2008;319(5870):1668–1672. https://doi.org/10.1126/science.1154584.
74. Kabashi E, Valdmanis PN, Dion P, et al. TARDBP mutations in individuals with sporadic and familial amyotrophic lateral sclerosis. *Nat Genet*. 2008;40(5):572–574. https://doi.org/10.1038/ng.132.
75. DeJesus-Hernandez M, Mackenzie IR, Boeve BF, et al. Expanded GGGGCC hexanucleotide repeat in noncoding region of C9ORF72 causes chromosome 9p-linked FTD and ALS. *Neuron*. 2011;72(2):245–256. PMC3202986. https://doi.org/10.1016/j.neuron.2011.09.011.
76. Renton AE, Majounie E, Waite A, et al. A hexanucleotide repeat expansion in C9ORF72 is the cause of chromosome 9p21-linked ALS-FTD. *Neuron*. 2011; 72(2):257–268. PMC3200438. https://doi.org/10.1016/j.neuron.2011.09.010.
77. Maruyama H, Morino H, Ito H, et al. Mutations of optineurin in amyotrophic lateral sclerosis. *Nature*. 2010;465(7295):223–226. https://doi.org/10.1038/nature08971.
78. Puls I, Jonnakuty C, LaMonte BH, et al. Mutant dynactin in motor neuron disease. *Nat Genet*. 2003;33(4):455–456. https://doi.org/10.1038/ng1123.
79. Johnson JO, Mandrioli J, Benatar M, et al. Exome sequencing reveals VCP mutations as a cause of familial ALS. *Neuron*. 2010;68(5):857–864. PMC3032425. https://doi.org/10.1016/j.neuron.2010.11.036.
80. Kim HJ, Kim NC, Wang Y-D, et al. Mutations in prion-like domains in hnRNPA2B1 and hnRNPA1 cause multisystem proteinopathy and ALS. *Nature*. 2013; 495(7442):467–473. PMC3756911. https://doi.org/10.1038/nature11922.
81. Liu Q, Shu S, Wang RR, et al. Whole-exome sequencing identifies a missense mutation in hnRNPA1 in a family with flail arm ALS. *Neurology*. 2016;87(17):1763–1769. https://doi.org/10.1212/wnl.0000000000003256.
82. Mackenzie IR, Nicholson AM, Sarkar M, et al. TIA1 mutations in amyotrophic lateral sclerosis and frontotemporal dementia promote phase separation and alter stress granule dynamics. *Neuron*. 2017;95(4):808–816.e9. PMC5576574. https://doi.org/10.1016/j.neuron.2017.07.025.
83. Johnson JO, Pioro EP, Boehringer A, et al. Mutations in the Matrin 3 gene cause familial amyotrophic lateral sclerosis. *Nat Neurosci*. 2014;17(5):664–666. PMC4000579. https://doi.org/10.1038/nn.3688.
84. Couthouis J, Hart MP, Shorter J, et al. A yeast functional screen predicts new candidate ALS disease genes. *Proc Natl Acad Sci U S A*. 2011;108(52):20881–20890. PMC3248518. https://doi.org/10.1073/pnas.1109434108.
85. Smith BN, Topp SD, Fallini C, et al. Mutations in the vesicular trafficking protein annexin A11 are associated with amyotrophic lateral sclerosis. *Sci Transl Med*. 2017; 9(388):eaad9157. PMC6599403. https://doi.org/10.1126/scitranslmed.aad9157.
86. Farhan SMK, Howrigan DP, Abbott LE, et al. Exome sequencing in amyotrophic lateral sclerosis implicates a novel gene, DNAJC7, encoding a heat-shock protein. *Nat Neurosci*. 2019;22(12):1966–1974. PMC6919277. https://doi.org/10.1038/s41593-019-0530-0.

87. Deng H-X, Chen W, Hong S-T, et al. Mutations in UBQLN2 cause dominant X-linked juvenile and adult-onset ALS and ALS/dementia. *Nature*. 2011;477(7363): 211–215. PMC3169705. https://doi.org/10.1038/nature10353.
88. Greenway MJ, Andersen PM, Russ C, et al. ANG mutations segregate with familial and 'sporadic' amyotrophic lateral sclerosis. *Nat Genet*. 2006;38(4):411–413. https://doi.org/10.1038/ng1742.
89. Wu C-H, Fallini C, Ticozzi N, et al. Mutations in the profilin 1 gene cause familial amyotrophic lateral sclerosis. *Nature*. 2012;488(7412):499–503. PMC3575525. https://doi.org/10.1038/nature11280.
90. Smith BN, Ticozzi N, Fallini C, et al. Exome-wide rare variant analysis identifies TUBA4A mutations associated with familial ALS. *Neuron*. 2014;84(2):324–331. PMC4521390. https://doi.org/10.1016/j.neuron.2014.09.027.
91. Freischmidt A, Wieland T, Richter B, et al. Haploinsufficiency of TBK1 causes familial ALS and fronto-temporal dementia. *Nat Neurosci*. 2015;18(5):631–636. https://doi.org/10.1038/nn.4000.
92. Kapeli K, Martinez FJ, Yeo GW. Genetic mutations in RNA-binding proteins and their roles in ALS. *Hum Genet*. 2017;136(9):1193. PMC5602095. https://doi.org/10.1007/s00439-017-1830-7.
93. Kato M, Han TW, Xie S, et al. Cell-free formation of RNA granules: low complexity sequence domains form dynamic fibers within hydrogels. *Cell*. 2012;149(4):753–767. PMC6347373. https://doi.org/10.1016/j.cell.2012.04.017.
94. Banani SF, Lee HO, Hyman AA, Rosen MK. Biomolecular condensates: organizers of cellular biochemistry. *Nat Rev Mol Cell Biol*. 2017;18(5):285–298. https://doi.org/10.1038/nrm.2017.7.
95. Murakami T, Qamar S, Lin JQ, et al. ALS/FTD mutation-induced phase transition of FUS liquid droplets and reversible hydrogels into irreversible hydrogels impairs RNP granule function. *Neuron*. 2015;88(4):678–690. PMC4660210. https://doi.org/10.1016/j.neuron.2015.10.030.
96. Patel A, Lee HO, Jawerth L, et al. A liquid-to-solid phase transition of the ALS protein FUS accelerated by disease mutation. *Cell*. 2015;162(5):1066–1077. https://doi.org/10.1016/j.cell.2015.07.047.
97. Lee K-H, Zhang P, Kim HJ, et al. C9orf72 dipeptide repeats impair the assembly, dynamics, and function of membrane-less organelles. *Cell*. 2016;167(3):774–788.e17. PMC5079111. https://doi.org/10.1016/j.cell.2016.10.002.
98. Lin Y, Mori E, Kato M, et al. Toxic PR poly-dipeptides encoded by the C9orf72 repeat expansion target LC domain polymers. *Cell*. 2016;167(3):789–802.e12. PMC5076566. https://doi.org/10.1016/j.cell.2016.10.003.
99. Martinez FJ, Pratt GA, Van Nostrand EL, et al. Protein-RNA networks regulated by normal and ALS-associated mutant HNRNPA2B1 in the nervous system. *Neuron*. 2016;92(4):780–795. PMC5123850. https://doi.org/10.1016/j.neuron.2016.09.050.
100. Boeynaems S, Bogaert E, Kovacs D, et al. Phase separation of C9orf72 dipeptide repeats perturbs stress granule dynamics. *Mol Cell*. 2017;65(6):1044–1055.e5. PMC5364369. https://doi.org/10.1016/j.molcel.2017.02.013. .
101. Molliex A, Temirov J, Lee J, et al. Phase separation by low complexity domains promotes stress granule assembly and drives pathological fibrillization. *Cell*. 2015;163(1): 123–133. PMC5149108. https://doi.org/10.1016/j.cell.2015.09.015.
102. Dao TP, Martyniak B, Canning AJ, et al. ALS-linked mutations affect UBQLN2 oligomerization and phase separation in a position- and amino acid-dependent manner. *Structure*. 2019;27(6):937–951.e5. https://doi.org/10.1016/j.str.2019.03.012.
103. Ciryam P, Kundra R, Morimoto RI, Dobson CM, Vendruscolo M. Supersaturation is a major driving force for protein aggregation in neurodegenerative diseases. *Trends Pharmacol Sci*. 2015;36(2):72–77. PMC4643722. https://doi.org/10.1016/j.tips.2014.12.004.

104. Eisenberg D, Jucker M. The amyloid state of proteins in human diseases. *Cell.* 2012;148(6):1188–1203. PMC3353745. https://doi.org/10.1016/j.cell.2012.02.022.
105. Knowles TPJ, Vendruscolo M, Dobson CM. The amyloid state and its association with protein misfolding diseases. *Nat Rev Mol Cell Biol.* 2014;15(6):384–396. https://doi.org/10.1038/nrm3810.
106. Abdolvahabi A, Shi Y, Chuprin A, Rasouli S, Shaw BF. Stochastic formation of fibrillar and amorphous superoxide dismutase oligomers linked to amyotrophic lateral sclerosis. *ACS Chem Nerosci.* 2016;7(6):799–810. https://doi.org/10.1021/acschemneuro.6b00048.
107. Arosio P, Knowles TPJ, Linse S. On the lag phase in amyloid fibril formation. *Phys Chem Phys.* 2015;17(12):7606–7618. https://doi.org/10.1039/c4cp05563b.
108. Knowles TPJ, Shu W, Devlin GL, et al. Kinetics and thermodynamics of amyloid formation from direct measurements of fluctuations in fibril mass. *Proc Natl Acad Sci U S A.* 2007;104(24):10016–10021. PMC1891240. https://doi.org/10.1073/pnas.0610659104.
109. Pellarin R, Caflisch A. Interpreting the aggregation kinetics of amyloid peptides. *J Mol Biol.* 2006;360(4):882–892. https://doi.org/10.1016/j.jmb.2006.05.033.
110. Shvadchak VV, Claessens MMAE, Subramaniam V. Fibril breaking accelerates α-synuclein fibrillization. *J Phys Chem B.* 2015;119(5):1912–1918. https://doi.org/10.1021/jp5111604.
111. Colby DW, Zhang Q, Wang S, et al. Prion detection by an amyloid seeding assay. *Proc Natl Acad Sci U S A.* 2007;104(52):20914–20919. PMC2409241. https://doi.org/10.1073/pnas.0710152105.
112. Stöhr J, Weinmann N, Wille H, et al. Mechanisms of prion protein assembly into amyloid. *Proc Natl Acad Sci U S A.* 2008;105(7):2409–2414. PMC2268150. https://doi.org/10.1073/pnas.0712036105.
113. Chattopadhyay M, Durazo A, Sohn SH, et al. Initiation and elongation in fibrillation of ALS-linked superoxide dismutase. *Proc Natl Acad Sci U S A.* 2008;105(48):18663–18668. PMC2585484. https://doi.org/10.1073/pnas.0807058105.
114. Chen AKH, Lin RYY, Hsieh EZJ, et al. Induction of amyloid fibrils by the C-terminal fragments of TDP-43 in amyotrophic lateral sclerosis. *J Am Chem Soc.* 2010;132(4):1186–1187. https://doi.org/10.1021/ja9066207.
115. Bigio EH, Wu JY, Deng H-X, et al. Inclusions in frontotemporal lobar degeneration with TDP-43 proteinopathy (FTLD-TDP) and amyotrophic lateral sclerosis (ALS), but not FTLD with FUS proteinopathy (FTLD-FUS), have properties of amyloid. *Acta Neuropathol.* 2013;125(3):463–465. https://doi.org/10.1007/s00401-013-1089-6.
116. Lowe J. New pathological findings in amyotrophic lateral sclerosis. *J Neurol Sci.* 1994;124:38–51. https://doi.org/10.1016/0022-510x(94)90175-9.
117. Page T, Gitcho MA, Mosaheb S, et al. FUS immunogold labeling TEM analysis of the neuronal cytoplasmic inclusions of neuronal intermediate filament inclusion disease: a frontotemporal lobar degeneration with FUS proteinopathy. *J Mol Neurosci.* 2011;45(3):409–421. https://doi.org/10.1007/s12031-011-9549-8.
118. Kato S, Takikawa M, Nakashima K, et al. New consensus research on neuropathological aspects of familial amyotrophic lateral sclerosis with superoxide dismutase 1 (SOD1) gene mutations: inclusions containing SOD1 in neurons and astrocytes. *Amyotroph Lateral Scler Other Motor Neuron Disord.* 2000;1(3):163–184. https://doi.org/10.1080/14660820050515160.
119. Kerman A, Liu H-N, Croul S, et al. Amyotrophic lateral sclerosis is a non-amyloid disease in which extensive misfolding of SOD1 is unique to the familial form. *Acta Neuropathol.* 2010;119(3):335–344. https://doi.org/10.1007/s00401-010-0646-5.
120. Abel O, Shatunov A, Jones AR, Andersen PM, Powell JF, Al-Chalabi A. Development of a smartphone app for a genetics website: the amyotrophic lateral sclerosis online genetics database (ALSoD). *JMIR Mhealth Uhealth.* 2013;1(2):e18. PMC4114449. https://doi.org/10.2196/mhealth.2706.

121. McCord JM, Fridovich I. Superoxide dismutase. An enzymic function for erythrocuprein (hemocuprein). *J Biol Chem.* 1969;244(22):6049–6055.
122. Wright GSA, Antonyuk SV, Hasnain SS. The biophysics of superoxide dismutase-1 and amyotrophic lateral sclerosis. *Q Rev Biophys.* 2019;52:e12. https://doi.org/10.1017/S003358351900012X.
123. Lelie HL, Liba A, Bourassa MW, et al. Copper and zinc metallation status of copper-zinc superoxide dismutase from amyotrophic lateral sclerosis transgenic mice. *J Biol Chem.* 2011;286(4):2795–2806. PMC3024775. https://doi.org/10.1074/jbc.M110.186999.
124. Tiwari A, Hayward LJ. Familial amyotrophic lateral sclerosis mutants of copper/zinc superoxide dismutase are susceptible to disulfide reduction. *J Biol Chem.* 2003; 278(8):5984–5992. https://doi.org/10.1074/jbc.M210419200.
125. McAlary L, Yerbury JJ, Aquilina JA. Glutathionylation potentiates benign superoxide dismutase 1 variants to the toxic forms associated with amyotrophic lateral sclerosis. *Sci Rep.* 2013;3:3275. PMC3834562. https://doi.org/10.1038/srep03275.
126. McAlary L, Andrew Aquilina J, Yerbury JJ. Susceptibility of mutant SOD1 to form a destabilized monomer predicts cellular aggregation and toxicity but not in vitro aggregation propensity. *Front Neurosci.* 2016;10:499. https://doi.org/10.3389/fnins.2016.00499.
127. Lindberg MJ, Byström R, Boknäs N, Andersen PM, Oliveberg M. Systematically perturbed folding patterns of amyotrophic lateral sclerosis (ALS)-associated SOD1 mutants. *Proc Natl Acad Sci U S A.* 2005;102(28):9754–9759. PMC1174986. https://doi.org/10.1073/pnas.0501957102.
128. Gurney M, Pu H, Chiu A, et al. Motor neuron degeneration in mice that express a human Cu, Zn superoxide dismutase mutation. *Science.* 1994;264(5166):1772–1775. https://doi.org/10.1126/science.8209258.
129. Reaume AG, Elliott JL, Hoffman EK, et al. Motor neurons in Cu/Zn superoxide dismutase-deficient mice develop normally but exhibit enhanced cell death after axonal injury. *Nat Genet.* 1996;13(1):43–47. https://doi.org/10.1038/ng0596-43.
130. Borchelt DR, Lee MK, Slunt HS, et al. Superoxide dismutase 1 with mutations linked to familial amyotrophic lateral sclerosis possesses significant activity. *Proc Natl Acad Sci U S A.* 1994;91(17):8292–8296. PMC44592. https://doi.org/10.1073/pnas.91.17.8292.
131. Lang L, Kurnik M, Danielsson J, Oliveberg M. Fibrillation precursor of superoxide dismutase 1 revealed by gradual tuning of the protein-folding equilibrium. *Proc Natl Acad Sci U S A.* 2012;109(44):17868–17873. PMC3497812. https://doi.org/10.1073/pnas.1201795109.
132. Chattopadhyay M, Nwadibia E, Strong CD, Gralla EB, Valentine JS, Whitelegge JP. The disulfide bond, but not zinc or dimerization, controls initiation and seeded growth in amyotrophic lateral sclerosis-linked Cu, Zn superoxide dismutase (SOD1) fibrillation. *J Biol Chem.* 2015;290(51):30624–30636. PMC4683282. https://doi.org/10.1074/jbc.M115.666503.
133. Fujiwara N, Nakano M, Kato S, et al. Oxidative modification to cysteine sulfonic acid of Cys111 in human copper-zinc superoxide dismutase. *J Biol Chem.* 2007;282(49): 35933–35944. https://doi.org/10.1074/jbc.M702941200.
134. Xu W-C, Liang J-Z, Li C, et al. Pathological hydrogen peroxide triggers the fibrillization of wild-type SOD1 via sulfenic acid modification of Cys-111. *Cell Death Dis.* 2018;9(2):67. PMC5833406. https://doi.org/10.1038/s41419-017-0106-4.
135. Anzai I, Tokuda E, Mukaiyama A, et al. A misfolded dimer of Cu/Zn-superoxide dismutase leading to pathological oligomerization in amyotrophic lateral sclerosis. *Protein Sci.* 2017;26(3):484–496. PMC5326558. https://doi.org/10.1002/pro.3094.
136. Pokrishevsky E, McAlary L, Farrawell NE, et al. Tryptophan 32-mediated SOD1 aggregation is attenuated by pyrimidine-like compounds in living cells. *Sci Rep.* 2018;8(1):15590. PMC6197196. https://doi.org/10.1038/s41598-018-32835-y.

137. Rasouli S, Abdolvahabi A, Croom CM, et al. Lysine acylation in superoxide dismutase-1 electrostatically inhibits formation of fibrils with prion-like seeding. *J Biol Chem*. 2017;292(47):19366–19380. PMC5702675. https://doi.org/10.1074/jbc.M117.805283.
138. Shi Y, Rhodes NR, Abdolvahabi A, et al. Deamidation of asparagine to aspartate destabilizes Cu, Zn superoxide dismutase, accelerates fibrillization, and mirrors ALS-linked mutations. *J Am Chem Soc*. 2013;135(42):15897–15908. https://doi.org/10.1021/ja407801x.
139. Rasouli S, Abdolvahabi A, Croom CM, Plewman DL, Shi Y, Shaw BF. Glycerolipid headgroups control rate and mechanism of superoxide dismutase-1 aggregation and accelerate fibrillization of slowly aggregating amyotrophic lateral sclerosis mutants. *ACS Chem Nerosci*. 2018;9(7):1743–1756. https://doi.org/10.1021/acschemneuro.8b00086.
140. Ivanova MI, Sievers SA, Guenther EL, et al. Aggregation-triggering segments of SOD1 fibril formation support a common pathway for familial and sporadic ALS. *Proc Natl Acad Sci U S A*. 2014;111(1):197–201. PMC3890817. https://doi.org/10.1073/pnas.1320786110.
141. Sangwan S, Zhao A, Adams KL, et al. Atomic structure of a toxic, oligomeric segment of SOD1 linked to amyotrophic lateral sclerosis (ALS). *Proc Natl Acad Sci U S A*. 2017;114(33):8770–8775. PMC5565441. https://doi.org/10.1073/pnas.1705091114.
142. Chia R, Howard Tattum M, Jones S, Collinge J, Fisher EMC, Jackson GS. Superoxide dismutase 1 and tgSOD1G93A mouse spinal cord seed fibrils, suggesting a propagative cell death mechanism in amyotrophic lateral sclerosis. *PLoS One*. 2010;5(5):e10627. https://doi.org/10.1371/journal.pone.0010627.
143. Turner BJ, Atkin JD, Farg MA, et al. Impaired extracellular secretion of mutant superoxide dismutase 1 associates with neurotoxicity in familial amyotrophic lateral sclerosis. *J Neurosci*. 2005;25(1):108–117. PMC6725218. https://doi.org/10.1523/JNEUROSCI.4253-04.2005.
144. Stevens JC, Chia R, Hendriks WT, et al. Modification of superoxide dismutase 1 (SOD1) properties by a GFP tag—implications for research into amyotrophic lateral sclerosis (ALS). *PLoS One*. 2010;5(3):e9541. https://doi.org/10.1371/journal.pone.0009541.
145. Crosby K, Crown AM, Roberts BL, Brown H, Ayers JI, Borchelt DR. Loss of charge mutations in solvent exposed Lys residues of superoxide dismutase 1 do not induce inclusion formation in cultured cell models. *PLoS One*. 2018;13(11):e0206751. https://doi.org/10.1371/journal.pone.0206751.
146. Prudencio M, Hart PJ, Borchelt DR, Andersen PM. Variation in aggregation propensities among ALS-associated variants of SOD1: correlation to human disease. *Hum Mol Genet*. 2009;18(17):3217–3226. PMC2722984. https://doi.org/10.1093/hmg/ddp260.
147. Pokrishevsky E, Hong RH, Mackenzie IR, Cashman NR. Spinal cord homogenates from SOD1 familial amyotrophic lateral sclerosis induce SOD1 aggregation in living cells. *PLoS One*. 2017;12(9):e0184384. PMC5587256. https://doi.org/10.1371/journal.pone.0184384.
148. Münch C, O'Brien J, Bertolotti A. Prion-like propagation of mutant superoxide dismutase-1 misfolding in neuronal cells. *Proc Natl Acad Sci U S A*. 2011;108(9):3548–3553. PMC3048161. https://doi.org/10.1073/pnas.1017275108.
149. Holmes BB, DeVos SL, Kfoury N, et al. Heparan sulfate proteoglycans mediate internalization and propagation of specific proteopathic seeds. *Proc Natl Acad Sci U S A*. 2013;110(33):E3138–E3147. PMC3746848. https://doi.org/10.1073/pnas.1301440110.
150. Grad LI, Yerbury JJ, Turner BJ, et al. Intercellular propagated misfolding of wild-type Cu/Zn superoxide dismutase occurs via exosome-dependent and -independent mechanisms. *Proc Natl Acad Sci U S A*. 2014;111(9):3620–3625. PMC3948312. https://doi.org/10.1073/pnas.1312245111.

151. Zeineddine R, Pundavela JF, Corcoran L, et al. SOD1 protein aggregates stimulate macropinocytosis in neurons to facilitate their propagation. *Mol Neurodegener.* 2015;10:57. PMC4628302. https://doi.org/10.1186/s13024-015-0053-4.
152. Bosco DA, Morfini G, Murat Karabacak N, et al. Wild-type and mutant SOD1 share an aberrant conformation and a common pathogenic pathway in ALS. *Nat Neurosci.* 2010;13(11):1396–1403. https://doi.org/10.1038/nn.2660.
153. Forsberg K, Jonsson PA, Andersen PM, et al. Novel antibodies reveal inclusions containing non-native SOD1 in sporadic ALS patients. *PLoS One.* 2010;5(7):e11552. PMC2904380. https://doi.org/10.1371/journal.pone.0011552.
154. Pokrishevsky E, Grad LI, Yousefi M, Wang J, Mackenzie IR, Cashman NR. Aberrant localization of FUS and TDP43 is associated with misfolding of SOD1 in amyotrophic lateral sclerosis. *PLoS One.* 2012;7(4):e35050. PMC3320864. https://doi.org/10.1371/journal.pone.0035050.
155. Paré B, Lehmann M, Beaudin M, et al. Misfolded SOD1 pathology in sporadic amyotrophic lateral sclerosis. *Sci Rep.* 2018;8(1):14223. PMC6155098. https://doi.org/10.1038/s41598-018-31773-z.
156. Forsberg K, Andersen PM, Marklund SL, Brännström T. Glial nuclear aggregates of superoxide dismutase-1 are regularly present in patients with amyotrophic lateral sclerosis. *Acta Neuropathol.* 2011;121(5):623–634. PMC3085063. https://doi.org/10.1007/s00401-011-0805-3.
157. Forsberg K, Graffmo K, Pakkenberg B, et al. Misfolded SOD1 inclusions in patients with mutations in C9orf72 and other ALS/FTD-associated genes. *J Neurol Neurosurg Psychiatry.* 2019;90(8):861–869. https://doi.org/10.1136/jnnp-2018-319386.
158. Deng H-X, Shi Y, Furukawa Y, et al. Conversion to the amyotrophic lateral sclerosis phenotype is associated with intermolecular linked insoluble aggregates of SOD1 in mitochondria. *Proc Natl Acad Sci U S A.* 2006;103(18):7142–7147. PMC1447523. https://doi.org/10.1073/pnas.0602046103.
159. Wang L, Deng H-X, Grisotti G, Zhai H, Siddique T, Roos RP. Wild-type SOD1 overexpression accelerates disease onset of a G85R SOD1 mouse. *Hum Mol Genet.* 2009;18(9):1642–1651. PMC2667291. https://doi.org/10.1093/hmg/ddp085.
160. Haidet-Phillips AM, Hester ME, Miranda CJ, et al. Astrocytes from familial and sporadic ALS patients are toxic to motor neurons. *Nat Biotechnol.* 2011;29(9):824–828. PMC3170425. https://doi.org/10.1038/nbt.1957.
161. Proctor EA, Mowrey DD, Dokholyan NV. β-Methylamino-L-alanine substitution of serine in SOD1 suggests a direct role in ALS etiology. *PLoS Comput Biol.* 2019;15(7):e1007225. PMC6668853. https://doi.org/10.1371/journal.pcbi.1007225.
162. Grad LI, Guest WC, Yanai A, et al. Intermolecular transmission of superoxide dismutase 1 misfolding in living cells. *Proc Natl Acad Sci U S A.* 2011;108(39):16398–16403. PMC3182705. https://doi.org/10.1073/pnas.1102645108.
163. Vande Velde C, Miller TM, Cashman NR, Cleveland DW. Selective association of misfolded ALS-linked mutant SOD1 with the cytoplasmic face of mitochondria. *Proc Natl Acad Sci U S A.* 2008;105(10):4022–4027. PMC2268797. https://doi.org/10.1073/pnas.0712209105.
164. Elam JS, Taylor AB, Strange R, et al. Amyloid-like filaments and water-filled nanotubes formed by SOD1 mutant proteins linked to familial ALS. *Nat Struct Biol.* 2003;10(6):461–467. https://doi.org/10.1038/nsb935.
165. Jonsson PA, Ernhill K, Andersen PM, et al. Minute quantities of misfolded mutant superoxide dismutase-1 cause amyotrophic lateral sclerosis. *Brain.* 2004;127(Pt. 1):73–88. https://doi.org/10.1093/brain/awh005.
166. Bergh J, Zetterström P, Andersen PM, et al. Structural and kinetic analysis of protein-aggregate strains in vivo using binary epitope mapping. *Proc Natl Acad Sci U S A.* 2015;112(14):4489–4494. PMC4394267. https://doi.org/10.1073/pnas.1419228112.

167. Taylor DM, Gibbs BF, Kabashi E, Minotti S, Durham HD, Agar JN. Tryptophan 32 potentiates aggregation and cytotoxicity of a copper/zinc superoxide dismutase mutant associated with familial amyotrophic lateral sclerosis. *J Biol Chem.* 2007;282(22):16329–16335. https://doi.org/10.1074/jbc.M610119200.
168. DuVal MG, Hinge VK, Snyder N, et al. Tryptophan 32 mediates SOD1 toxicity in a in vivo motor neuron model of ALS and is a promising target for small molecule therapeutics. *Neurobiol Dis.* 2019;124:297–310. https://doi.org/10.1016/j.nbd.2018.11.025.
169. Crown A, McAlary L, Fagerli E, et al. Tryptophan residue 32 in human Cu-Zn superoxide dismutase modulates prion-like propagation and strain selection. *PLoS One.* 2020;15(1):e0227655. https://doi.org/10.1371/journal.pone.0227655.
170. Béringue V, Vilotte J-L, Laude H. Prion agent diversity and species barrier. *Vet Res.* 2008;39(4):47. https://doi.org/10.1051/vetres:2008024.
171. Ayers JI, Fromholt S, Koch M, et al. Experimental transmissibility of mutant SOD1 motor neuron disease. *Acta Neuropathol.* 2014;128(6):791–803. https://doi.org/10.1007/s00401-014-1342-7.
172. Ayers JI, Diamond J, Sari A, et al. Distinct conformers of transmissible misfolded SOD1 distinguish human SOD1-FALS from other forms of familial and sporadic ALS. *Acta Neuropathol.* 2016;132(6):827–840. PMC5107152. https://doi.org/10.1007/s00401-016-1623-4.
173. Ayers JI, Fromholt SE, O'Neal VM, Diamond JH, Borchelt DR. Prion-like propagation of mutant SOD1 misfolding and motor neuron disease spread along neuroanatomical pathways. *Acta Neuropathol.* 2016;131(1):103–114. PMC4699876. https://doi.org/10.1007/s00401-015-1514-0.
174. Bidhendi EE, Bergh J, Zetterström P, Andersen PM, Marklund SL, Brännström T. Two superoxide dismutase prion strains transmit amyotrophic lateral sclerosis-like disease. *J Clin Invest.* 2016;126(6):2249–2253. PMC4887173. https://doi.org/10.1172/JCI84360.
175. Lang L, Zetterström P, Brännström T, Marklund SL, Danielsson J, Oliveberg M. SOD1 aggregation in ALS mice shows simplistic test tube behavior. *Proc Natl Acad Sci U S A.* 2015;112(32):9878–9883. PMC4538623. https://doi.org/10.1073/pnas.1503328112.
176. Chan PK, Chattopadhyay M, Sharma S, et al. Structural similarity of wild-type and ALS-mutant superoxide dismutase-1 fibrils using limited proteolysis and atomic force microscopy. *Proc Natl Acad Sci U S A.* 2013;110(27):10934–10939. PMC3704032. https://doi.org/10.1073/pnas.1309613110.
177. McCampbell A, Cole T, Wegener AJ, et al. Antisense oligonucleotides extend survival and reverse decrement in muscle response in ALS models. *J Clin Invest.* 2018;128(8):3558–3567. PMC6063493. https://doi.org/10.1172/JCI99081.
178. Bravo-Hernandez M, Tadokoro T, Navarro MR, et al. Spinal subpial delivery of AAV9 enables widespread gene silencing and blocks motoneuron degeneration in ALS. *Nat Med.* 2020;26(1):118–130. https://doi.org/10.1038/s41591-019-0674-1.
179. Capper MJ, Wright GSA, Barbieri L, et al. The cysteine-reactive small molecule ebselen facilitates effective SOD1 maturation. *Nat Commun.* 2018;9(1):1693. PMC5923229. https://doi.org/10.1038/s41467-018-04114-x.
180. Roberts BR, Lim NKH, McAllum EJ, et al. Oral treatment with CuII(atsm) increases mutant SOD1 in vivo but protects motor neurons and improves the phenotype of a transgenic mouse model of amyotrophic lateral sclerosis. *J Neurosci.* 2014;34(23):8021–8031. https://doi.org/10.1523/jneurosci.4196-13.2014.
181. Benatar M, Wuu J, Andersen PM, et al. Randomized, double-blind, placebo-controlled trial of arimoclomol in rapidly progressive ALS. *Neurology.* 2018;90(7):e565–e574. PMC5818014. https://doi.org/10.1212/WNL.0000000000004960.

182. McAlary L, Plotkin SS, Cashman NR. Emerging developments in targeting proteotoxicity in neurodegenerative diseases. *CNS Drugs*. 2019;33(9):883–904. PMC6776490. https://doi.org/10.1007/s40263-019-00657-9.
183. Buratti E, Dörk T, Zuccato E, Pagani F, Romano M, Baralle FE. Nuclear factor TDP-43 and SR proteins promote in vitro and in vivo CFTR exon 9 skipping. *EMBO J*. 2001;20(7):1774–1784. PMC145463. https://doi.org/10.1093/emboj/20.7.1774.
184. Buratti E, Brindisi A, Giombi M, Tisminetzky S, Ayala YM, Baralle FE. TDP-43 binds heterogeneous nuclear ribonucleoprotein A/B through its C-terminal tail: an important region for the inhibition of cystic fibrosis transmembrane conductance regulator exon 9 splicing. *J Biol Chem*. 2005;280(45):37572–37584. https://doi.org/10.1074/jbc.M505557200.
185. Hefferon TW, Groman JD, Yurk CE, Cutting GR. A variable dinucleotide repeat in the CFTR gene contributes to phenotype diversity by forming RNA secondary structures that alter splicing. *Proc Natl Acad Sci U S A*. 2004;101(10):3504–3509. PMC373492. https://doi.org/10.1073/pnas.0400182101.
186. Mercado PA, Ayala YM, Romano M, Buratti E, Baralle FE. Depletion of TDP 43 overrides the need for exonic and intronic splicing enhancers in the human apoA-II gene. *Nucleic Acids Res*. 2005;33(18):6000–6010. PMC1270946. https://doi.org/10.1093/nar/gki897.
187. Ling JP, Pletnikova O, Troncoso JC, Wong PC. TDP-43 repression of nonconserved cryptic exons is compromised in ALS-FTD. *Science*. 2015;349(6248):650–655. PMC4825810. https://doi.org/10.1126/science.aab0983.
188. Wu L-S, Cheng W-C, Hou S-C, Yan Y-T, Jiang S-T, Shen CKJ. TDP-43, a neuro-pathosignature factor, is essential for early mouse embryogenesis. *Genesis*. 2010;48(1):56–62. https://doi.org/10.1002/dvg.20584.
189. Sephton CF, Good SK, Atkin S, et al. TDP-43 is a developmentally regulated protein essential for early embryonic development. *J Biol Chem*. 2010;285(9):6826–6834. PMC2825476. https://doi.org/10.1074/jbc.M109.061846.
190. Kraemer BC, Schuck T, Wheeler JM, et al. Loss of murine TDP-43 disrupts motor function and plays an essential role in embryogenesis. *Acta Neuropathol*. 2010; 119(4):409–419. PMC2880609. https://doi.org/10.1007/s00401-010-0659-0.
191. Shiina Y, Arima K, Tabunoki H, Satoh J-I. TDP-43 dimerizes in human cells in culture. *Cell Mol Neurobiol*. 2010;30(4):641–652. https://doi.org/10.1007/s10571-009-9489-9.
192. Zhang Y-J, Caulfield T, Xu Y-F, et al. The dual functions of the extreme N-terminus of TDP-43 in regulating its biological activity and inclusion formation. *Hum Mol Genet*. 2013;22(15):3112–3122. PMC3699067. https://doi.org/10.1093/hmg/ddt166.
193. Afroz T, Hock E-M, Ernst P, et al. Functional and dynamic polymerization of the ALS-linked protein TDP-43 antagonizes its pathologic aggregation. *Nat Commun*. 2017;8(1):45. PMC5491494. https://doi.org/10.1038/s41467-017-00062-0.
194. Wang A, Conicella AE, Schmidt HB, et al. A single N-terminal phosphomimic disrupts TDP-43 polymerization, phase separation, and RNA splicing. *EMBO J*. 2018;37(5). : e97452https://doi.org/10.15252/embj.201797452.
195. Sun Y, Arslan PE, Won A, Yip CM, Chakrabartty A. Binding of TDP-43 to the 3′UTR of Its cognate mRNA enhances its solubility. *Biochemistry*. 2014;53(37): 5885–5894. https://doi.org/10.1021/bi500617x.
196. Lukavsky PJ, Daujotyte D, Tollervey JR, et al. Molecular basis of UG-rich RNA recognition by the human splicing factor TDP-43. *Nat Struct Mol Biol*. 2013;20(12): 1443–1449. https://doi.org/10.1038/nsmb.2698.
197. Bhardwaj A, Myers MP, Buratti E, Baralle FE. Characterizing TDP-43 interaction with its RNA targets. *Nucleic Acids Res*. 2013;41(9):5062–5074. PMC3643599. https://doi.org/10.1093/nar/gkt189.

198. Buratti E, Baralle FE. Characterization and functional implications of the RNA binding properties of nuclear factor TDP-43, a novel splicing regulator of CFTR exon 9. *J Biol Chem*. 2001;276(39):36337–36343. https://doi.org/10.1074/jbc.M104236200.
199. Conicella AE, Zerze GH, Mittal J, Fawzi NL. ALS mutations disrupt phase separation mediated by α-helical structure in the TDP-43 low-complexity C-terminal domain. *Structure*. 2016;24(9):1537–1549. PMC5014597. https://doi.org/10.1016/j.str.2016.07.007.
200. Li H-R, Chiang W-C, Chou P-C, Wang W-J, Huang J-R. TAR DNA-binding protein 43 (TDP-43) liquid-liquid phase separation is mediated by just a few aromatic residues. *J Biol Chem*. 2018;293(16):6090–6098. PMC5912450. https://doi.org/10.1074/jbc.AC117.001037.
201. Berning BA, Walker AK. The pathobiology of TDP-43 C-terminal fragments in ALS and FTLD. *Front Neurosci*. 2019;13:335. PMC6470282. https://doi.org/10.3389/fnins.2019.00335.
202. Guenther EL, Cao Q, Trinh H, et al. Atomic structures of TDP-43 LCD segments and insights into reversible or pathogenic aggregation. *Nat Struct Mol Biol*. 2018;25(6):463–471. https://doi.org/10.1038/s41594-018-0064-2.
203. Guo W, Chen Y, Zhou X, et al. An ALS-associated mutation affecting TDP-43 enhances protein aggregation, fibril formation and neurotoxicity. *Nat Struct Mol Biol*. 2011;18(7):822–830. PMC3357956. https://doi.org/10.1038/nsmb.2053.
204. Jiang L-L, Zhao J, Yin X-F, et al. Two mutations G335D and Q343R within the amyloidogenic core region of TDP-43 influence its aggregation and inclusion formation. *Sci Rep*. 2016;6:23928. PMC4814915. https://doi.org/10.1038/srep23928.
205. Mompeán M, Buratti E, Guarnaccia C, et al. Structural characterization of the minimal segment of TDP-43 competent for aggregation. *Arch Biochem Biophys*. 2014;545:53–62. https://doi.org/10.1016/j.abb.2014.01.007.
206. Cao Q, Boyer DR, Sawaya MR, Ge P, Eisenberg DS. Cryo-EM structures of four polymorphic TDP-43 amyloid cores. *Nat Struct Mol Biol*. 2019;26(7):619–627. https://doi.org/10.1038/s41594-019-0248-4.
207. Igaz LM, Kwong LK, Chen-Plotkin A, et al. Expression of TDP-43 C-terminal fragments in vitro recapitulates pathological features of TDP-43 proteinopathies. *J Biol Chem*. 2009;284(13):8516–8524. PMC2659210. https://doi.org/10.1074/jbc.M809462200.
208. Zhang Y-J, Xu Y-F, Cook C, et al. Aberrant cleavage of TDP-43 enhances aggregation and cellular toxicity. *Proc Natl Acad Sci U S A*. 2009;106(18):7607–7612. PMC2671323. https://doi.org/10.1073/pnas.0900688106.
209. Yang C, Tan W, Whittle C, et al. The C-terminal TDP-43 fragments have a high aggregation propensity and harm neurons by a dominant-negative mechanism. *PLoS One*. 2010;5(12):e15878. PMC3013128. https://doi.org/10.1371/journal.pone.0015878.
210. Furukawa Y, Kaneko K, Watanabe S, Yamanaka K, Nukina N. A seeding reaction recapitulates intracellular formation of Sarkosyl-insoluble transactivation response element (TAR) DNA-binding protein-43 inclusions. *J Biol Chem*. 2011;286(21):18664–18672. PMC3099683. https://doi.org/10.1074/jbc.M111.231209.
211. Laferrière F, Maniecka Z, Pérez-Berlanga M, et al. TDP-43 extracted from frontotemporal lobar degeneration subject brains displays distinct aggregate assemblies and neurotoxic effects reflecting disease progression rates. *Nat Neurosci*. 2019;22(1):65–77. https://doi.org/10.1038/s41593-018-0294-y.
212. Smethurst P, Newcombe J, Troakes C, et al. In vitro prion-like behaviour of TDP-43 in ALS. *Neurobiol Dis*. 2016;96:236–247. PMC5113659. https://doi.org/10.1016/j.nbd.2016.08.007.

213. Zeineddine R, Whiten DR, Farrawell NE, et al. Flow cytometric measurement of the cellular propagation of TDP-43 aggregation. *Prion*. 2017;11(3):195–204. PMC5480386. https://doi.org/10.1080/19336896.2017.1314426.
214. Feiler MS, Strobel B, Freischmidt A, et al. TDP-43 is intercellularly transmitted across axon terminals. *J Cell Biol*. 2015;211(4):897–911. PMC4657165. https://doi.org/10.1083/jcb.201504057.
215. McAlary L, Plotkin SS, Yerbury JJ, Cashman NR. Prion-like propagation of protein misfolding and aggregation in amyotrophic lateral sclerosis. *Front Mol Neurosci*. 2019;12:262. https://doi.org/10.3389/fnmol.2019.00262.
216. Shimonaka S, Nonaka T, Suzuki G, Hisanaga S-I, Hasegawa M. Templated aggregation of TAR DNA-binding protein of 43 kDa (TDP-43) by seeding with TDP-43 peptide fibrils. *J Biol Chem*. 2016;291(17):8896–8907. PMC4861459. https://doi.org/10.1074/jbc.M115.713552.
217. Igaz LM, Kwong LK, Lee EB, et al. Dysregulation of the ALS-associated gene TDP-43 leads to neuronal death and degeneration in mice. *J Clin Invest*. 2011;121(2):726–738. PMC3026736. https://doi.org/10.1172/JCI44867.
218. Walker AK, Spiller KJ, Ge G, et al. Functional recovery in new mouse models of ALS/FTLD after clearance of pathological cytoplasmic TDP-43. *Acta Neuropathol*. 2015;130(5):643–660. PMC5127391. https://doi.org/10.1007/s00401-015-1460-x.
219. Walker AK, Tripathy K, Restrepo CR, et al. An insoluble frontotemporal lobar degeneration-associated TDP-43 C-terminal fragment causes neurodegeneration and hippocampus pathology in transgenic mice. *Hum Mol Genet*. 2015;24(25): 7241–7254. PMC5007600. https://doi.org/10.1093/hmg/ddv424.
220. Salado IG, Redondo M, Bello ML, et al. Protein kinase CK-1 inhibitors as new potential drugs for amyotrophic lateral sclerosis. *J Med Chem*. 2014;57(6):2755–2772. PMC3969104. https://doi.org/10.1021/jm500065f.
221. Boyd JD, Lee P, Feiler MS, et al. A high-content screen identifies novel compounds that inhibit stress-induced TDP-43 cellular aggregation and associated cytotoxicity. *J Biomol Screen*. 2014;19(1):44–56. PMC3913261. https://doi.org/10.1177/1087057113501553.
222. Pozzi S, Thammisetty SS, Codron P, et al. Virus-mediated delivery of antibody targeting TAR DNA-binding protein-43 mitigates associated neuropathology. *J Clin Invest*. 2019;129(4):1581–1595. PMC6436898. https://doi.org/10.1172/JCI123931.
223. Belly A, Moreau-Gachelin F, Sadoul R, Goldberg Y. Delocalization of the multifunctional RNA splicing factor TLS/FUS in hippocampal neurones: exclusion from the nucleus and accumulation in dendritic granules and spine heads. *Neurosci Lett*. 2005;379(3):152–157. https://doi.org/10.1016/j.neulet.2004.12.071.
224. Wang X, Arai S, Song X, et al. Induced ncRNAs allosterically modify RNA-binding proteins in cis to inhibit transcription. *Nature*. 2008;454(7200):126–130. PMC2823488. https://doi.org/10.1038/nature06992.
225. Iko Y, Kodama TS, Kasai N, et al. Domain architectures and characterization of an RNA-binding protein, TLS. *J Biol Chem*. 2004;279(43):44834–44840. https://doi.org/10.1074/jbc.M408552200.
226. Nomura T, Watanabe S, Kaneko K, Yamanaka K, Nukina N, Furukawa Y. Intranuclear aggregation of mutant FUS/TLS as a molecular pathomechanism of amyotrophic lateral sclerosis. *J Biol Chem*. 2014;289(2):1192–1202. PMC3887186. https://doi.org/10.1074/jbc.M113.516492.
227. Luo F, Gui X, Zhou H, et al. Atomic structures of FUS LC domain segments reveal bases for reversible amyloid fibril formation. *Nat Struct Mol Biol*. 2018;25(4):341–346. https://doi.org/10.1038/s41594-018-0050-8.

228. Mackenzie IRA, Ansorge O, Strong M, et al. Pathological heterogeneity in amyotrophic lateral sclerosis with FUS mutations: two distinct patterns correlating with disease severity and mutation. *Acta Neuropathol.* 2011;122(1):87–98. PMC3319073. https://doi.org/10.1007/s00401-011-0838-7.
229. King A, Troakes C, Smith B, et al. ALS-FUS pathology revisited: singleton FUS mutations and an unusual case with both a FUS and TARDBP mutation. *Acta Neuropathol Commun.* 2015;3:62. PMC4600255. https://doi.org/10.1186/s40478-015-0235-x.
230. Falcon B, Zhang W, Schweighauser M, et al. Tau filaments from multiple cases of sporadic and inherited Alzheimer's disease adopt a common fold. *Acta Neuropathol.* 2018;136(5):699–708. PMC6208733. https://doi.org/10.1007/s00401-018-1914-z.
231. Guerrero-Ferreira R, Taylor NM, Mona D, et al. Cryvo-EM structure of alpha-synuclein fibrils. *Elife.* 2018;7:e36402. PMC6092118. https://doi.org/10.7554/eLife.36402.

CHAPTER TWELVE

The role of the cellular prion protein in the uptake and toxic signaling of pathological neurodegenerative aggregates

Carlo Scialò and Giuseppe Legname*
Laboratory of Prion Biology, Department of Neuroscience, Scuola Internazionale Superiore Di Studi Avanzati (SISSA), Trieste, Italy
*Corresponding author: e-mail address: legname@sissa.it

Contents

Abstract

Neurodegenerative disorders are invariably associated with intra- or extra-cellular deposition of aggregates composed of misfolded insoluble proteins. These deposits composed of tau, amyloid-β or α-synuclein spread from cell to cell, in a prion-like

Progress in Molecular Biology and Translational Science, Volume 175
ISSN 1877-1173
https://doi.org/10.1016/bs.pmbts.2020.08.008

manner. Emerging evidence suggests that the circulating soluble species of these misfolded proteins (usually referred as oligomers) could play a major role in pathology, while insoluble aggregates would represent their protective less toxic counterparts. Convincing data support the hypothesis that the cellular prion protein, PrP^C, act as a toxicity-transducing receptor for amyloid-β oligomers. As a consequence, several studies extended investigations to the role played by PrP^C in binding aggregates of proteins other than Aβ, such as tau and α-synuclein, for its possible common role in mediating toxic signaling. A better characterization of the biological relevance of PrP^C as key ligand and potential mediator of toxicity for multiple proteinaceous aggregated species, prions or PrP^{Sc} included, would bring relevant therapeutic implications. Here we will first describe the structure of the prion protein and the hypothesized interplay with its pathological counterpart PrP^{Sc} and then we will recapitulate the most relevant discoveries regarding the role of PrP^C in the interaction with aggregated forms of other neurodegeneration-associated proteins.

1. Introduction

Neurodegenerative disorders (NDs) are characterized by a progressive and selective failure of specific subsets of neurons associated with intra- or extra-cellular deposition of insoluble protein aggregates.[1] Extracellular deposits comprise aggregates with immunoreactivity for amyloid-β (Aβ) or prion protein (PrP), while proteins found in intracellular deposits include, among others, tau and α-synuclein (α-syn). Alzheimer's disease (AD) is characterized by the progressive accumulation of pathological plaques of Aβ, associated with the presence of neurofibrillary tangles composed of hyperphosphorylated and cleaved forms of the microtubule-associated protein tau.[2,3] Besides AD, where tau is deposited in association with Aβ, tauopathies include frontotemporal dementia linked to chromosome 17 (FTLD-17), progressive supranuclear palsy (PSP), argyrophilic grain disease (AGD), corticobasal degeneration (CBD) and Pick's disease.[4] Lewy bodies (LB) and Lewy neurites (LN) composed of α-syn are characteristic of Parkinson's disease (PD) and dementia with Lewy bodies (DLB), while α-syn oligodendroglial and neuronal deposits are present in multiple system atrophy (MSA).[5,6] Prion diseases, or Transmissible Spongiform Encephalopathies (TSEs), are unique neurodegenerative disorders showing an infectious transmissible protein species capable of recapitulating clinical disease[7,8] with no clear evidence to naturally occurring human-to-human transmission of other neurodegenerative condition.[9,10] Despite this relevant distinction, clinical, cellular, molecular and biochemical studies provided evidence supporting a common modality for the spreading and propagation of the neurodegenerative process, i.e., the "prion-like" hypothesis.[11–14] The "prion-like" term is used to underline, on one hand, the similarities with the prion replication and propagation process and on the other one, the lack of infectivity. The canonical model for the replication of misfolded proteins lies in the prion paradigm.[15] In prion disorders, the critical pathological step is the conversion of the normal cellular prion protein (PrP^C) into its β-sheet enriched pathological conformer PrP-scrapie (PrP^{Sc}).[16] According to the "prion-like" hypothesis, misfolded protein assemblies in NDs other than prion disorders, act similarly as seeds of aggregation that can sequester their native isoforms and convert them into pathological molecules, thereby growing in size. The term "seed" indicates the smallest amount of a misfolded protein able to template the pathological conversion onto native molecules and to spread throughout nearby or connected brain areas, after its release in the extra-cellular space.[13] Due to their macroscopic features and their relative abundance in NDs affected tissues, aggregates composed of misfolded fibrillary structures have occupied a central role in the interpretation of disease etiology. Pathology investigations, genetic discoveries, animal and biophysics prediction models all supported a strict connection between pathologic protein aggregates and NDs.[17] Nevertheless, emerging evidence suggests that circulating soluble species of misfolded proteins (usually referred as oligomers) could play a major role in the pathology, while insoluble aggregates would represent their protective less toxic counterparts.[4,17–27] Accordingly, the severity of cognitive deficits in AD correlates more strongly with the levels of soluble forms of Aβ than with insoluble amyloid plaque load[26] and impairment of hippocampal synaptic plasticity can be detected before the formation of insoluble Aβ plaques in amyloid-β precursor protein (APP) transgenic mouse models of AD.[28] Moreover, soluble Aβ oligomers (Aβ-o) showed detrimental effects on hippocampal long term potentiation (LTP) *in vitro* and *in vivo*.[26] Similar detrimental effects have been shown for tau oligomers (tau-o) extracted from AD diseased brains or obtained by *in vitro* fibrillization[21,29] and for *in vitro* obtained α-syn oligomers (*syn*-o).[30–32] The discovery of circulating soluble oligomeric species and of their toxic effects was paralleled by the research for putative ligands and downstream effectors of their detrimental activity. Aβ-o were the most extensively investigated pathological oligomeric species with multiples ligands identified.[27] Among these, the prion protein, PrP^C, occupies a relevant position[33–38] and was recently recognized as the highest affinity binding partner for Aβ-o.[39] Many studies revealed that the binding between PrP^C and Aβ-o is mediated by high affinity interaction (in the sub-nanomolar range) and that it produces at least some of the Aβ-o-induced neurotoxic effects.[19,34,35,37,40–46] Conversely, other studies showed the presence of PrP-independent Aβ-o deleterious effects.[47–50] This apparent discrepancy has several potential explanations that will be discussed later below in another section. The tantalizing hypothesis that PrP^C, already proved to be central in the pathogenesis of prion disorders, could represent a common acceptor for multiple neurodegeneration-associated amyloid species, stimulated further investigations to validate this possibility. As a consequence, in recent years the interplay between PrP^C and oligomers of proteins other than Aβ, such as tau and α-syn, has been increasingly investigated.[19,22,23,51–55] The possibility of an interaction between tau and the prion protein was already suggested by neuropathological examinations of Gerstmann-Straüssler-Scheinker syndrome (GSS) cases, a subset of familial forms of TSEs. In fact, GSS-linked mutations of the *PRNP* gene, encoding for the cellular prion protein, display to co-deposition of PrP^{Sc} with hyperphosphorylated forms of tau.[56–63] The nature of this co-deposition is still a matter of debate and the study of the molecular interplay between PrP^C and tau could shed new light in the understanding of this phenomenon. Furthermore, a better characterization of the role of PrP^C as a key ligand and potential mediator of toxicity for multiple pathologic oligomeric species (included PrP^{Sc}), would bring relevant therapeutic implications. Here we will first describe the structure of the prion protein and the hypothesized interplay with its pathological counterpart, PrP^{Sc}, and then we will recapitulate the most relevant discoveries regarding the role of PrP^C in the interaction with aggregated forms of several neurodegeneration-associated proteins.

2. The prion protein structure and PrP^C-mediated PrP^{Sc} toxicity

2.1 Prion protein structure

The cellular prion protein, PrP^C, is a glycosylphosphatidylinositol (GPI)-anchored protein of 231 aa encoded by the *PRNP* gene located on chromosome 20, in humans.[64–68] It is composed of two structurally well-defined regions: a long, N-terminal flexible tail and a globular C-terminal domain containing 3 α-helices and 2 short β-strands flanking the first α-helix. The flexible tail is further divided in a small charged cluster (CC1), an octarepeat (OR) region and a central domain (CD), which comprises a second charge cluster (CC2) and a hydrophobic domain (HD).

The pre-pro-protein is translocated to the endoplasmic reticulum (ER) where it is subjected to several post-translational modifications including N-linked glycosylation at residues N181 and N197, formation of a single disulfide bond at position C179 and C214, cleavage of the C-terminal signal peptide and subsequent attachment of the GPI anchor at position 231.[69] PrP^C is widely expressed in the CNS during early development, in adult neurons and glial cells. In addition, expression of PrP^C has been reported in several tissues including lymphoid organs and the heart[67,70,71] and at lower levels in the kidney and liver.[71,72] Several putative functions have been suggested for PrP^C, including ion balance homeostasis, neuritogenesis, neuronal homeostasis, cell signaling, cell adhesion and a protective role against stress.[73] PrP^C could serve as a dynamic platform for the assembly of signaling modules at the cell surface, acting with the properties of a cell surface scaffold protein.[74] According to this model, PrP^C shares several features with intracellular scaffold proteins, as for instance the ability to recruit spatially organized groups of binding partners, mediation of signaling pathways, mutual allosteric regulation with binding molecules and oligomerization-dependent effect on signaling.[74] Since PrP^C does not span the plasma membrane, accessory molecules are required to transduce signals into the cytosol. Several PrP^C binding partners have been identified so far. Among them the most studied are the metabotropic glutamate receptors (mGluR1 and mGluR5) *via* laminin γ1 chain interaction,[75] the α7 type of nicotinic acetylcholine receptor (α7nAChR) following the binding of the cochaperone hop/STI1,[76] the neural cell adhesion molecule (NCAM)[77] and the Laminin Receptor Precursor/Laminin Receptor (LRP/LR).[78]

2.2 PrP^C is essential for prion replication

Several studies unambiguously proved that the expression of PrP^C is essential for prion propagation. Knock-out mice for the *Prnp* gene are resistant to prion diseases and to the propagation of the scrapie infectious agent.[79] In a seminal study, neural tissue overexpressing PrP^C was grafted into the brain of PrP-deficient mice which were later inoculated with infectious prions. PrP^C deficient neurons, exposed to PrP^{Sc} material produced by PrP^C over-expressing neuronal grafts, were devoid of neuropathological alterations.[80] In a subsequent study, Mallucci et al. showed that depleting endogenous neuronal PrP^C in mice with established prion infection was able to revert early spongiform changes and to prevent neuronal loss and progression to clinical disease. Interestingly, this occurred despite the continuous production of infectious PrP^{Sc} by surrounding astrocytes, leading to levels of PrP^{Sc} deposition comparable to terminally ill wild-type mice.[81] The expression of physiological levels of a form of PrP^{C} devoid of the GPI anchor (ΔGPI-PrP) was permissive for PrP^{Sc} replication but produced a clinically silent phenotype.[82] When ΔGPI-PrP was expressed at higher levels it produced a late onset spontaneous phenotype associated with the deposition of large amyloid plaques of PrP^{Sc}. Interestingly, disease onset was accelerated by co-expression of wild-type full-length PrP^{C}.[83]

2.3 PrP^{C} mediates PrP^{Sc} toxicity

All the above-mentioned studies, besides confirming the essential role of PrP^{C} as the indispensable substrate for prion replication, pointed to PrP^{C} as the cellular mediator of at least part, if not the majority, of PrP^{Sc} neurotoxic effects. Moreover, different studies suggested that PrP^{C} can acquire a neurotoxic potential also in the absence of PrP^{Sc}/prion propagation.[84,85] Mice models expressing PrPΔHD, a PrP^{C} mutant with a deletion of its hydrophobic domain (HD), demonstrated that neurotoxicity of PrPΔHD is linked to a PrP^{C}-dependent signaling pathway.[85–87] Importantly, these phenotypes were suppressed in a dose-dependent manner by co-expression of wild-type PrP^{C}, suggesting that the wild-type and mutated molecules interact with each other, or compete for binding to a common molecular target that mediates both physiological and pathological effects. These toxic mutants of PrP^{C} induced spontaneous ionic currents, recordable by patch clamping techniques, when expressed in cell lines[88] and in primary neurons.[89] Notably, also these currents were silenced by co-expression of wild-type PrP^{C} in the same cells. Further studies to entangle the relationship between PrPΔHD mutants and their toxicity suggested the presence of an auto-inhibitory mechanism that regulates the functional activity of PrP^{C}, mediated by an intramolecular docking between N- and C-terminal domains of PrP^{C}.[90] According to this proposed model, the toxic activity of the flexible N-terminal domain is regulated in *cis* by the globular structured C-terminal domain and anti-PrP antibodies directed against the C-terminal domain would produce a neurotoxic effect by disrupting this interaction. Interestingly enough, the authors suggest that the same effect could be mediated by PrP pathological ligands, as PrP^{Sc} or circulating pathological oligomers (Fig. 1).[90] Even if this model provides an interesting mechanistic explanation for oligomers-induced PrP^{C} mediated neurotoxicity, the fact that mice expressing N-terminally truncated PrP^{C} remain susceptible to prion diseases[95,96] argues against this, as a primary

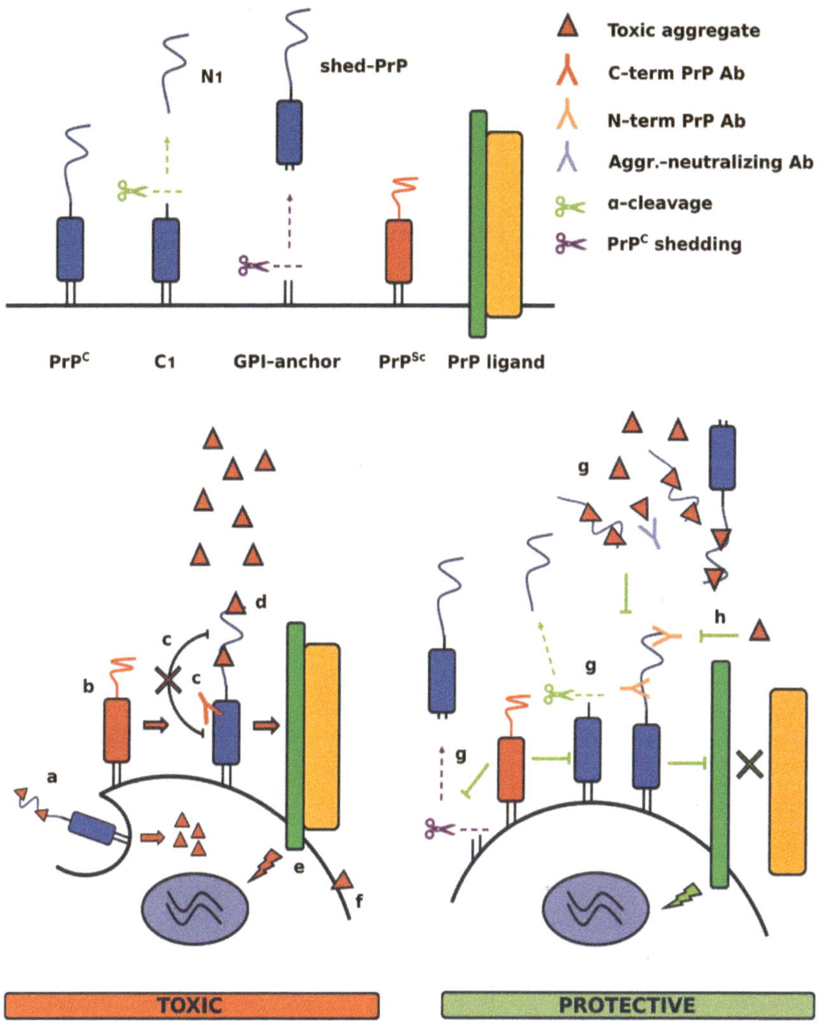

Fig. 1 Schematic representation of the major features involved in the complex interplay between PrP^C and toxic pathological aggregates (a) PrP^C is involved in the uptake and internalization of tau and α-syn fibrils.[51,52] (b) The interaction between PrP^{Sc} and PrP^C is necessary for prion replication[79] and is also the driver of PrP^{Sc}-induced toxic signaling,[43] mediated by PrP^C *via* several putative binding partners (e).[91] (c) An auto-inhibitory mechanism regulates the functional activity of PrP^C, mediated by an intra-molecular docking between N- and C-terminal domains of PrP^C.[90] Anti-PrP^C antibodies directed against PrP^C C-terminal domain (c) or PrP^C toxic ligands (d) produce a neurotoxic effect by disrupting this intra-molecular interaction.[90] (d) PrP^C mediates Aβ, tau and α-syn aggregates neurotoxic effects *via* two binding sites located in its N-terminus[19,40,43,46] (e) PrP^C mediates cell signaling acting as an extracellular scaffolding protein, able to *(Continued)*

pathogenic mechanism. Rambold et al., showed that PrP^{Sc} produced and released by ScN2a chronically infected cells induced apoptosis in SH-SY5Y cells, only if cells were transiently transfected to express PrP^{C}.[84] The scrapie prion-induced cell death was paralleled by the activation of the Jun N-terminal kinase (JNK).[84] It should be stressed that the authors did not characterize whether the toxic PrP^{Sc} molecules were a PK-resistant intermediate generated during the conversion process or oligomers or fibrils. In the attempt to provide evidence that this phenomenon was not related to prion replication and accumulation in SH-SY5Y cells exposed to PrP^{Sc}, but that was instead specifically linked to the transmission of a toxic signal induced by PrP^{Sc} *via* PrP^{C}, the same group repeated the experiment expressing heterologous PrP^{C} molecules, less susceptible to prion conversion due to species barrier. Heterologous PrP^{C}, be it of hamster, human, cervid or bovine origin, efficiently mediated toxic signaling of mouse PrP^{Sc}.[43] Furthermore, they showed that PrP^{Sc} neurotoxic activity in SH-SY5Y cells expressing PrP^{C} can be reduced by the use of the oligomer-specific antibody A11.[43] A similar PrP^{C}-mediated neurotoxic effect was observed in the same experimental settings also with Aβ-o,[43] and will be discussed in details in the dedicated paragraph.

2.4 PrP^{C} proteolytic cleavage and its role in pathology

The prion protein is subject to post-translational proteolytic processing. These cleavage events have been shown to be involved in the regulation of its physiological functions, to produce biologically active fragments and to potentially influence pathology. Four cleavages appear to be conserved and currently represent the best investigated processing events on PrP^{C}.[93,97] The first studied proteolytic processing of PrP^{C} takes place in a

Fig. 1—Cont'd organize multiprotein complexes at the cell surface[74] (f) Toxic Aβ oligomers could bind directly to the cell membrane and disrupt its normal function, leading to cytotoxicity.[92] N1 could bind to these species, or to the membrane surface, in a manner that prevents the formation of cell membrane damage (not shown). (g) PrP^{C} α-cleavage and shedding protective role: both the released N1 fragment and shed-PrP contain the N-terminal part of the protein harboring binding sites for toxic aggregates, so, they could bind and neutralize them. The same effect could be induced by aggregates-directed neutralizing antibodies (in light blue).[43] Cleavage events decrease the presence of PrP^{C} at the cell membrane impeding the transmission of aggregates-induced toxic response(s). In prion disorders, cleavage events would reduce the amount of PrP^{C} which could be converted in PrP^{Sc}.[93] (h) Antibodies directed against the N-terminus of PrP^{C} showed a protective role against PrP^{C}-mediated neurotoxic signaling, probably competing with toxic aggregates on their binding sites.[22,23,35,94]

region termed "α-cleavage site" (located at positions K110/H111 or H111/M112 in the human sequence) and releases the N-terminal flexible part of PrP^{C} (N1 fragment of ~11 kDa) while leaving the globular C-terminal domain (C1 fragment of ~16 kDa) attached to the membrane *via* its GPI anchor.[93,97,98] Another minor proteolytic event termed "β-cleavage" occurs at residues 90–91. In consequence of this cleavage, an N-terminal fragment of ~9 kDa (termed N2) is released whereas a corresponding C2 fragment of 18–20 kDa (depending on the glycosylation status) remains bound to the cellular membrane. In contrast to α-cleavage, β-cleavage is mainly observed under pathological conditions and much less present or even absent in normal conditions. A third physiological cleavage of PrP^{C}, called shedding, occurs in close proximity to its GPI-anchor and results in the release of the almost full-length protein (shed-PrP) from the plasma membrane.[93,97,99] Recently, another proteolytic event of PrP^{C}, i.e., γ-cleavage, has been identified and described,[100] but the specific cutting site and its patho- physiological relevance need further investigations. Several reports highlighted the importance of α-cleavage to regulate many PrP^{C} physiological functions,[93,97] a detailed description of which goes beyond the scopes of this chapter. We will briefly discuss only those which showed to have a potential influence in the interaction between PrP^{C} and pathological aggregates of proteinaceous species, PrP^{Sc} included. At least two modes of how proteolytic processing of PrP^{C} could play a role in disease have been suggested. First, both the released N1 fragment and shed-PrP contain the N-terminal part of the protein which is thought to interact with pathological assemblies, so, once released, they could bind and neutralize them. Second, cleavage events would decrease the presence of PrP^{C} at the cell membrane, thus impeding the transmission of aggregates-induced toxic response(s) as well as reducing the substrate for the conversion of PrP^{C} in PrP^{Sc}, in prion disorders (Fig. 1).[93] Accordingly, the N1 fragment showed neuroprotective properties against PrP^{Sc}[43] and Aβ-mediated synaptic damage[34,92,101] and both PrP^{C} shedding and α-cleavage are increased during AD[102] (see next paragraph).

3. PrP^{C}-mediated Aβ toxicity

3.1 Aβ oligomers and AD heterogeneity

It is now recognized that Aβ pathological deposits encompass a heterogeneous mixture of peptides with different solubility, stability and biological and toxic properties[27] which can be observed either in different areas of a single AD affected brain or in patients suffering from various subtypes of AD.[103,104]

Part of the heterogeneity in the length of Aβ peptides is related to different γ-secretase C-terminal cleavage positions in the amyloid precursor protein (APP). In fact, γ-secretase generates several peptides, of which Aβ43, Aβ42, Aβ40, Aβ38 and Aβ37 variants are the most studied (Aβ numbering indicates the last C-terminal amino acid residue).[27] Additional heterogeneity is generated by enzymatic processes mediated by aminopeptidases, glutaminylcyclase or isomerases, resulting in more than 20 Aβ peptides that participate in putative Aβ functions in the normal brain and oligomerization and fibrillization in the AD affected brain.[27] It is well accepted that Aβ monomers do not interrupt normal synaptic function, whereas small (low-n Aβ oligomers, Aβ-o) and large Aβ aggregates (protofibrils) impair synaptic plasticity.[26] These different pathologic Aβ oligomeric circulating species likely contribute to the heterogeneity of AD clinical manifestations.[105–108] Aβ-o can be obtained by *in vitro* aggregation of Aβ monomers, by collection from the culture medium of specific cell lines or from the soluble fraction of AD diseased brains.[26]

3.2 PrP^C as Aβ oligomers receptor

Aβ-o have been shown to act as potent neurotoxins[24,26] with many putative ligands/receptors proposed so far.[27] Among the list of potential receptors, PrP^C has been extensively studied and has been recently recognized as the highest affinity binding partner for Aβ-o.[39] The first striking demonstration of PrP^C and Aβ-o interaction was obtained using an unbiased expression cloning screening approach to identify potential ligands for a specific class of Aβ42 synthetic oligomers, namely, Aβ-derived diffusible ligands (ADDLs).[37] ADDLs binding affinity to PrP^C was in the nanomolar scale and the ADDLs-mediated blockade of LTP was absent in hippocampal slices from young adult PrP null mice. Furthermore, anti-PrP antibodies prevented ADDLs binding to PrP^C and were able to rescue synaptic plasticity in hippocampal slices. The deletion of 11 amino acids resulting in the Δ95–105 variant, significantly reduced this binding (~80% of reduction).[37] After this first discovery, many other studies confirmed PrP^C-mediated Aβ-o neurotoxic effects[19,34,35,40–46,91,94,109] while others demonstrated Aβ-o deleterious effect which did not require PrP^C expression.[47–50]

3.3 Interpretation of controversies

The first study against PrP^C involvement in mediating Aβ-o toxicity, despite confirming an interaction between Aβ42-o and PrP^C, observed comparable detrimental effects on memory impairment in wild-type and *PRNP* null mice.[47] Another study showed that Aβ42-o-induced synaptic depression, loss of dendritic spines and blockade of LTP were present in hippocampal slices prepared from PrP^{C}-deficient animals.[50] To unravel the impact of PrP^{C} in AD, mice lacking or overexpressing membrane-anchored or secreted PrP^{C} were crossed with *APPPS1*$^{+}$ mice co-expressing mutant *APP* (*APP*$^{KM670/671NL}$) and mutant presenilin-1 which suffer from Aβ-dependent cognitive deficits. The ablation or overexpression of PrP^{C} had no effect on the impairment of hippocampal synaptic plasticity.[48] Finally, PrP^{C} expression was irrelevant in modifying Aβ-induced neurological impairment in another transgenic AD mouse model, the J20 mouse line.[49] These apparent discrepancies were interpreted as the result of the use of heterogeneous Aβ preparations in different studies and also justified by the likely hypothesis that not all Aβ-o detrimental effects should be PrP^{C}-mediated. In fact, since AD is a clinicopathological syndrome with multiple genetic and environmental causative factors and since pathological Aβ is not represented by a unique molecule, it seems not plausible that a single receptor could be responsible for several toxic effects mediated by a *plethora* of Aβ toxic entities. In response to Kessels et al. which claimed the use of the same Aβ42 oligomeric species,[50] Lauren et al. questioned the similarity of the Aβ preparations that were only characterized by SDS-PAGE and suggested that PrP^{C} dependency may also vary with dosing of the administered Aβ42 oligomers.[110] The contrasting findings obtained with transgenic mice could also be related to differences in the soluble Aβ species circulating at the ages when animals were sacrificed (*APPPS1*$^{+}$ mice were examined at 2–4 months of age while J20 mice were 6–8 months old). In support of this hypothesis, a later study in the *APPPS1*$^{+}$ mice showed some PrP-dependent effects at 14 months when soluble Aβ assemblies become abundant.[94]

3.4 PrP^{C}-mediated Aβ oligomer toxicity

In the last 10 years further investigations, published also by authors who initially described PrP-independent Aβ toxicity, supported the role of PrP^{C} in mediating at least part of Aβ-o detrimental effects, suggesting that a consensus may be forming. Transgenic mice expressing familial AD transgenes encoding *APPswe* and *PSen1ΔE9* on a *PRNP* null background presented no detectable impairment of spatial learning and memory, despite displaying pathological levels of Aβ accumulation.[109] The cluster of basic amino acids 23–27 (KKRPK) at the very N-terminus of PrP^{C} was identified as a relevant Aβ42-o binding site in addition to the previously characterized

92–110 region.[40] PrP^C expression was necessary for detrimental effects of *ex vivo* material extracted from human AD brain in hippocampal slices and the anti-PrP antibodies ICSM-35 (which recognizes an epitope within the residues 93–102) and ICSM-18 (recognizing residues 143–153) were both able to block the Aβ-mediated inhibition of LTP *in vivo* and *in vitro*.[35] Resenberger et al., co-cultivated SH-SY5Y cells (expressing PrP^C or not) with a stably transfected Chinese hamster ovary cell line (CHO-7PA2) that expresses the familial AD mutation V717F in the *APP* gene and secretes Aβ toxic species in the culture medium. A significant increase in apoptotic cell death was observed in SH-SY5Y cells transiently transfected to express PrP^C, while control SH-SY5Y cells expressing GPI-anchored GFP displayed no adverse effects on cell viability.[43] Interestingly, SH-SY5Y cells expressing PrP^C did not show increased apoptosis when co-cultured with control CHO cells or with CHO-7PA2 cells pre-treated with a γ-secretase inhibitor, showing that the detrimental effect was effectively driven by the presence of Aβ toxic species in the culture medium. In the same study, they observed that toxic signaling *via* PrP^C required its intrinsically disordered N-terminal domain and the GPI anchor. Interestingly, a secreted version of PrP N-terminal domain was able to associate with Aβ conformers and to antagonize their toxic signaling.[43] Larson et al. showed that soluble Aβ extracted from AD brains was able to bind to PrP^C at neuronal dendritic spines *in vivo* and *in vitro* where it formed a complex with Fyn, resulting in the activation of the kinase. Furthermore, using the antibody 6D11 to prevent Aβ-o binding to PrP^C, they abolished Fyn activation and Fyn-dependent downstream effects.[94] In the attempt to identify and characterize active PrP-dependent, PrP-independent and benign Aβ oligomeric assemblies, Nicoll and colleagues showed that Aβ assemblies that trigger PrP^C-mediated toxicity are protofibrils with a defined tridimensional structure.[42] More in details, the examined *in vitro* produced Aβ42 protofibrils presented an elaborated nanotube architecture with a triple helix structure wound around an hollow core.[42]

3.5 PrP^C proteolytic cleavage and its role in AD pathology

Full-length PrP^C is able to inhibit *in vitro* Aβ40 and Aβ42 fiber formation and to cause disassembly of pre-formed Aβ mature fibrils, trapping Aβ in an oligomeric form enriched in antiparallel β-sheet.[46] Aβ recognition sites mapped to specific residues in the natively unstructured N-terminal half of PrP^C, confirming the observations of previous studies pointing at residues

95–110 and 23–27 as primary and secondary binding enhancement regions. According to the authors, this interaction would maintain Aβ in a more toxic oligomeric conformation and thus favor pathology.[46] Partially in contrast with these results, several studies demonstrated an *in vitro* and *in vivo* protective role for the released full-length recombinant PrP and its N-terminal fragment.[34,45,92,101] PrPN1, the main physiological cleavage fragment of PrP^C, protects primary cultured neurons against toxicity and cell death triggered by oligomer-enriched conditioned medium collected from a CHO cell line overexpressing the London mutated β*APP* (APP_{LDN}) gene.[101] PrPN1 also protects neurons against oligomers extracted from AD affected brain tissues.[101] Fluharty and colleagues have shown that PrPN1 binds early Aβ42 oligomeric intermediates during Aβ42 polymerization into amyloid fibrils[34] confirming that its binding affinity is determined by positively charged residues in the two previously identified Aβ42-o binding sites (positions 23–31 and 95–105). Interestingly, N1 strongly suppresses Aβ42-o toxicity in cultured murine hippocampal neurons, in a *Caenorhabditis elegans*-based assay, and *in vivo* in a mouse model of Aβ-induced memory dysfunction.[34] These results suggest that PrP^C cleavage processing could play a relevant role in modulating AD pathology. This hypothesis was verified in the *APPswe/PSen1ΔE9* mouse model of AD, which exhibited altered PrP^C processing in brain cortex characterized by increased levels of both α- and β-cleavage.[102] Since the released N1 fragment contains the N-terminal part of the protein which is thought to interact with pathological oligomers, the observed increase in α-processing could be interpreted as a protective response with two potential beneficial functions: (i) to decrease PrP^C-mediated toxic signaling reducing the presence of PrP^C on the cell surface; (ii) to neutralize Aβ-o pathological species before their binding to down-stream receptors.[93] On the other hand, the increase in β-cleavage, which has been previously associated to an excess of reactive oxygen species (ROS),[111] could reflect distinct processes such as lysosomal failure to clear Aβ aggregates and/or increased ROS production induced by stress, which is a common feature in AD mouse models.[102] Interestingly, other reports evidenced a protective role against cytotoxic Aβ42-o of soluble (membrane anchor-free) recombinant human prion protein (recHuPrP) and of the N1 fragment, also independently of the presence of endogenous PrP^C.[92] More in details, both recHuPrP and N1 blocked Aβ42-o mediated inhibition of LTP in hippocampal slices and in wild-type primary hippocampal neurons. Furthermore, they reduced Aβ42-o-induced cytotoxic effects when the experiment was performed with hippocampal slices and neurons derived from *PRNP* null mice.[92] The authors explained these results suggesting a double protective action of recHuPrP and N1. The first mechanism would specifically target the PrP^C-dependent LTP impairment, likely through a competitive inhibition of Aβ42-o binding to PrP^C on the membrane surface. The second, less specific protective role against cytotoxicity, might be due to recHuPrP and N1 interference in the interaction between Aβ and the cell membrane: recHuPrP and N1 would bind to Aβ42-o species, or to the membrane surface, in a manner that prevents the formation of cell membrane leakage through defects and/or pores in the lipid bilayer.[92] These hypothesized mechanisms were supported by another study which analyzed chemically cross-linked small oligomeric species of Aβ42, finding that recHuPrP inhibits the cytotoxicity of these species in neurons as well as their membrane permeation effects in liposomes.[45] Another recent study also supported a protective role of the N1 fragment and investigated the potential beneficial effect of monoclonal antibodies grafted with one of the two PrP-Aβ-o binding sites (i.e., aa 19–33 or 87–112).[41] PrP-grafted antibodies bound a significant portion of aggregated Aβ in aqueous AD extracts, but when these antibodies were co-incubated with neurons treated with brain extracts, they did not reduce toxicity. By contrast, PrP fragment N1 did protect against Aβ.[41] Since conversely to N1, the two grafted antibodies harbored only one binding sequence each, this suggests that the presence of both Aβ-o binding sites on the molecule is a necessary requisite for the PrP-mediated protective role.[41]

3.6 PrP^C mediators of Aβ oligomers induced toxic signaling

As stated before, it has been suggested that PrP^C mediates cell signaling acting as an extracellular scaffolding protein, able to organize multiprotein complexes at the cell surface.[74] Several studies suggested that this scaffolding can bring neurotoxic effects. For instance, PrP^C binding to Aβ-o seems to produce a maladaptive signaling *via* metabotropic glutamate receptor 5 (mGluR5),[112] which in turn, initiates multiple changes in synaptic homeostasis, leading to excitotoxicity, endoplasmic reticulum stress, and eventually to synaptic degradation and neuronal cell death.[37,94,102,109,113–116] Furthermore, PrP^C was found to be enriched in post-synaptic densities (PSD) where the binding with *in vitro* and *ex vivo* derived Aβ-o species activates the nonreceptor tyrosine kinase Fyn, leading to NR2B phosphorylation and altered NMDAR localization and, ultimately, to destabilization of dendritic spines.[44] Another group investigated the role of a peptide mimicking the binding site of laminin γ1 onto PrP^C (Ln-γ1) observing that once bound to PrP^C, it induced intracellular Ca^{2+} increase in neurons *via* the complex PrP^C-mGluR5.[91] Even if Ln-γ1 promoted the internalization of PrP^C and mGluR5 and transiently decreased Aβ-o biding to neurons, the peptide did not impact Aβ-o toxicity.[91] Since mGluR5 showed a key role in the transmission of toxic signaling also in prion diseases, the same group tested whether mGluR5 knock-out mice would be susceptible to prion infection, observing mild, but significant, effects on disease progression, without affecting survival of infected mice.[91]

4. PrP^C-mediated tau toxicity

4.1 Tau neurotoxic species

Tau is a microtuble-associated protein that under physiological conditions regulates microtubules assembly, dynamic behavior and spatial organization.[117] The gene encoding tau, *MAPT*, is located on chromosome 17q21.3, spans approximately 150 kb and consists of 16 exons, only some of them constitutively translated.[118] Exons 2, 3 and 10 are alternatively spliced in the adult brain, giving rise to different tau isoforms, that range from 37 to 46 kDa.[119] The interaction between tau and tubulin is mediated by four imperfect repeat domains (encompassing 31–32 residues) encoded by exons 9–12.[120] Alternative splicing of exon 10 results in the production of isoforms containing either 3 or 4 binding domains (3R and 4R tau).[121] Neurodegenerative conditions known as tauopathies have as a common feature the accumulation of insoluble tau inclusions called neurofibrillary tangles (NFTs) in the cell bodies of neurons and glia. Comparative biochemistry of tau pathological aggregates shows that they differ in both phosphorylation and content of tau isoforms, which enabled a molecular classification of tauopathies.[4] Recent evidence implicated soluble, diffusible tau oligomers as important drivers of synaptotoxicity.[21,29,117,122–124] With the use of a tau oligomer-specific antibody, T22, Lasagna-Reeves et al. found that tau-o in human AD brain samples were fourfold higher than those in control brains and that tau-o were present in early stages of AD cytopathology.[122] *Post-mortem* analysis of brains from subjects affected by mild cognitive impairment showed that cognitive decline was related with pre-fibrillar tau rather than NFTs[125,126] and several tau transgenic animal models showed behavioral deficits, synaptic dysfunction, and cell death in the absence of NFTs formation.[124] Exogenous application of *in vitro* obtained soluble tau aggregates and tau oligomers extracted from AD brains impaired hippocampal LTP *in vitro,* and memory in mice.[21,122] Interestingly, tau-o could also act in combination with Aβ-o to produce these detrimental effects, since sub-toxic doses of the two species, administered in combination, led to LTP and memory impairment.[21] The same group observed that *APP* expression was required for an efficient intra-neuronal uptake of Aβ-o and tau-o.[29] Furthermore, *APP*-null mice were resistant to Aβ-o and tau-o induced impairment in memory and LTP, suggesting that the toxic effect was dependent upon *APP* expression.[29]

4.2 PrP^C-mediated tau-o toxicity

Given that PrP^C has been identified as a receptor for PrP^{Sc} and Aβ oligomers (see above), in the last few years several groups expanded this field of investigation to the interaction between PrP^C and other prion-like aggregated species. Since only a little number of studies have been published so far regarding the potential crosstalk between PrP^C and tau aggregates,[19,23,52,53,127] these reports should be interpreted with caution and further studies are needed to provide definitive information. The possibility of an interaction between tau and the prion protein was already suggested by neuropathological examination of GSS cases which displayed the co-deposition of PrP^{Sc} together with hyperphosphorylated forms of tau, even if the nature of this phenomenon remains still elusive.[56–63] Furthermore, full-length recombinant tau has been reported to bind to recombinant PrP^C, *in vitro*.[128] More in details, the N-terminus (amino acids 1–91) and tandem repeats region (amino acids 186–283) of tau were responsible for the interaction, which involved the octapeptide repeat domain on PrP^C. The P102L GSS-related mutant and other PrP^C mutants characterized by the presence of two or seven extra octapeptide repeats (associated with familial forms of prion disorders) showed more tau binding affinity than wild-type PrP^C.[128] The first study showing a PrP^C-mediated causality of tau oligomers, investigated the neurotoxic effect of secretomes of human induced pluripotent stem cell (iPSC)-derived models of AD.[53] In this study they observed that the secretome of iPSC-derived neurons harboring trisomy of chromosome 21 (Ts21secretome) was enriched in a complex mixture of extracellular tau species capable of inducing *in vivo* LTP impairment. Furthermore, they observed that this detrimental effect was prevented by tau immunodepletion of the Ts21 secretome or by pre-injection of the anti-PrP 6D11 antibody.[53] They presented comparable results using Aβ-o enriched secretomes of other iPSC-derived neuronal cell lines harboring mutations in *APP* or *PS1*.[53] Intracerebroventricular injection of soluble aggregates formed from either wild-type or mutant P301S human recombinant tau potently inhibited hippocampal LTP *in vivo*, while tau monomers and end-stage fibrils appeared inactive.[23] Moreover, AD brain soluble extracts inhibited LTP in a tau-dependent manner. In fact, their effect on LTP was abolished by tau immunodepletion or co-injection of a mid-region anti-tau monoclonal antibody, Tau5.[23] Antibodies against the mid-region (6D11) and the N-terminus (MI-0131) of PrP^C, also prevented the LTP inhibition induced by both recombinant and brain-derived tau, confirming a PrP^C-mediated tau toxic effect.[23] Recently, Corbett and colleagues provided evidence that soluble aggregates of tau (and also of Aβ and α-syn, see next paragraph) bind to PrP^C *in vitro* and on mouse cortical neurons, *via* at least one of the previously described N-terminal PrP^C Aβ-o binding sites.[19] Tau soluble aggregates (produced *in vitro* and extracted from diseased brains) caused both functional (impairment of LTP) and structural (neuritic dystrophy) effects prevented by PrP^C ablation or pre-treatment with anti-PrP^C blocking antibodies.[19]

4.3 PrP^C mediates tau uptake and complex tau-induced effects

Ondrejack and colleagues observed that, conversely to Aβ-o, soluble full-length tau aggregates injection *in vivo*, increased the threshold for long-term depression (LTD) induction.[127] This effect was mediated by PrP^C and was reduced by systemic treatment with Ro 25-6981, a selective antagonist of GluN2B subunit-containing NMDA receptors.[127] Interestingly, the same molecule was not able to prevent tau-induced and PrP^C-mediated inhibition of LTP. When co-administered with Aβ-o, tau soluble aggregates exhibited complex effects toward Aβ-o mediated LTD and LTP, completely blocking Aβ-facilitated LTD, while facilitating Aβ-induced inhibition of LTP.[127] Taken together, these results suggest that both tau and Aβ oligomeric species interact with PrP^C to elicit, sometimes opposed, effects on synaptic plasticity, probably mediated by different effectors.[127] A complex interplay between PrP^C and tau fibrils was recently evidenced also by De Cecco and colleagues, which observed that fibrils of tau K18 (i.e., a truncated form of human tau containing only the 4 microtubule binding repeats) are actively internalized in N2a cells *via* PrP^C.[52] The uptake of tau K18 fibrils was reduced in PrP^C-knockout N2a cells and by pre-treatment with antibodies against the N-terminus of PrP^C. Surprisingly, exposure of ScN2a cells (i.e., chronically prion-infected N2a cells) to tau K18 fibrils, reduced PrP^{Sc} levels and increased PrP^C α-cleavage.[52] The authors proposed that tau-induced clearance of PrP^{Sc} could be related to either an increase in PrP^C processing or to a direct binding of tau fibrils to PrP^C, ultimately impeding its conversion in PrP^{Sc}. If confirmed, this data suggest a bidirectional interaction between the two proteins.[52]

5. PrP^C-mediated α-synuclein toxicity

5.1 α-Synuclein neurotoxic species

Despite being all defined by the abnormal aggregation of the same protein, α-synucleinopathies show a variety of symptoms and clinical manifestations ranging from motor alterations to neuropsychiatric disturbances, cognitive decline and memory impairment.[129] LB pathology is not confined to the *substantia nigra*, but extends to multiple brain areas including hippocampus and cortex, explaining the presence of non-motor symptoms.[130,131] Recent studies suggested that, as for other NDs, α-syn oligomers (α-*syn*-o) may represent the most toxic α-syn species which, once released from neuronal cells, could contribute to the spread of pathology and, ultimately, to the major pathological features of α-synucleinopathies.[32] Extracellular α-syn-o, but not monomers or fibers, showed a detrimental effect on hippocampal LTP and increased basal synaptic transmission, through a mechanism dependent on NMDA receptor activation.[30] Furthermore, extracellular α-syn released from neuronal cells, binds to the Toll-like receptor 2 (TLR2), which, in turns, activates inflammatory responses in microglia, leading, eventually, to a neuroinflammatory response.[31] Interestingly, only specific types of α-*syn*-o were able to interact with and activate TLR2, suggesting that this α-syn-mediated TLR2 activation was conformation-sensitive.[31] Finally, Mao and colleagues suggested that the transmembrane protein, lymphocyte-activation gene 3 (LAG3/CD223) could be a neuronal receptor for synthetic α-synuclein preformed fibrils (PFF). Depletion of LAG3, or antibodies against LAG3, substantially reduced the transmission and consequent neurotoxicity of pathologic α-syn PFF.[132] In the last years several studies described, sometimes with conflicting results,[55] a number of α-syn toxic effects and mechanisms of inter-neuronal transmission, mediated, at least in part, by the cellular prion protein,[19,22,51,133] which will be discussed in the next paragraph.

5.2 PrP^C-mediated α-syn toxicity and uptake

The first study which investigated the participation of PrP^C in α-syn propagation and toxicity, compared the spreading and the extent of α-syn pathology in $PRNP^{+/+}$, $PRNP^{0/0}$ and Tga20 (i.e., overexpressing PrP^C) mice injected, in the striatum, with α-syn PFF.[133] *PRNP* expression was not mandatory for α-syn transport in the mouse brain, although $PRNP^{+/+}$ and Tga20 mice displayed increased levels of α-syn transport. Tga20 mice also showed increased Lewy body-like pathology in motor pyramidal neurons.[133] The same year, another independent study confirmed that PrP^C overexpression enhanced α-syn spreading and its pathologic deposition *in vitro* and *in vivo*.[51] Several studies suggested that α-syn binding sites on PrP^C[19,22,54] seemingly correspond to the previously described PrP^C Aβ-o high affinity binding regions.[40,46] HEK293 cells expressing a PrP^C mutant (PrPΔCC) devoid of its central domain (i.e., aa 90–110 or aa 91–115) presented significantly lower binding affinity for α-syn PFF[133] and α-*syn*-o mediated detrimental effect on LTP was abolished in hippocampal slices pre-treated with 6D11 anti-PrP antibody.[22] Very recently, Corbett et al. demonstrated that Aβ, tau and α-syn protofibrils (also defined as soluble protein aggregates, or SPAs), all induce neurotoxic effects *via* PrP^C and, even if with relative differences according to the specific protein aggregate, their interaction with PrP^C was mediated by the two previously described Aβ-o binding site I (aa 23–33) and II (aa 88–111).[19] Ferreira et al. investigated putative effectors of α-*syn*-o-induced impairment of LTP and demonstrated that α-syn-o physically interact with PrP^C, inducing the phosphorylation of Fyn kinase *via* mGluR5 with the consequent activation of the NMDAR subunit 2B and altered calcium homeostasis.[22] Notably, mGluR5 was already described as a PrP^C binding partner in the transmission of Aβ-o-induced toxic signaling.[91,112] As for tau, also α-syn fibrils presented a surprising effect on prion clearance in an *in vitro* cellular system.[51] When fibrils of α-syn were added to N2a cells chronically infected with prions, they induced a reduction in the levels of the PK-resistant PrP^{Sc} material and an increase in PrP^C α-cleavage processing, confirming that the interplay between pathological proteinaceous aggregates and PrP^C could involve bidirectional effects.[51,52] Conversely to the above mentioned studies, La Vitola and colleagues, found that PrP^C was not mandatory to mediate α-*syn*-o detrimental effects *in vitro* and *in vivo*.[55] They observed that α-syn-o toxicity was comparable in $PRNP^{+/+}$ and $PRNP^{0/0}$ neurons and both $PRNP^{+/+}$ and $PRNP^{0/0}$ mice injected with α-syn-o presented memory deficits and hippocampal gliosis. Furthermore, they demonstrated by means of surface plasmon resonance that there was no interaction between PrP^C and their preparation of α-syn-o.[55] Taken together, these results suggest that further experiments are needed to clearly address the relationship between PrP^C and α-syn pathological species. As for Aβ-o, it is possible that different pathological conformers dramatically influence the binding with PrP^C and also drive PrP^C-independent toxic effects. Therefore, to ultimately clarify this point and to unambiguously evaluate the significance of PrP^C in α-synucleinopathies, it is of utmost importance to increase the characterization and to standardize the preparation protocols of tested aggregates, as recent studies are correctly already doing.[19]

6. Conclusions and future perspectives

After initial controversies, the interaction between PrP^C and Aβ toxic oligomeric species seems now widely accepted. As a consequence, several studies extended their focus to validate the hypothesis that PrP^C mediates toxic signaling induced by aggregates of proteins other than Aβ, as tau and α-synuclein. PrP^C interaction with these toxic species seemingly involves complex mechanisms of signal transduction and is likely influenced by PrP^C physio-pathological cleavage processing. Moreover, initial reports suggest that the interplay between PrP^C and pathological aggregates could lead to bidirectional effects. Fig. 1 schematically illustrates the major proposed PrP^C-mediated effects discussed in this chapter. Lessons learned from the first studies on PrP^C and Aβ interaction could be used as a guide to design future investigations and to correctly interpret resulting data. Several points seem to be crucial to obtain definite information. First of all, it is now clear that the preparation protocol used to obtain synthetic aggregates is a key element to increase the reproducibility of results and to help in the characterization of the putative pathological conformer responsible for PrP^C-mediated toxic effects. Related to this point, it will be also relevant to compare the composition of the *in vitro* obtained conformers to pathological aggregates extracted from diseased tissues. Since patient-derived Aβ, tau and α-syn enriched samples are enormously precious, it is reasonable to start collecting data on *in vitro* obtained aggregates and to subsequently verify their biological relevance as a second step. A second crucial point is that authors in the field should be encouraged to report both positive and negative results and detailed explanations of the mechanistic processes involved in the transmission of toxicity. Rigorous presentation of collected data would bring to a faster understanding of the biological significance of the PrP^C molecule for many neurodegenerative disorders, with relevant therapeutic implications. In fact, if the nature of the interaction between PrP^C and neurodegenerative pathological aggregates will be confirmed,

PrP^C could be the target of several therapeutic approaches, as for instance the use of antisense oligonucleotides to knock-down the *PRNP* gene *via* RNA interference, the administration of PrP^C derivatives (i.e., the full-length recombinant protein or its N-terminal fragment) or anti-PrP^C antibodies. In this regard it should be mentioned that anti-PrP^C antibodies should be handled with extreme attention to avoid undesired detrimental effects due to the disruption of physiologically relevant and protective PrP^C inter-domain interactions.

References

1. Kovacs GG. Molecular pathological classification of neurodegenerative diseases: turning towards precision medicine. *Int J Mol Sci*. 2016;17:189.
2. Grundke-Iqbal I, Iqbal K, Quinlan M, et al. Microtubule-associated protein tau. A component of Alzheimer paired helical filaments. *J Biol Chem*. 1986;261:6084–6089.
3. Masters CL, Simms G, Weinman NA, et al. Amyloid plaque core protein in Alzheimer disease and Down syndrome. *Proc Natl Acad Sci USA*. 1985;82:4245–4249.
4. Sergeant N, Delacourte A, Buee L. Tau protein as a differential biomarker of tauopathies. *Biochim Biophys Acta*. 2005;1739:179–197.
5. Braak H, Del Tredici K. Neuroanatomy and pathology of sporadic Parkinson's disease. *Adv Anat Embryol Cell Biol*. 2009;201:1–119.
6. Goedert M, Spillantini MG, Del Tredici K, Braak H. 100 Years of Lewy pathology. *Nat Rev Neurol*. 2013;9:13–24.
7. Moore RA, Taubner LM, Priola SA. Prion protein misfolding and disease. *Curr Opin Struct Biol*. 2009;19:14–22.
8. Smethurst P, Sidle KC, Hardy J. Review: prion-like mechanisms of transactive response DNA binding protein of 43 kDa (TDP-43) in amyotrophic lateral sclerosis (ALS). *Neuropathol Appl Neurobiol*. 2015;41:578–597.
9. Holmes BB, Diamond MI. Amyotrophic lateral sclerosis and organ donation: is there risk of disease transmission? *Ann Neurol*. 2012;72:832–836.
10. Jucker M, Walker LC. Pathogenic protein seeding in Alzheimer disease and other neurodegenerative disorders. *Ann Neurol*. 2011;70:532–540.
11. Aguzzi A. Cell biology: beyond the prion principle. *Nature*. 2009;459:924–925.
12. Aguzzi A, Lakkaraju AKK. Cell biology of prions and prionoids: a status report. *Trends Cell Biol*. 2016;26:40–51.
13. Maniecka Z, Polymenidou M. From nucleation to widespread propagation: a prion-like concept for ALS. *Virus Res*. 2015;207:94–105.
14. Walker LC, Jucker M. Neurodegenerative diseases: expanding the prion concept. *Annu Rev Neurosci*. 2015;38:87–103.
15. Prusiner SB. Biology and genetics of prions causing neurodegeneration. *Annu Rev Genet*. 2013;47:601–623.
16. Prusiner SB. Novel proteinaceous infectious particles cause scrapie. *Science*. 1982; 216:136–144.
17. Caughey B, Lansbury PT. Protofibrils, pores, fibrils, and neurodegeneration: separating the responsible protein aggregates from the innocent bystanders. *Annu Rev Neurosci*. 2003;26:267–298.
18. Conway KA, Harper JD, Lansbury PT. Accelerated in vitro fibril formation by a mutant alpha-synuclein linked to early-onset Parkinson disease. *Nat Med*. 1998; 4:1318–1320.

19. Corbett GT, Wang Z, Hong W, et al. PrP is a central player in toxicity mediated by soluble aggregates of neurodegeneration-causing proteins. *Acta Neuropathol.* 2020;139: 503–526.
20. El-Agnaf OM, Jakes R, Curran MD, et al. Aggregates from mutant and wild-type alpha-synuclein proteins and NAC peptide induce apoptotic cell death in human neuroblastoma cells by formation of beta-sheet and amyloid-like filaments. *FEBS Lett.* 1998;440:71–75.
21. Fa M, Puzzo D, Piacentini R, et al. Extracellular tau oligomers produce an immediate impairment of LTP and memory. *Sci Rep.* 2016;6:19393.
22. Ferreira DG, Temido-Ferreira M, Miranda HV, et al. Alpha-synuclein interacts with PrP(C) to induce cognitive impairment through mGluR5 and NMDAR2B. *Nat Neurosci.* 2017;20:1569–1579.
23. Ondrejcak T, Klyubin I, Corbett GT, et al. Cellular prion protein mediates the disruption of hippocampal synaptic plasticity by soluble tau in vivo. *J Neurosci.* 2018; 38:10595–10606.
24. Selkoe DJ, Hardy J. The amyloid hypothesis of Alzheimer's disease at 25 years. *EMBO Mol Med.* 2016;8:595–608.
25. Usenovic M, Niroomand S, Drolet RE, et al. Internalized tau oligomers cause neurodegeneration by inducing accumulation of pathogenic tau in human neurons derived from induced pluripotent stem cells. *J Neurosci.* 2015;35:14234–14250.
26. Li S, Selkoe DJ. A mechanistic hypothesis for the impairment of synaptic plasticity by soluble Aβ oligomers from Alzheimer's brain. *J Neurochem.* 2020;1–15. https://doi.org/10.1111/jnc.15007.
27. Benilova I, Karran E, De Strooper B. The toxic Abeta oligomer and Alzheimer's disease: an emperor in need of clothes. *Nat Neurosci.* 2012;15:349–357.
28. Oddo S, Caccamo A, Tran L, et al. Temporal profile of amyloid-beta (Abeta) oligomerization in an in vivo model of Alzheimer disease. A link between Abeta and tau pathology. *J Biol Chem.* 2006;281:1599–1604.
29. Puzzo D, Piacentini R, Fa M, et al. LTP and memory impairment caused by extracellular Abeta and Tau oligomers is APP-dependent. *Elife.* 2017;6:e26991.
30. Diogenes MJ, Dias RB, Rombo DM, et al. Extracellular alpha-synuclein oligomers modulate synaptic transmission and impair LTP via NMDA-receptor activation. *J Neurosci.* 2012;32:11750–11762.
31. Kim C, Ho DH, Suk JE, et al. Neuron-released oligomeric alpha-synuclein is an endogenous agonist of TLR2 for paracrine activation of microglia. *Nat Commun.* 2013;4:1562.
32. Lee HJ, Bae EJ, Lee SJ. Extracellular alpha—synuclein-a novel and crucial factor in Lewy body diseases. *Nat Rev Neurol.* 2014;10:92–98.
33. Dohler F, Sepulveda-Falla D, Krasemann S, et al. High molecular mass assemblies of amyloid-beta oligomers bind prion protein in patients with Alzheimer's disease. *Brain.* 2014;137:873–886.
34. Fluharty BR, Biasini E, Stravalaci M, et al. An N-terminal fragment of the prion protein binds to amyloid-beta oligomers and inhibits their neurotoxicity in vivo. *J Biol Chem.* 2013;288:7857–7866.
35. Freir DB, Nicoll AJ, Klyubin I, et al. Interaction between prion protein and toxic amyloid beta assemblies can be therapeutically targeted at multiple sites. *Nat Commun.* 2011;2:336.
36. Ganzinger KA, Narayan P, Qamar SS, et al. Single-molecule imaging reveals that small amyloid-beta1-42 oligomers interact with the cellular prion protein (PrP(C)). *Chembiochem.* 2014;15:2515–2521.
37. Lauren J, Gimbel DA, Nygaard HB, Gilbert JW, Strittmatter SM. Cellular prion protein mediates impairment of synaptic plasticity by amyloid-beta oligomers. *Nature.* 2009;457:1128–1132.

38. Zou WQ, Xiao X, Yuan J, et al. Amyloid-beta42 interacts mainly with insoluble prion protein in the Alzheimer brain. *J Biol Chem*. 2011;286:15095–15105.
39. Smith LM, Kostylev MA, Lee S, Strittmatter SM. Systematic and standardized comparison of reported amyloid-beta receptors for sufficiency, affinity, and Alzheimer's disease relevance. *J Biol Chem*. 2019;294:6042–6053.
40. Chen S, Yadav SP, Surewicz WK. Interaction between human prion protein and amyloid-beta (Abeta) oligomers: role of N-terminal residues. *J Biol Chem*. 2010;285: 26377–26383.
41. Mengel D, Hong W, Corbett GT, et al. PrP-grafted antibodies bind certain amyloid beta-protein aggregates, but do not prevent toxicity. *Brain Res*. 2019;1710:125–135.
42. Nicoll AJ, Panico S, Freir DB, et al. Amyloid-beta nanotubes are associated with prion protein-dependent synaptotoxicity. *Nat Commun*. 2013;4:2416.
43. Resenberger UK, Harmeier A, Woerner AC, et al. The cellular prion protein mediates neurotoxic signalling of beta-sheet-rich conformers independent of prion replication. *EMBO J*. 2011;30:2057–2070.
44. Um JW, Nygaard HB, Heiss JK, et al. Alzheimer amyloid-beta oligomer bound to postsynaptic prion protein activates Fyn to impair neurons. *Nat Neurosci*. 2012;15: 1227–1235.
45. Williams TL, Choi JK, Surewicz K, Surewicz WK. Soluble prion protein binds isolated low molecular weight amyloid-beta oligomers causing cytotoxicity inhibition. *ACS Chem Nerosci*. 2015;6:1972–1980.
46. Younan ND, Sarell CJ, Davies P, Brown DR, Viles JH. The cellular prion protein traps Alzheimer's Abeta in an oligomeric form and disassembles amyloid fibers. *FASEB J*. 2013;27:1847–1858.
47. Balducci C, Beeg M, Stravalaci M, et al. Synthetic amyloid-beta oligomers impair long-term memory independently of cellular prion protein. *Proc Natl Acad Sci USA*. 2010;107:2295–2300.
48. Calella AM, Farinelli M, Nuvolone M, et al. Prion protein and Abeta-related synaptic toxicity impairment. *EMBO Mol Med*. 2010;2:306–314.
49. Cisse M, Sanchez PE, Kim DH, et al. Ablation of cellular prion protein does not ameliorate abnormal neural network activity or cognitive dysfunction in the J20 line of human amyloid precursor protein transgenic mice. *J Neurosci*. 2011;31:10427–10431.
50. Kessels HW, Nguyen LN, Nabavi S, Malinow R. The prion protein as a receptor for amyloid-beta. *Nature*. 2010;466:E3–E4; discussion E4–5.
51. Aulic S, Masperone L, Narkiewicz J, et al. Alpha-synuclein amyloids hijack prion protein to gain cell entry, facilitate cell-to-cell spreading and block prion replication. *Sci Rep*. 2017;7:10050.
52. De Cecco E, Celauro L, Vanni S, et al. The uptake of tau amyloid fibrils is facilitated by the cellular prion protein and hampers prion propagation in cultured cells. *J Neurochem*. 2020;1–15. https://doi.org/10.1111/jnc.15040.
53. Hu NW, Corbett GT, Moore S, et al. Extracellular forms of Abeta and tau from iPSC models of Alzheimer's disease disrupt synaptic plasticity. *Cell Rep*. 2018;23:1932–1938.
54. Urrea L, Segura-Feliu M, Masuda-Suzukake M, et al. Involvement of cellular prion protein in alpha-synuclein transport in neurons. *Mol Neurobiol*. 2018;55:1847–1860.
55. La Vitola P, Beeg M, Balducci C, et al. Cellular prion protein neither binds to alpha-synuclein oligomers nor mediates their detrimental effects. *Brain*. 2019;142:249–254.
56. Alzualde A, Indakoetxea B, Ferrer I, et al. A novel PRNP Y218N mutation in Gerstmann-Straussler-Scheinker disease with neurofibrillary degeneration. *J Neuropathol Exp Neurol*. 2010;69:789–800.
57. Ghetti B, Tagliavini F, Giaccone G, et al. Familial Gerstmann-Straussler-Scheinker disease with neurofibrillary tangles. *Mol Neurobiol*. 1994;8:41–48.
58. Hsiao K, Dlouhy SR, Farlow MR, et al. Mutant prion proteins in Gerstmann-Straussler-Scheinker disease with neurofibrillary tangles. *Nat Genet*. 1992;1:68–71.

59. Ishizawa K, Komori T, Shimazu T, et al. Hyperphosphorylated tau deposition parallels prion protein burden in a case of Gerstmann-Straussler-Scheinker syndrome P102L mutation complicated with dementia. *Acta Neuropathol.* 2002;104:342–350.
60. Simpson M, Johanssen V, Boyd A, et al. Unusual clinical and molecular-pathological profile of Gerstmann-Straussler-Scheinker disease associated with a novel PRNP mutation (V176G). *JAMA Neurol.* 2013;70:1180–1185.
61. Tranchant C, Sergeant N, Wattez A, et al. Neurofibrillary tangles in Gerstmann-Straussler-Scheinker syndrome with the A117V prion gene mutation. *J Neurol Neurosurg Psychiatry.* 1997;63:240–246.
62. Woulfe J, Kertesz A, Frohn I, et al. Gerstmann-Straussler-Scheinker disease with the Q217R mutation mimicking frontotemporal dementia. *Acta Neuropathol.* 2005;110: 317–319.
63. Yamazaki M, Oyanagi K, Mori O, et al. Variant Gerstmann-Straussler syndrome with the P105L prion gene mutation: an unusual case with nigral degeneration and widespread neurofibrillary tangles. *Acta Neuropathol.* 1999;98:506–511.
64. Aguzzi A, Miele G. Recent advances in prion biology. *Curr Opin Neurol.* 2004;17: 337–342.
65. Del Rio JA, Ferrer I, Gavin R. Role of cellular prion protein in interneuronal amyloid transmission. *Prog Neurobiol.* 2018;165–167:87–102.
66. del Rio JA, Gavin R. Functions of the cellular prion protein, the end of Moore's law, and Ockham's razor theory. *Prion.* 2016;10:25–40.
67. Linden R, Martins VR, Prado MA, et al. Physiology of the prion protein. *Physiol Rev.* 2008;88:673–728.
68. Prusiner SB, Scott MR, DeArmond SJ, Cohen FE. Prion protein biology. *Cell.* 1998;93:337–348.
69. Stahl N, Borchelt DR, Hsiao K, Prusiner SB. Scrapie prion protein contains a phosphatidylinositol glycolipid. *Cell.* 1987;51:229–240.
70. Ford MJ, Burton LJ, Morris RJ, Hall SM. Selective expression of prion protein in peripheral tissues of the adult mouse. *Neuroscience.* 2002;113:177–192.
71. Miele G, Alejo Blanco AR, Baybutt H, et al. Embryonic activation and developmental expression of the murine prion protein gene. *Gene Expr.* 2003;11:1–12.
72. Tichopad A, Pfaffl MW, Didier A. Tissue-specific expression pattern of bovine prion gene: quantification using real-time RT-PCR. *Mol Cell Probes.* 2003;17:5–10.
73. Legname G, Moda F. The prion concept and synthetic prions. *Prog Mol Biol Transl Sci.* 2017;150:147–156.
74. Linden R. The biological function of the prion protein: a cell surface scaffold of signaling modules. *Front Mol Neurosci.* 2017;10:77.
75. Beraldo FH, Arantes CP, Santos TG, et al. Metabotropic glutamate receptors transduce signals for neurite outgrowth after binding of the prion protein to laminin gamma1 chain. *FASEB J.* 2011;25:265–279.
76. Beraldo FH, Arantes CP, Santos TG, et al. Role of alpha7 nicotinic acetylcholine receptor in calcium signaling induced by prion protein interaction with stress-inducible protein 1. *J Biol Chem.* 2010;285:36542–36550.
77. Slapsak U, Salzano G, Amin L, et al. The N terminus of the prion protein mediates functional interactions with the neuronal cell adhesion molecule (NCAM) fibronectin domain. *J Biol Chem.* 2016;291:21857–21868.
78. Rieger R, Edenhofer F, Lasmezas CI, Weiss S. The human 37-kDa laminin receptor precursor interacts with the prion protein in eukaryotic cells. *Nat Med.* 1997;3: 1383–1388.
79. Bueler H, Aguzzi A, Sailer A, et al. Mice devoid of PrP are resistant to scrapie. *Cell.* 1993; 73:1339–1347.
80. Brandner S, Isenmann S, Raeber A, et al. Normal host prion protein necessary for scrapie-induced neurotoxicity. *Nature.* 1996;379:339–343.

81. Mallucci G, Dickinson A, Linehan J, et al. Depleting neuronal PrP in prion infection prevents disease and reverses spongiosis. *Science*. 2003;302:871–874.
82. Chesebro B, Trifilo M, Race R, et al. Anchorless prion protein results in infectious amyloid disease without clinical scrapie. *Science*. 2005;308:1435–1439.
83. Stohr J, Watts JC, Legname G, et al. Spontaneous generation of anchorless prions in transgenic mice. *Proc Natl Acad Sci USA*. 2011;108:21223–21228.
84. Rambold AS, Muller V, Ron U, et al. Stress-protective signalling of prion protein is corrupted by scrapie prions. *EMBO J*. 2008;27:1974–1984.
85. Winklhofer KF, Tatzelt J, Haass C. The two faces of protein misfolding: gain- and loss-of-function in neurodegenerative diseases. *EMBO J*. 2008;27:336–349.
86. Baumann F, Tolnay M, Brabeck C, et al. Lethal recessive myelin toxicity of prion protein lacking its central domain. *EMBO J*. 2007;26:538–547.
87. Li A, Christensen HM, Stewart LR, et al. Neonatal lethality in transgenic mice expressing prion protein with a deletion of residues 105–125. *EMBO J*. 2007;26: 548–558.
88. Solomon IH, Huettner JE, Harris DA. Neurotoxic mutants of the prion protein induce spontaneous ionic currents in cultured cells. *J Biol Chem*. 2010;285:26719–26726.
89. Biasini E, Unterberger U, Solomon IH, et al. A mutant prion protein sensitizes neurons to glutamate-induced excitotoxicity. *J Neurosci*. 2013;33:2408–2418.
90. Wu B, McDonald AJ, Markham K, et al. The N-terminus of the prion protein is a toxic effector regulated by the C-terminus. *Elife*. 2017;6:e23473.
91. Beraldo FH, Ostapchenko VG, Caetano FA, et al. Regulation of amyloid beta oligomer binding to neurons and neurotoxicity by the prion protein-mGluR5 complex. *J Biol Chem*. 2016;291:21945–21955.
92. Scott-McKean JJ, Surewicz K, Choi JK, et al. Soluble prion protein and its N-terminal fragment prevent impairment of synaptic plasticity by Abeta oligomers: implications for novel therapeutic strategy in Alzheimer's disease. *Neurobiol Dis*. 2016;91:124–131.
93. Linsenmeier L, Altmeppen HC, Wetzel S, et al. Diverse functions of the prion protein—does proteolytic processing hold the key? *Biochim Biophys Acta Mol Cell Res*. 2017;1864:2128–2137.
94. Larson M, Sherman MA, Amar F, et al. The complex PrP(c)-Fyn couples human oligomeric Abeta with pathological tau changes in Alzheimer's disease. *J Neurosci*. 2012;32:16857–16871.
95. Supattapone S, Muramoto T, Legname G, et al. Identification of two prion protein regions that modify scrapie incubation time. *J Virol*. 2001;75:1408–1413.
96. Turnbaugh JA, Unterberger U, Saa P, et al. The N-terminal, polybasic region of PrP(C) dictates the efficiency of prion propagation by binding to PrP(Sc). *J Neurosci*. 2012;32:8817–8830.
97. Altmeppen HC, Puig B, Dohler F, et al. Proteolytic processing of the prion protein in health and disease. *Am J Neurodegener Dis*. 2012;1:15–31.
98. Chen SG, Teplow DB, Parchi P, et al. Truncated forms of the human prion protein in normal brain and in prion diseases. *J Biol Chem*. 1995;270:19173–19180.
99. Borchelt DR, Rogers M, Stahl N, Telling G, Prusiner SB. Release of the cellular prion protein from cultured cells after loss of its glycoinositol phospholipid anchor. *Glycobiology*. 1993;3:319–329.
100. Lewis V, Johanssen VA, Crouch PJ, et al. Prion protein "gamma-cleavage": characterizing a novel endoproteolytic processing event. *Cell Mol Life Sci*. 2016;73:667–683.
101. Guillot-Sestier MV, Sunyach C, Ferreira ST, et al. alpha-Secretase-derived fragment of cellular prion, N1, protects against monomeric and oligomeric amyloid beta (Abeta)-associated cell death. *J Biol Chem*. 2012;287:5021–5032.
102. Ostapchenko VG, Beraldo FH, Guimaraes AL, et al. Increased prion protein processing and expression of metabotropic glutamate receptor 1 in a mouse model of Alzheimer's disease. *J Neurochem*. 2013;127:415–425.

103. Condello C, Lemmin T, Stohr J, et al. Structural heterogeneity and intersubject variability of Abeta in familial and sporadic Alzheimer's disease. *Proc Natl Acad Sci USA*. 2018;115:E782–E791.
104. Rasmussen J, Mahler J, Beschorner N, et al. Amyloid polymorphisms constitute distinct clouds of conformational variants in different etiological subtypes of Alzheimer's disease. *Proc Natl Acad Sci USA*. 2017;114:13018–13023.
105. Karantzoulis S, Galvin JE. Distinguishing Alzheimer's disease from other major forms of dementia. *Expert Rev Neurother*. 2011;11:1579–1591.
106. Lam B, Masellis M, Freedman M, Stuss DT, Black SE. Clinical, imaging, and pathological heterogeneity of the Alzheimer's disease syndrome. *Alzheimers Res Ther*. 2013;5:1.
107. Cohen M, Appleby B, Safar JG. Distinct prion-like strains of amyloid beta implicated in phenotypic diversity of Alzheimer's disease. *Prion*. 2016;10:9–17.
108. Scialo C, De Cecco E, Manganotti P, Legname G. Prion and prion-like protein strains: deciphering the molecular basis of heterogeneity in neurodegeneration. *Viruses*. 2019;11:261.
109. Gimbel DA, Nygaard HB, Coffey EE, et al. Memory impairment in transgenic Alzheimer mice requires cellular prion protein. *J Neurosci*. 2010;30:6367–6374.
110. Lauren J, Gimbel DA, Nygaard HB, Gilbert JW, Strittmatter SM. Laurén et al. reply. *Nature*. 2010;466:E4–E5.
111. McMahon HE, Mange A, Nishida N, et al. Cleavage of the amino terminus of the prion protein by reactive oxygen species. *J Biol Chem*. 2001;276:2286–2291.
112. Um JW, Kaufman AC, Kostylev M, et al. Metabotropic glutamate receptor 5 is a coreceptor for Alzheimer abeta oligomer bound to cellular prion protein. *Neuron*. 2013;79:887–902.
113. Bate C, Williams A. Amyloid-beta-induced synapse damage is mediated via cross-linkage of cellular prion proteins. *J Biol Chem*. 2011;286:37955–37963.
114. Haas LT, Kostylev MA, Strittmatter SM. Therapeutic molecules and endogenous ligands regulate the interaction between brain cellular prion protein (PrPC) and metabotropic glutamate receptor 5 (mGluR5). *J Biol Chem*. 2014;289:28460–28477.
115. Hu NW, Nicoll AJ, Zhang D, et al. mGlu5 receptors and cellular prion protein mediate amyloid-beta-facilitated synaptic long-term depression in vivo. *Nat Commun*. 2014; 5:3374.
116. Kudo W, Lee HP, Zou WQ, et al. Cellular prion protein is essential for oligomeric amyloid-beta-induced neuronal cell death. *Hum Mol Genet*. 2012;21:1138–1144.
117. Medina M, Avila J. The role of extracellular Tau in the spreading of neurofibrillary pathology. *Front Cell Neurosci*. 2014;8:113.
118. Buee L, Bussiere T, Buee-Scherrer V, Delacourte A, Hof PR. Tau protein isoforms, phosphorylation and role in neurodegenerative disorders. *Brain Res Brain Res Rev*. 2000;33:95–130.
119. Andreadis A, Brown WM, Kosik KS. Structure and novel exons of the human tau gene. *Biochemistry*. 1992;31:10626–10633.
120. Lee G, Neve RL, Kosik KS. The microtubule binding domain of tau protein. *Neuron*. 1989;2:1615–1624.
121. Himmler A, Drechsel D, Kirschner MW, Martin Jr DW. Tau consists of a set of proteins with repeated C-terminal microtubule-binding domains and variable N-terminal domains. *Mol Cell Biol*. 1989;9:1381–1388.
122. Lasagna-Reeves CA, Castillo-Carranza DL, Sengupta U, et al. Identification of oligomers at early stages of tau aggregation in Alzheimer's disease. *FASEB J*. 2012;26: 1946–1959.
123. Lasagna-Reeves CA, Castillo-Carranza DL, Jackson GR, Kayed R. Tau oligomers as potential targets for immunotherapy for Alzheimer's disease and tauopathies. *Curr Alzheimer Res*. 2011;8:659–665.

124. Guerrero-Munoz MJ, Gerson J, Castillo-Carranza DL. Tau oligomers: the toxic player at synapses in Alzheimer's disease. *Front Cell Neurosci.* 2015;9:464.
125. Vana L, Kanaan NM, Ugwu IC, et al. Progression of tau pathology in cholinergic Basal forebrain neurons in mild cognitive impairment and Alzheimer's disease. *Am J Pathol.* 2011;179:2533–2550.
126. Mufson EJ, Ward S, Binder L. Prefibrillar tau oligomers in mild cognitive impairment and Alzheimer's disease. *Neurodegener Dis.* 2014;13:151–153.
127. Ondrejcak T, Hu NW, Qi Y, et al. Soluble tau aggregates inhibit synaptic long-term depression and amyloid beta-facilitated LTD in vivo. *Neurobiol Dis.* 2019;127:582–590.
128. Wang XF, Dong CF, Zhang J, et al. Human tau protein forms complex with PrP and some GSS- and fCJD-related PrP mutants possess stronger binding activities with tau in vitro. *Mol Cell Biochem.* 2008;310:49–55.
129. Marti MJ, Tolosa E, Campdelacreu J. Clinical overview of the synucleinopathies. *Mov Disord.* 2003;18(suppl 6):S21–S27.
130. Spillantini MG, Schmidt ML, Lee VM, et al. Alpha-synuclein in Lewy bodies. *Nature.* 1997;388:839–840.
131. Braak H, Ghebremedhin E, Rub U, Bratzke H, Del Tredici K. Stages in the development of Parkinson's disease-related pathology. *Cell Tissue Res.* 2004;318:121–134.
132. Mao X, Ou MT, Karuppagounder SS, et al. Pathological alpha-synuclein transmission initiated by binding lymphocyte-activation gene 3. *Science.* 2016;353:aah3374.
133. Urrea L, Ferrer I, Gavin R, Del Rio JA. The cellular prion protein (PrP(C)) as neuronal receptor for alpha-synuclein. *Prion.* 2017;11:226–233.

CHAPTER THIRTEEN

Cell-free amplification of prions: Where do we stand?

Federico Angelo Cazzaniga[a], Chiara Maria Giulia De Luca[a], Edoardo Bistaffa[a], Alessandra Consonni[b], Giuseppe Legname[c], Giorgio Giaccone[a], and Fabio Moda[a,*]

[a]Fondazione IRCCS Istituto Neurologico Carlo Besta, Unit of Neurology 5 and Neuropathology, Milan, Italy
[b]Fondazione IRCCS Istituto Neurologico Carlo Besta, Neurology IV—Neuroimmunology and Neuromuscular Diseases Unit, Milan, Italy
[c]Laboratory of Prion Biology, Department of Neuroscience, Scuola Internazionale Superiore Di Studi Avanzati (SISSA), Trieste, Italy
*Corresponding author: e-mail address: fabio.moda@istituto-besta.it

Contents

Abstract

Neurodegenerative diseases (NDs) such as Alzheimer's disease (AD), Parkinson's disease (PD), atypical parkinsonisms, frontotemporal dementia (FTLD) and prion diseases are characterized by the accumulation of misfolded proteins in the central nervous system (CNS). Although the cause for the initiation of protein aggregation is not well understood, these aggregates are disease-specific. For instance, AD is characterized by the intraneuronal accumulation of tau and extracellular deposition of amyloid-β (Aβ), PD is marked by the intraneuronal accumulation of α-synuclein, many FTLD are associated with the accumulation of TDP-43 while prion diseases show aggregates of misfolded prion protein. Hence, misfolded proteins are considered disease-specific biomarkers and their identification and localization in the CNS, collected *postmortem*, is required

Progress in Molecular Biology and Translational Science, Volume 175
ISSN 1877-1173
https://doi.org/10.1016/bs.pmbts.2020.08.005

for a definitive diagnosis. With the development of two innovative cell-free amplification techniques named Protein Misfolding Cyclic Amplification (PMCA) and Real-Time Quaking-Induced Conversion (RT-QuIC), traces of disease-specific biomarkers were found in CSF and other peripheral tissues (e.g., urine, blood, and olfactory mucosa) of patients with different NDs. These techniques exploit an important feature shared by many misfolded proteins, that is their ability to interact with their normally folded counterparts and force them to undergo similar structural rearrangements. Essentially, RT-QuIC and PMCA mimic *in vitro* the same pathological processes of protein misfolding which occur *in vivo* in a very rapid manner. For this reason, they have been employed for studying different aspects of protein misfolding but, overall, they seem to be very promising for the *premortem* diagnosis of NDs.

1. Neurodegenerative diseases and protein misfolding

Neurodegenerative diseases (NDs), such as Alzheimer's disease (AD), Parkinson's disease (PD), atypical parkinsonisms, dementia with Lewy bodies (DLB), frontotemporal dementia (FTLD) and prion diseases affect almost 50 million people worldwide but this figure is expected to increase further since the world population is rapidly aging.[1]

Although phenotypically heterogeneous, NDs share a common pathological mechanism that involves the intra- or extracellular accumulation of protein aggregates in specific brain areas. Each disease is characterized by its own subset of protein aggregates whose formation results from conformational changes in the secondary structure of the native proteins. Once misfolded, these proteins become prone to self-assemble and to form fibrillar aggregates. According to (i) the type of protein that undergoes misfolding and aggregation, (ii) the type of cells involved (e.g., neurons, astrocytes, oligodendrocytes) and (iii) the neuroanatomical distribution of these protein aggregates it is possible to specifically identify each ND (Table 1). For this reason, misfolded proteins are considered disease-specific biomarkers and their exact identification and localization is a requirement for formulating a definitive diagnosis of NDs that, at present, is possible only at *postmortem*, after an extensive neuropathological analysis of the brain tissue collected at autopsy. Among disease-specific misfolded proteins, α-synuclein (αS) is found to accumulate in a group of pathologies named synucleinopathies which includes Parkinson's disease (PD), dementia with Lewy bodies (DLB), multiple system atrophy (MSA) and pure autonomic failure (PAF). Particularly, in PD, αS aggregates in neuronal cell bodies and processes (known as Lewy bodies and Lewy neuritis, respectively) mainly located in the brainstem.[2] MSA is characterized by αS deposition in oligodendrocytes

Table 1 List of the most common neurodegenerative diseases with the characteristic misfolded proteins that aggregate in the brain.

Classification	Disease	Misfolded protein
Synucleinopathies	Parkinson's disease (PD)	α-Synuclein
	Dementia with Lewy bodies (DLB)	α-Synuclein
	Multiple system atrophy (MSA)	α-Synuclein
	Pure autonomic failure (PAF)	α-Synuclein
Primary tauopathies	Corticobasal degeneration (CBD)	Tau (4R)
	Progressive supranuclear palsy (PSP)	Tau (4R)
	Frontotemporal dementia with parkinsonism linked to chromosome-17 (FTLD-17)	Tau (4R)
	Pick's disease (PiD)	Tau (3R)
Secondary tauopathies	Alzheimer's disease (AD)	Aβ + Tau (3R + 4R)
Other proteinopathies	Cerebral amyloid angiopathy	Aβ
	Prion diseases	PrP^{Sc}
	Frontotemporal dementia (FTLD)	TDP-43 (or tau)
	Amyotrophic lateral sclerosis (ALS)	TDP-43

of cerebellum, basal ganglia and pons while in DLB Lewy bodies and Lewy neuritis are found throughout the neocortex and brainstem.[3] Finally, PAF shows αS aggregates in autonomic ganglia and nerves.[4] Aggregates of misfolded tau are typically found in a group of pathologies named tauopathies that includes, progressive supranuclear palsy (PSP), corticobasal degeneration (CBD), Pick's disease (PiD), and frontotemporal dementia with parkinsonism linked to chromosome-17 (FTLD-17). PSP and CBD show aggregates made up of 4-repeat (4R) tau in neuronal and glial cells. Especially, PSP shows the characteristic tufted astrocytes which are mainly localized in basal ganglia and brainstem while astrocytic plaques are typically found in CBD and prevalently localized in neocortex and basal ganglia.[5] In many cases of FTLD-17, splicing mutations occur in intron 10 thus leading to the overproduction and aggregation of 4R tau in multiple brain regions, including the hippocampus, neocortex and substantia nigra.[6] On the other hand, PiD shows intraneuronal aggregates made up of 3-repeat (3R) tau (in the form of round argyrophilic inclusions named Pick bodies) in the cerebral cortex and some subcortical structures.[7] PSP, CBD, FTLD-17, and PiD are considered primary tauopathies since tau is believed to be the major player of the neurodegenerative process. Tau aggregates can be found also in Alzheimer's disease (AD). However, in this case, 4R and 3R tau co-deposit along with another protein named amyloid-β (Aβ) which forms extracellular amyloid plaques.[8] AD is considered a secondary tauopathy and the exact relationship between tau and Aβ aggregates is still a matter of debate. Aggregates of Aβ are also found in the walls of the cerebral arteries of patients with cerebral amyloid angiopathy (CAA),[9] in some forms of DLB[10] and in skeletal muscles of patients with a muscle disease named inclusion body myositis (IBM).[11] TAR DNA-binding protein 43 (TDP-43) is another disease-specific misfolded protein that accumulates in some variants of FTLD or amyotrophic lateral sclerosis (ALS).[12] Finally, a group of disorders named transmissible spongiform encephalopathies (TSEs) or prion diseases, are caused by a misfolded form of the prion protein, named prion, or PrP^{Sc}. PrP^{Sc} is found to intra- or extracellularly accumulate in a variety of TSEs, including, Creutzfeldt-Jakob disease (CJD), variably protease-sensitive prionopathy (VPSPr), Gerstmann-Sträussler-Scheinker disease (GSS), and Fatal Familial Insomnia (FFI).[13,14]

The clinical (*antemortem*) diagnoses of NDs are extremely challenging because symptoms may overlap between pathologies, especially at early stages of diseases. Moreover, such diagnoses rely on criteria that are based on clinical, instrumental and neuropsychological evaluations where less specific biomarkers are investigated. For this reason, the diagnostic accuracy of NDs is limited and varies from about 62% (as in the case of MSA) to 95% (as in the case of PSP).[15–21]

However, recent evidence seems to indicate that disease-specific biomarkers are not confined to the CNS but can appear in peripheral tissues and body fluids, including cerebrospinal fluid (CSF), blood, urine and olfactory mucosa (OM) but at concentrations which are below the detection limits of the most common diagnostic techniques.[22–27] Thanks to the advent of two ultrasensitive techniques named Protein Misfolding Cyclic Amplification (PMCA) and Real-Time Quaking-Induced Conversion (RT-QuIC), originally developed in the field of prion diseases and subsequently extended to other NDs, it is now possible to show the presence of peripheral disease-specific biomarkers in CSF, urine, blood and olfactory mucosa of patients with several NDs (see Sections 4.3 and 5.3).

The exact mechanisms through which disease-specific biomarkers spread in the body are still unknown.[28–31] However, it has been recently shown that blood- or urine-derived exosomes carry many misfolded proteins involved in NDs as cargo.[32] Therefore, these vesicles might effectively contribute to disease-specific biomarkers spreading in peripheral tissues. Overall, disease-specific biomarkers seem to appear in peripheral tissues in the early stages of the disease. Thus, the collection of peripheral tissues with non-invasive procedures and analyses with ultrasensitive techniques may revolutionize the field of NDs finally leading to early diagnosis, patients' stratification and targeted therapies.

2. Transmission of protein misfolding: The example of prions

One of the most fascinating aspects of misfolded proteins is the ability to template their structure to the native counterparts. This phenomenon has been originally observed in prion diseases where PrP^{Sc} is the infectious agent able to transmit the disease by interacting with PrP^{C} and forcing it to misfold and acquire a similar pathological conformation. In this way, PrP^{Sc} can spread throughout the CNS by recruiting and converting new PrP^{C} molecules in an autocatalytic manner (Fig. 1).

This is the first example of protein-based infectious propagation of pathology, whose efficiency depends on many other endogenous factors and mechanisms. PrP^{C} is a glycoprotein attached to the cell surface *via* a glycophosphatidylinositol (GPI)-anchor. In humans, PrP^{C} is composed of 253 amino acids encoded by *PRNP* gene which is located in the short arm on chromosome 20 (20q13). Normally, PrP^{C} is characterized by a high content in α-helical structures while PrP^{Sc} is largely composed of β-sheet structures.[33,34] Thus, conversion causes an increase in the amount of β-sheet structure of PrP^{Sc} which renders it partially resistant to

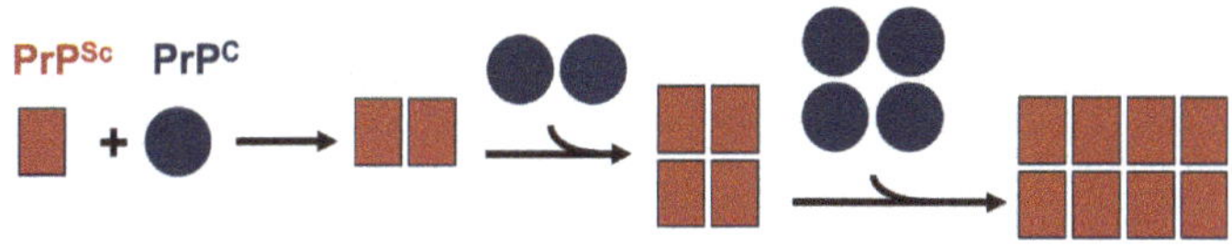

Fig. 1 Mechanism of autocatalytic conversion of PrP^{C} into PrP^{Sc}. PrP^{C} (blue circle) interacts with PrP^{Sc} (red rectangle) and undergoes structural rearrangements which lead to the formation of new PrP^{Sc}.

proteolysis (e.g., treatment with proteinase K) and partly insoluble in mild detergents.[35,36] These features are exploited to discriminate PrP^{Sc} and PrP^{C}.[37,38] Indeed, the latter is completely degraded by proteinase K (PK) and readily soluble in detergents. PrP^{Sc} does not induce humoral or innate immune responses in the host and its spreading is sustained by an autocatalytic mechanism in which PrP^{C} is converted into the misfolded *infectious* form by PrP^{Sc} (Fig. 1).[39–41] Basically, PrP^{Sc} acts as a template for the conversion of PrP^{C}.[42–44] For this reason, the availability of PrP^{C} is indispensable to promote PrP^{Sc} propagation in the CNS and to sustain disease progression.[45] Mouse and hamster are the most commonly used laboratory animals for studying prion diseases. Upon animal inoculations, prions are able to generate typical neuropathological alterations mainly represented by spongiform changes, astrocytic activation and PrP^{Sc} deposition.[46] Prion inoculation in PrP^{C} knock-out mice ($PrP^{0/0}$) failed to induce disease, thus demonstrating that PrP^{C} is necessary to promote PrP^{Sc} spreading and disease progression.[47] These observations confirmed that PrP^{Sc}-PrP^{C} interaction is required to induce the profound conformational rearrangement of PrP^{C} into the pathogenic form. The mechanism which induces the formation of PrP^{Sc} is still unknown but, once accumulating, PrP^{Sc} is able to (i) spread in the CNS and, occasionally, to other tissues of the same host, (ii) spread between hosts belonging to the same species and, sometimes, (iii) spread between hosts of different species.[48–51] Prion transmission between hosts may occur but is characterized by extremely low efficiency due to a phenomenon known as species barrier (as in the case of bovine prions that were transmitted to humans causing the new variant of Creutzfeldt-Jakob disease).[52,53] Likewise, a transmission barrier is known to limit the efficiency of prion transmission between hosts belonging to the same species.[54]

A puzzling aspect of TSEs lies in the fact that the same protein, PrP^{Sc}, can cause a variety of diseases that are phenotypically heterogeneous. For instance, the sporadic forms of CJD are currently classified in six subgroups and every form has its own peculiar features, in terms of clinical manifestations and neuropathological alterations.[55] Such differences are currently thought to be associated with different conformational variants of PrP^{Sc} (referred to as *strains*) that confer the protein with specific pathological features (Fig. 2). The term *strains* indicates faithfully self-replicating PrP^{Sc} that produces unique patterns of disease if serially transmitted to susceptible animals.[56–58] There are many factors that contribute to modulating the abnormal structure acquired by PrP^{Sc}. Some of them are known, such as the polymorphism at codon 129 of *PRNP* gene,[59–61] while others still need to be identified.

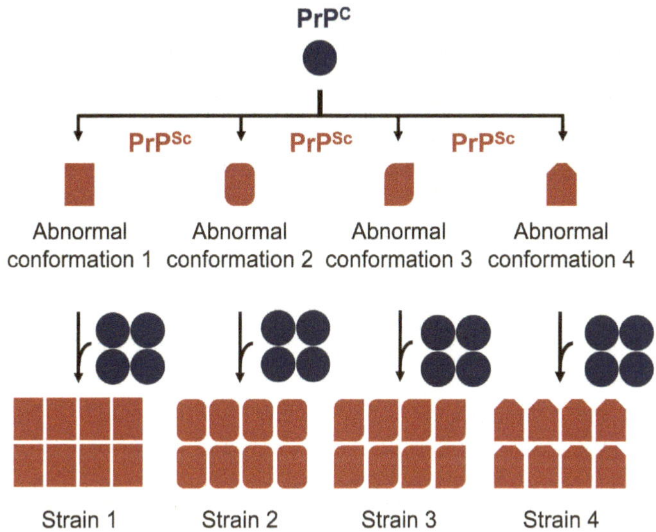

Fig. 2 Generation of different prion strains. Native PrP^C (blue circle) can undergo misfolding and acquire slightly different abnormal conformations which give rise to specific prion "strains" (red rectangular shapes). Each strain is able to self-propagate and is characterized by its own biological properties (neurotoxicity, efficiency to spread in the CNS, cell tropism) which lead to unique pathological phenotypes of the disease.

Recent evidence suggests that other misfolded proteins, including αS, tau, Aβ and TDP-43 might exist in different conformations that are able to act like PrP^{Sc}, therefore, promoting the conformational rearrangement of their native counterparts and contributing to disease progression.

3. Evidence that αS, tau, Aβ and TDP-43 behave like PrP^{Sc}

A growing body of evidence strongly supports the idea that the pathogenesis of other NDs is initiated and sustained by the endogenous misfolding, aggregation and spreading of αS, tau, Aβ, and TDP-43, as observed in prion diseases.[62] Cell and animal experiments have been fundamental to evaluate the prion-based mechanism of spreading of these proteins. For instance, prion-like transmission of αS was originally observed in several patients with PD who received intra-striatal grafts of fetal dopaminergic neurons with the aim of ameliorating their clinical symptoms. When analyzed at autopsy, transplanted neurons revealed the presence of αS aggregates in the form of Lewy bodies. This finding showed that pathological αS had transferred from the host to grafted neurons, probably in a prion-like fashion.[63] Subsequently, treatment of cells with different types of αS aggregates often resulted in the conversion of endogenous αS into insoluble, ubiquitinated and hyperphosphorylated pathological αS assemblies.[24,64–69] Similarly, inoculation of αS aggregates (either artificially produced in laboratory or contained in the CNS of patients with synucleinopathies) in genetically modified or wild-type animals resulted in misfolding and aggregation of endogenous αS.[70–73]

Prion-like transmission of tau has been shown either in cells or animal models upon inoculation with aggregates artificially produced or contained in the brain of patients with primary or secondary tauopathies.[30,65,74–79] One of the most interesting aspects of these studies was the fact that the type of tau aggregates and the cells affected in mice were similar to those found in the brain of human patients.[79]

In 1993, it has been reported that Aβ load was increased in the brain of marmosets after intracerebral injection of AD brain homogenates.[80] More recently, with the generation of genetically modified mouse models of cerebral β-amyloidosis, it has been shown that intracerebral inoculation of AD brains in these mice induced significant Aβ lesions, while injection of control brains did not.[81–85] Studies performed in cells and organotypic slice cultures helped to elucidate some mechanisms through which exogenous Aβ aggregates could efficiently promote endogenous Aβ aggregation.[86] Very recent data seem to suggest that Aβ might be transmissible between humans. In particular, aggregates of Aβ were found in the brain parenchyma and blood vessels of patients treated with the cadaveric human growth hormones supposed to be contaminated with pathological Aβ.[87,88]

Finally, an efficient prion-like spreading of TDP-43 has been observed in cells and more recently in susceptible animal models.[89–91]

Taken together these results indicate that, in a self-perpetuating and cyclic process αS, tau, Aβ, and TDP-43 spread to interconnected neurons and glial cells, following a well-defined spatial-temporal pattern and contributing to the progression of the pathology. However, whether the ultimate neuronal loss which occurs in NDs is caused by protein aggregates still remains to be clearly elucidated.

Recently, the existence of αS, tau, Aβ, and TDP-43 strains has been documented. For example, different strains of αS account for at least four different synucleinopathies, including PD, MSA, DLB, and PAF (see Table 1). Even the phenotypical heterogeneity that can be found within each group of diseases is thought to be associated with minimal variations in the structure of the protein. For instance, different structural variants of αS can be responsible for different pathological phenotypes of PD,[92,93] DLB[94,95] or even MSA. This latter might be further classified in parkinsonian

(MSA-P) and cerebellar (MSA-C) phenotype, thus supporting a possible contribution of different αS strains.[96] Similarly, different strains of tau can cause either different tauopathies, including CBD, PSP, PiD, FTLD-17, and AD, or different pathological phenotypes of the same disorder.[97–101] Finally, different strains of Aβ and TDP-43 have been associated with diverse pathological phenotypes of AD and FTLD, respectively.[102,103]

All these protein aggregates can acquire different β-sheet structures as recently shown by high-resolution structural studies.[104–106] This is not entirely surprising since most misfolded proteins are intrinsically disordered under the physiological condition and can adopt a heterogeneous ensemble of conformations.[107–110] Therefore, in some cases, misfolded proteins can adopt misfolded conformations that confer protein pathological properties that are retained during serial transmissions either performed in cells or in animals. For these reasons, the concept of strain can be extended to αS, tau, Aβ, and TDP-43. For instance, tau extracted from *postmortem* brains of AD (AD-tau), progressive supranuclear palsy (PSP-tau), and corticobasal degeneration (CBD-tau) patients were intracerebrally inoculated in mice and efficiently induced the aggregation of endogenous tau while recapitulating the diversity of neuropathology found in these human tauopathies.[79] Similarly, structurally different Aβ strains collected from patients with different AD phenotypes were able to maintain their pathological features when inoculated in animal models of AD pathology.[102,111,112] For this reason, animal models can be successfully employed for characterizing misfolded proteins strains, understanding their mechanisms of spreading and developing targeted therapeutic agents that can inhibit their specific propagation. During the self-propagation process, each strain interacts with its soluble protein counterpart and promote its structural rearrangement in a similar aberrant conformation.[113]

In some cases, misfolded proteins can promote misfolding and aggregation of different proteins through a phenomenon known as *cross-seeding*. This occurrence often results in the co-existence of multiple misfolded proteins in the CNS of the same patient. For instance, co-deposition of αS and tau was found in AD-related disorders (ADRD).[114–116] αS aggregates were found in more than 50% of AD[117,118]; while tau aggregates were found in patients with Parkinson's disease dementia.[115,119] These observations support the fact that misfolded proteins can synergistically interact while promoting their mutual aggregation and causing strong overlapping between pathologies, often aggravating the clinical and neuropathological course of the diseases.

Thanks to the development of the PMCA and RT-QuIC technologies, which utilize the same principle but slightly different technical procedures, it is now possible to reproduce *in vitro* the process of protein misfolding which occurs *in vivo*, in a very rapid manner. These technologies are used to study different aspects of misfolded proteins, including the molecular mechanisms which lead to their generation and propagation. Above all, both technologies are currently being successfully exploited in the clinical diagnosis of NDs.

4. Protein Misfolding Cyclic Amplification (PMCA)

4.1 Brief description of PMCA

In 2001, Soto and co-workers developed a technique, named Protein Misfolding Cyclic Amplification (PMCA), that was able to mimic *in vitro* the process of prion conversion that occurs *in vivo*.[120] The reaction requires the incubation of a substrate (generally represented by brain homogenates of healthy wild-type or genetically modified mice), containing a high amount of PrP^{C}, with a small amount of PrP^{Sc}. During the incubation phase, PrP^{Sc} induces a structural rearrangement of the PrP^{C} present in the substrate into PrP^{Sc}. The newly formed PrP^{Sc} molecules start to aggregate while the sonication phase promotes the fragmentation of these aggregates into small polymers that are able to convert new PrP^{C} molecules. Both phases are alternated in a cyclic manner (Fig. 3).[121] At the end of the reaction, which generally requires 48–96 h, an aliquot of the amplified material is collected, diluted in freshly prepared brain homogenate and subjected to additional PMCA cycles (referred to as rounds of amplification). The final reaction product is then collected and analyzed by means of Western blotting after PK treatment.[122] The technique is highly sensitive and PrP^{C} conversion can occur even in the presence of a single molecule of oligomeric PrP^{Sc}. In this way, minute amounts of PrP^{Sc} can be amplified and finally detected by Western blot. PMCA efficiently amplifies different prion strains, including those adapted to hamsters (e.g., 263K, hyper, drowsy),[121,123,124] cattle (e.g., bovine spongiform encephalopathy—BSE),[125] cervids (e.g., chronic wasting disease—CWD),[126] ovine (e.g., scrapie),[127,128] mice (e.g., RML and Me7)[129] and humans (variant form of CJD—vCJD) and Fatal Familial Insomnia (FFI).[130–132]

One of the most important aspects of PMCA is the fact that amplified products seem to maintain the biochemical and infectious features of the original PrP^{Sc}.[133] For this reason, it represents a great tool for studying prion

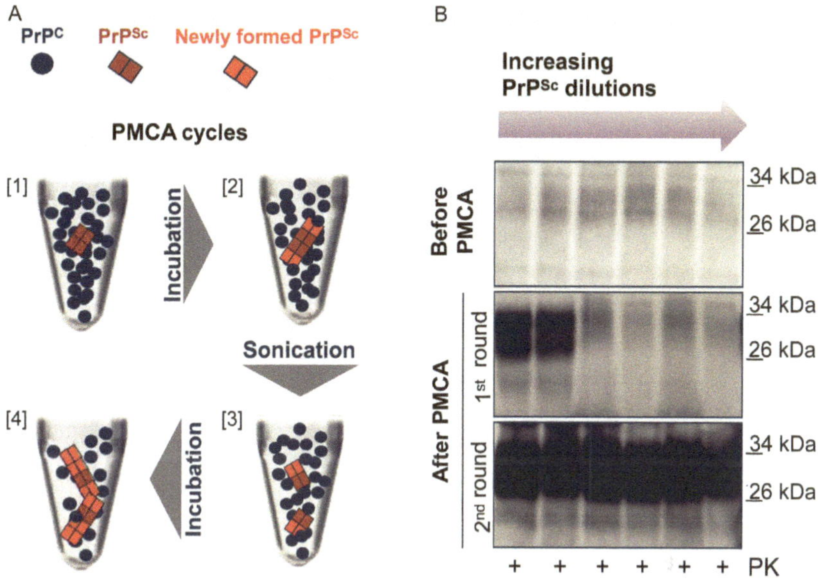

Fig. 3 Schematic representation of PMCA. PMCA alternates phases of incubation to phases of sonication. During the incubation phase, PrP^{Sc} forces PrP^{C} to acquire the pathologic conformation and to form new PrP^{Sc} species which aggregate. During the sonication phase, PrP^{Sc} aggregates are fragmented into small seeds of PrP^{Sc} which induces further conversion of PrP^{C} molecules (A). PrP^{Sc} is serially diluted in brain homogenates of healthy mice (containing high amounts of PrP^{C}). Before PMCA, none of the PrP^{Sc} dilutions is detectable. After one round of PMCA, two dilutions are detectable by Western blot, while after two rounds of amplification PrP^{Sc} is detected in all dilutions (B).

biology, the mechanism of prion transmission, the species barrier which limits prion transmission between species and the recently discovered phenomenon of prion strains selection/adaptation.[134] Given its high sensitivity, PMCA has been successfully applied in the field of prion diagnosis (see Section 4.3).

4.2 PMCA applications in prion research

From its development in 2001, PMCA has been used to study different aspects of prion biology. In particular, the technology has been employed to evaluate the species barrier which influences prion transmissibility between individuals of the same or different species.[135] It was originally employed to evaluate the zoonotic potential of some animal prion strains, including those responsible for BSE, scrapie and CWD. Collected results confirmed the extraordinary ability of PMCA to faithfully reflect the phenomenon of the species barrier observed in animals genetically modified to express human PrP^{C} and inoculated with these animal prion strains.[135–139] PMCA is currently exploited for evaluating an important aspect of prion behavior that has recently emerged. Indeed, several reports have shown that prions can undergo a process of selection and adaptation according to specific environmental changes.[140–144] As a consequence, novel PrP^{Sc} conformations, characterized by a higher propensity to propagate in the new environment, arise and contributed to sustaining disease progression. Several data already indicated that pharmacological treatments of prion-infected cell cultures or animal models resulted in the formation of drug-resistant PrP^{Sc} variants.[145–149] Probably, the treatment has significantly modified the environment of prion replication, thus leading to the selection of a drug-resistant PrP^{Sc} able to sustain the pathological process in the new replication context. Of course, this finding has a great implication from a therapeutic point of view and might explain why no effective therapies for these devastating disorders have been developed so far (see Section 4.3). PMCA helps to understand the mechanism of prion selection and adaptation in response to well-controlled modifications of the replication environment.[143] Particularly, through PMCA the effects of different therapeutics to induce prion strain selection and adaptation are currently being assessed and new pharmacological treatments can be evaluated.[150] Moreover, the technology offers a great opportunity to selectively modify the replication environment, thus allowing the evaluation of the role of specific molecules (e.g., polyanions, RNA, amyloidogenic proteins) to influence PrP^{C}-PrP^{Sc} conversion.[141,151–154] Finally, given its high sensitivity in amplifying minute amounts of prions, PMCA has been used for a rapid assessment of the efficacy of specific sterilization procedures (chemical or physical) to effectively eliminate prion infectivity, especially from surgical instruments.[155]

4.3 PMCA applications in prion diseases diagnosis

In the last few years, PMCA has been optimized for the analysis of peripheral tissues collected from patients with different forms of prion diseases with the aim of verifying whether they could contain traces of PrP^{Sc} undetectable with the current diagnostic techniques. These analyses revealed that PrP^{Sc} could be detected with high sensitivity and specificity in CSF,[156] urine[157] and blood of patients with vCJD.[158] Interestingly, PrP^{Sc} was detected in the blood of vCJD patients even at the pre-clinical stage of the disease.[159] This finding is of utmost importance, especially in light that there may be asymptomatic carriers of vCJD who potentially could infect other individuals, especially through blood transfusions. This point has raised important concerns for the public health, especially after the discovery that vCJD was present in subjects transfused with blood (or blood derivatives) collected from asymptomatic individuals silently incubating vCJD.[160,161] More recently, two asymptomatic cases of vCJD were described in which PrP^{Sc} was detected in the spleen but not the brain. The spleen of one of these subjects was inoculated in mice and efficiently transmitted the disease, thus demonstrating the potential for asymptomatic individuals to transmit vCJD.[162,163] Therefore, PMCA analysis of blood samples, other than supporting the clinical diagnosis of vCJD, represents an optimal strategy to specifically reduce the risk of further disease transmission between humans. At present, there is no treatment or cure for vCJD (or any other prion diseases), thus it is essential to prevent its spreading through containment measures. Interestingly, under specific experimental conditions, PMCA products are able to retain the infectious features of the original PrP^{Sc}.[164] Hence, PMCA offers a great opportunity to test the efficacy of therapeutic compounds to interfere with vCJD replication *in vitro*.

Unfortunately, while vCJD-PrP^{Sc} can be efficiently amplified by means of PMCA, other human prion strains, especially those associated with the most common sporadic CJD (sCJD), appear more difficult to amplify. Increasing efforts have been made in an attempt to modify the PMCA experimental conditions to overcome this problem. The discovery of new substrates (brain homogenates of bank voles) look promising to increase the efficiency of PMCA. In particular, with this substrate human PrP^{Sc} strains associated with a genetic form of prion disease (FFI), iatrogenic forms of CJD due to *dura mater* transplant (iCJD) and some sporadic CJD forms (sCJD) could be amplified although with limited efficiency.[132,165] These modifications led to the discovery that the OM of patients with FFI contains infectious PrP^{Sc}.[132] Preliminary data that are currently being obtained from our laboratory in collaboration with the laboratory of prof. Juan Maria Torres, suggest that the amplified OM-PrP^{Sc} maintains infectious features when inoculated in genetically modified mice (personal communication).

The possibility to amplify different human and animal prion strains while maintaining their infectious features in cell-free systems makes the PMCA an excellent tool for rapid high-throughput screening of therapeutic compounds.[136,164,166–168]

In vivo, PrP^{C}-PrP^{Sc} conversion begins long before disease symptoms onset (even decades). Unfortunately, once symptoms appear, irreversible brain damages and extensive PrP^{Sc} deposition have already occurred. Through the use of PMCA, it would be possible to evaluate the efficacy of anti-prion compounds in the very early phases of prion conversion, thus virtually representing the early stages of the disease. This will enable rapid identification and selection of the most effective molecules able to interfere with the first moment of PrP^{C}-PrP^{Sc} conversion. Up to now, several anti-prion compounds have been identified and used with the aim of blocking the mechanism of PrP^{C}-PrP^{Sc} conversion. Their efficacy has been evaluated using different *in vitro* and *in vivo* models.[169–174] However, none of these compounds have been able to definitely block disease progression *in vivo*. This failure might be due to prions that can undergo conformational changes which determine the emergence of new drug-resistant strains.[140,143,146] Thus, PMCA can test large libraries of compounds and evaluate the appearance of drug-resistant conformers, in a cheaper and faster way than cell and animal assays.

To date, few reports have described the application of PMCA for studying the aggregation of αS[175] and tau[176] while there are still no reports of PMCA analyses of Aβ, TDP-43 or other misfolded proteins.

5. Real-Time Quaking-Induced Conversion (RT-QuIC)

5.1 Brief description of RT-QuIC

In 2011, Atarashi and collaborators have developed the Real-Time Quaking-Induced Conversion (RT-QuIC) assay for studying prion diseases.[177–180] This technique uses recombinant prion protein (recPrP) as a substrate of reaction which is subjected to phases of incubation and shaking which are cyclically repeated. Under these experimental conditions, recPrP is forced to misfold, aggregate and produce large amyloid fibrils. Fibrils growth is monitored with the use of the Thioflavin-T (ThT) fluorescent dye which intercalates into amyloid structures and emits a fluorescence signal. The increase in fluorescence intensities is directly proportional to the formation of prion amyloids (Fig. 4).

The addition of PrP^{Sc} to the reaction mix significantly accelerates the kinetics of recPrP aggregation. This effect can be generally observed even when attograms of PrP^{Sc} are added to the reaction, therefore showing the extraordinary sensitivity of the technology.[179] From its development, the experimental settings of the RT-QuIC underwent several optimizations with the aim of identifying a unique recPrP capable of aggregating in the presence of any type of PrP^{Sc}, regardless of the strain and the animal species.

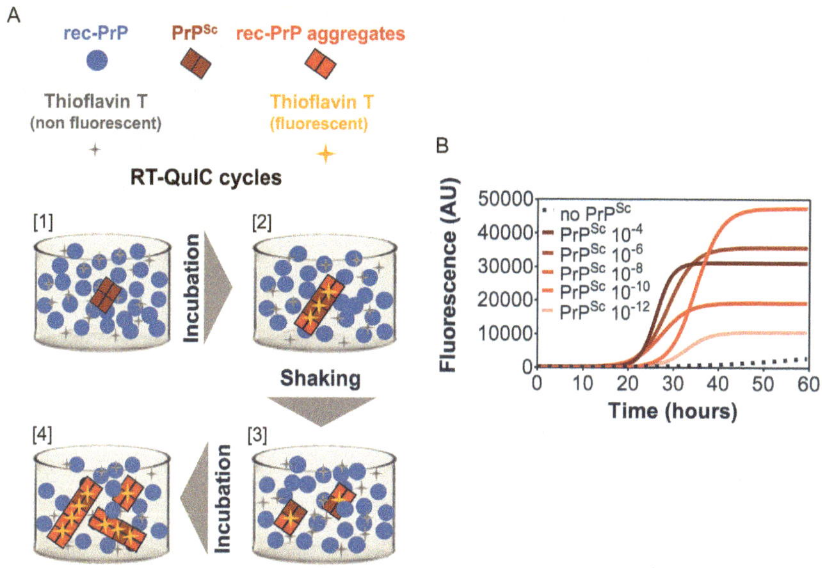

Fig. 4 Schematic representation of RT-QuIC. RT-QuIC alternates phases of incubation to phases of shaking. During the incubation phase, PrP^{Sc} promotes recPrP misfolding and aggregation. During the shaking phase, recPrP aggregates are fragmented into small seeds which induce further misfolding and aggregation of recPrP. Fibrils formation is monitored with the use of ThT whose fluorescence values are directly proportional to the growth of recPrP aggregates (A). PrP^{Sc} is serially diluted in a reaction mix containing recPrP and ThT. All PrP^{Sc} dilutions are able to seed recPrP aggregation within 30 h (B).

RecPrP with the amino acid sequence of the bank vole was found to be an excellent acceptor of prions (either from human or animal origins), capable of aggregating in response to almost all PrP strains.[181] However, in contrast to PMCA, it is still unclear whether recPrP aggregates acquire some infectious or pathological features upon inoculation in animal models.[182,183] Unlike PMCA, this assay works well in detecting very low amounts of PrP^{Sc} associated with the sporadic forms of CJD (sCJD). For this reason, the technology represents valid support for the diagnosis of prion diseases (see Section 5.2) and has been recently introduced among the Italian diagnostic criteria for these disorders.[184]

5.2 RT-QuIC applications in prion research

RT-QuIC has been used to study some aspects of prion biology. In particular, it has been applied for evaluating the phenomenon of the species barrier in highly pure systems, without the presence of cellular cofactors.

However, the lack of important protein-protein interactions or other environmental components prevents the technique to faithfully mimic this phenomenon, as it occurs *in vivo*, therefore requiring relevant interpretations of the results. Overall, the RT-QuIC experimental settings used can drastically modify the outcomes of the analyses, sometimes leading to opposite results.[185] Thus, studying the species barrier with RT-QuIC is possible but the results should be carefully interpreted. Thanks to RT-QuIC analyses, it has been recently shown that the amino terminal part of the prion protein might play an important role in influencing the species barrier.[186] Additionally, since the technology is characterized by a sensitivity which is similar or sometimes higher than that of animal bioassays, it has been used to (i) re-assess prion distribution in different human tissues (where PrP^{Sc} was not detectable by more conventional techniques such as Western blotting and immunohistochemistry),[187] (ii) evaluate prion infectivity titration (classically performed by bioassays),[188] (iii) support prion strain typing (since different human prion strains show distinct responses by RT-QuIC),[181,189] (iv) test the efficacy of prion disinfectants (including sodium hydroxide, bleach, and hypochlorous acid) and (v) analyze the effects of new anti-prion compounds.[190,191] In 2016, the group of Inga Zerr exploited RT-QuIC for assessing the efficacy of doxycycline to inhibit recPrP aggregation after the addition of PrP^{Sc} collected from the brain or CSF of patients with sCJD.[192] The study showed that the compound was able to efficiently interfere with recPrP aggregation in a dose- and time-dependent manner. Therefore, the time at which the treatment is administered and its concentration might be extremely important in determining the efficiency in blocking prion propagation *in vivo*. In 2017, Hyeon and collaborators optimized the RT-QuIC for screening the effect of acridine, dextran sulfate sodium (DSS), and tannic acid to interfere with recPrP aggregation induced by PrP^{Sc} contained in brain homogenates and CSF of patients with sCJD. The study showed that all compounds efficiently blocked recPrP aggregation in a specific and reliable manner.[193] In a more recent study, the group of Kanthasamy combined the RT-QuIC with organotypic slice culture assays (OSCAR) thus creating an innovative screening platform for different anti-prion therapeutics.[194] Even if RT-QuIC can be used to evaluate the efficacy of anti-prion compounds, it has several limitations that need to be considered. For instance, many therapeutics which target the CNS are generally dissolved in lipophilic solvents (e.g., dimethyl sulfoxide (DMSO)).[195] In this way, such compounds are able to cross the blood-brain barrier and reach the CNS. However, many of these solvents are not suitable for RT-QuIC analyses since they can unspecifically trigger the aggregation of recPrP even in the absence of PrP^{Sc}. Finally, the technique has been used to screen the effects of inhibitors (including chemical stabilizers of the prion protein) to halt the process of recPrP aggregation.[192,194,196,197] Remarkably, all RT-QuIC evaluations are currently based on fluorescence emission. Since many therapeutic compounds or inhibitors may quench the fluorescence signal of the ThT, any negative results should be further verified by other techniques (e.g., Western blotting after PK digestion of the reaction's products, Transmission Electron Microscopy analyses, etc.), in order to effectively demonstrate the lack of recPrP aggregates, and hence that of PrP^{Sc}.

5.3 RT-QuIC applications in prion diseases diagnosis

As previously discussed, RT-QuIC represents an optimal assay for the detection of peripherally circulating PrP^{Sc}, especially in the case of patients with sCJD (for whom PMCA does not efficiently work, at present). In particular, a first study published in 2011, showed the ability of the RT-QuIC to detect PrP^{Sc} in the CSF of a series of Japanese patients with sCJD and then in 30 CSF of Australian sCJD patients with a sensitivity of 80% and specificity of 100%. This finding demonstrated the diagnostic utility of RT-QuIC to support the clinical diagnosis of patients with sCJD.[178] Further studies were then performed with the aim of increasing the sensitivity of the technology in detecting PrP^{Sc} in CSF.[183,198,199] Thanks to the results gathered from International ring trial experiments, the reproducibility of the technique has been verified and its reliability and robustness in the clinical practice confirmed.[200,201] Subsequently, RT-QuIC was employed for the analysis of CSF collected from patients with some genetic forms of prion diseases and demonstrated that also, in this case, the assay was able to efficiently detect PrP^{Sc}.[202] We and others have then contributed to optimizing such technology while greatly enhancing its sensitivity and specificity for prion detection in the CSF of patients with different forms of the disease.[183,188,202–204] In 2014, it was shown that the RT-QuIC could detect PrP^{Sc} in samples of OM collected from patients with sCJD. In particular, OM elicited stronger and faster RT-QuIC responses than those promoted by CSF samples of the same patients. Moreover, RT-QuIC of OM samples were characterized by higher sensitivity and specificity compared to those of CSF.[205] Our group contributed in showing that PrP^{Sc} could be also detected in OM samples of patients with FFI.[132] Taken together these results showed that CSF and OM are important sources of PrP^{Sc} that can be efficiently analyzed by means of RT-QuIC thus providing fundamental support for the clinical diagnosis of prion diseases. Recent reports demonstrated that RT-QuIC could efficiently detect PrP^{Sc} in the skin and eye samples from patients with prion diseases, however, this has not yet been replicated.[206,207]

5.4 RT-QuIC applications in the diagnosis of other NDs

Following the success of RT-QuIC in detecting low amounts of PrP^{Sc} in several tissues (e.g., brain homogenates, CSF, OM, skin, and eyes) of patients with prion diseases, the technology has been extended to the analysis of other NDs. Indeed, since misfolded αS, tau, Aβ, and TDP-43 possess prion-like features, the technology has been adapted for the analysis of tissues collected from patients with NDs characterized by their accumulation. Recombinant proteins (mainly expressed in *E. coli*), including αS (recαS), tau (rectau) and TDP-43 or synthetically produced proteins, including Aβ peptides ($A\beta_{1-40}$, $A\beta_{1-42}$) were used as substrates in many RT-QuIC experiments:

- *Aβ*: in 2014 Claudio Soto showed that traces of misfolded Aβ oligomers were detectable in the CSF of patients with AD by RT-QuIC. In particular, the ability of misfolded Aβ to promote the polymerization of monomeric $A\beta_{1-42}$ peptide (used as a substrate of reaction) was exploited. The results of this study showed that the CSF samples of AD patients significantly increased the kinetics of $A\beta_{1-42}$ aggregation compared to those of CSF collected from patients with other neurodegenerative or neurological disorders. Therefore, it was possible to distinguish AD patients from other subjects with an overall sensitivity of 90% and specificity of 92%.[208]
- *αS*: in 2016, the group of Alison Green showed that the CSF of patients with PD and DLB contained misfolded αS able to promote the aggregation of human recαS used as RT-QuIC reaction's substrate. The study included patients with a diagnosis of (i) DLB, (ii) PD, (iii) PSP, (iv) CBD, (v) DLB with AD pathology, (vi) AD with incidental LBs and (vii) pure AD that were neuropathologically confirmed. PD and DLB patients were identified with 95% and 92% sensitivity, respectively, and with 100% specificity. A sensitivity of 65% was observed for patients with DLB with AD pathology or AD with incidental Lewy bodies. Negative results were obtained from patients with PSP, CBD, pure AD and controls.[209] In 2017, Claudio Soto published another work showing that misfolded αS present in the CSF of patients with PD efficiently

promoted the aggregation of human recαS, while samples collected from control patients (affected by other neurodegenerative or neurological disorders) did not. In conclusion, this test correctly identified patients with PD with an overall sensitivity of 88.5% and specificity of 96.9%.[210] One year later, Byron Caughey described an improved RT-QuIC assay for αS with similar sensitivity and specificity to the prior assays, but the analysis could be completed within 1–2 days (while those of Green and Soto's groups required from 5 to 13 days). In this case, CSF from PD ($n=12$), DLB ($n=17$) and controls ($n=31$, including 16 AD) yielded 93% diagnostic sensitivity and 100% specificity. In particular, almost all CSF belonging to PD and DLB patients (27 out of 29) significantly increased the kinetics of aggregation of the mutated K23Q recαS[211] while controls did not. In 2019, Anumantha Kanthasamy's group confirmed the ability of RT-QuIC to detect misfolded αS in CSF of patients with PD ($n=15$) compared to age-matched healthy control ($n=11$). In this case, wild type human recαS was used as reaction's substrate and its kinetics of aggregation was specifically accelerated only by CSF samples of PD patients. Therefore, the test had 100% sensitivity and specificity.[212] During the same year, the group of Marcel Verbeek demonstrated the diagnostic utility of RT-QuIC to distinguish α-synucleinopathies from non-α-synucleinopathies and controls in a group of patients with parkinsonism and uncertain clinical diagnoses upon inclusion. In particular, the RT-QuIC analysis of CSF samples collected from these subjects yielded to the diagnostic accuracy of 84% with a sensitivity of 75% and specificity of 94%.[213] Another paper was then published where the CSF collected from Leucine-rich kinase 2 (LRRK2)-linked PD and idiopathic PD patients were analyzed by means of RT-QuIC for assessing their ability to promote recαS. RT-QuIC analysis revealed that misfolded αS was detected in idiopathic PD patients with 90% sensitivity and 80% specificity, while it was detectable with lower efficiency in CSF of patients with LRRK2-linked PD.[214] Gianluigi Zanusso and his group applied the RT-QuIC technology for detecting misfolded αS in the CSF of patients with a clinical diagnosis of CJD that were re-classified at neuropathological evaluation as DLB ($n=7$), other neurodegenerative diseases with αS mixed pathology ($n=20$) or without Lewy-related pathology ($n=49$). The test was able to discriminate α-synucleinopathies from non-α-synucleinopathies with 92.9% sensitivity and 95.9% specificity. The test was then repeated using the CSF of patients with DLB ($n=36$) and was positive in 17/20 patients with probable DLB and negative in patients with possible DLB or AD. Therefore, in this case, RT-QuIC appeared to be an optimal test for DLB diagnosis.[215] In 2020, the group of Claudio Soto showed that RT-QuIC could detect misfolded αS in the CSF of patients with PD and MSA with an overall sensitivity of 95.4% and specificity of 100%. Through biochemical and biophysical analyses they showed that the final reaction products acquired peculiar features useful to distinguish between PD and MSA.[216] Analogous results were reported in 2019 by our group which demonstrated that, other than CSF, OM samples collected from patients with PD and MSA efficiently promoted the aggregation of human recαS by means of RT-QuIC. Also in this case, the final reaction products acquired biochemical and morphological features useful to distinguish between PD and MSA. As controls, OM samples collected from patients with PSP and CBD ($n=18$) were used. This latter did not induce recαS aggregation, except for 3 samples. Considering the limited number of samples analyzed we decided not to calculate the sensitivity and specificity of the test. However, while almost all RT-QuIC experiments have been performed using CSF as a source of peripheral disease-specific biomarkers, here we demonstrate that, as for prion diseases, OM samples may represent optimal tissues for RT-QuIC analysis. Indeed, in contrast to CSF, these tissues can be periodically collected with non-invasive procedures.[217] Finally, the group of Anumantha Kanthasamy demonstrated that the submandibular gland tissues collected from PD patients were able to promote recαS aggregation by means of RT-QuIC with 100% sensitivity and 94% specificity when challenged with samples collected from patients with incidental DLB and controls.[218] This study further contributes to demonstrating that peripherally accessible tissues can be successfully exploited for αS RT-QuIC analyses.

- *Tau*: in 2017, Byron Caughey has developed an RT-QuIC assay where 3R-tau (cysteine-free with a point mutation at residue 332 cysteine to serine) was used as a substrate of reaction for the analysis of CSF samples collected from patients with PiD and controls (patients with other neurodegenerative diseases or healthy subjects). The results of these analyses demonstrated that CSF samples collected from patients with PiD were able to promote 3R tau aggregation with higher efficiency compared to control samples.[219] Two years later, Byron Caughey developed another RT-QuIC assay where 4R-tau (cysteine-free with mutations at residues 291 and 322 cysteine to serine and a poly-histidine tag) was used as a substrate of reaction for testing the ability of CSF collected from PSP, CBD, and unaffected controls to promote its aggregation. CSF samples were collected either from patients whose diagnosis was neuropathologically confirmed or from living patients whose diagnoses remained clinically verified. CSF samples from all neuropathologically confirmed PSP and CBD cases induced efficient 4R tau aggregation while controls did not. Notably, RT-QuIC analyses of CSF collected from living subjects had weaker 4R tau seeding activities.[220]

- *TDP-43*: there are no RT-QuIC assays optimized for the analysis of CSF samples (or other peripheral tissues) collected from patients with FTLD or ALS, at present.

6. Conclusions

While the presence of aggregated αS, tau, Aβ and TDP-43 in NDs has been known for sometime, their exact role in the disease etiology remained for a long time uncertain. Only recently, they were found to be the triggering factors of NDs that can adopt diverse structural conformations which are able to self-propagate while causing the phenotypic heterogeneity of these diseases. The existence of misfolded filamentous aggregates characterized by diverse molecular conformations (strains) has been recently confirmed by high-resolution techniques, including cryogenic Electron Microscopy (cryo-EM) and solid-state nuclear magnetic resonance (ssNMR).[101,110,221–223]

Therefore, since RT-QuIC and PMCA can mimic the process of protein misfolding which occurs *in vivo*, they represent optimal tools for understanding the molecular events involved in the initiation and progression of NDs, the molecular alterations which lead to the formation of different αS, Aβ, tau and TDP-43 strains, while providing more detailed insight into the relationship between protein structure and function of the conformationally different misfolded strains. Characterizing various strains is fundamental for understanding their mechanisms of spreading and for developing targeted therapies aimed at blocking this event and hopefully curing the individual's disease phenotype. These strains are indeed able to faithfully template their structure to the recombinant/synthetic proteins used as RT-QuIC reactions substrates, therefore suggesting that they possess a conformational memory that can be transmitted, more or less efficiently, to their normal counterpart, even *in vitro*. However, as detailed in Section 3, misfolded proteins can promote the aggregation of a distinct aggregation-prone protein *in vivo*.[39] This phenomenon is known as cross-seeding and can occur even *in vitro* thus complicating the interpretation of the RT-QuIC and PMCA results and requiring support by other analytical techniques.[224–227]

Finally, this new generation of cell-free amplification techniques have been shown to reach sensitivity levels matching that of bioassays[122,188] and they are proving to be extremely useful to enhance the accuracy of the *premortem* diagnosis of many NDs. Current diagnostic methods rely upon clinical examinations, imaging techniques and analyses of less specific biomarkers which limit the diagnostic accuracy of NDs. The possibility of exploiting these technologies for detecting peripheral disease-specific biomarkers (collectible with non-invasive procedures) in the early stages of the disease, represents a great revolution in the field of NDs diagnosis. At present, both technologies have been optimized for the diagnosis of prion diseases. Particularly, RT-QuIC underwent extensive processes of harmonization and validation at the international level and now represents a reliable and robust test which is used to support the clinical diagnosis of many prion diseases worldwide. In contrast, huge efforts still need to be carried out to harmonize the RT-QuIC procedures for the analysis of peripheral tissues collected from NDs due to αS, Aβ, tau, and TDP-43 which are currently lab-specific. This process is fundamental to strengthen the RT-QuIC diagnostic utility for AD, PD, DLB, atypical parkinsonisms, and frontotemporal dementia, especially using peripheral tissues other than CSF.

Considering the fact that these technologies can detect peripheral disease-specific biomarkers long before the onset of the clinical signs of the disease, they would allow early diagnosis, before the CNS is severely compromised.[201,202,228,229] Notably, other than detecting peripheral disease-specific biomarkers in the early stages of the disease, these technologies can even reveal minimal variations in misfolded proteins' structure. This is a crucial observation that has been made possible thanks to the combination of RT-QuIC and PMCA assays with highly specialized biophysical techniques, including among the others, atomic force microscopy (AFM), transmission electron microscopy (TEM), cryo-EM, ssNMR, Raman and Fourier Transform Infrared spectroscopy, X-ray scattering and electrospray ionization mass spectrometry (ESI-MS), that revealed the presence of structural variations (at subatomic resolution) in the final amplification products.[101,144,217,230,231] Thus, each strain is able to promote the formation of protein aggregates *in vitro* which are characterized by peculiar structural features, identifiable only by means of high-resolution techniques. The identification of such structural differences would enable patients stratification which is pivotal for improving their enrollment in clinical trials and treatments (when available).

Therefore, without neglecting their limits and stressing the fact that they will not replace cell or animal bioassays, RT-QuIC and PMCA represent simple but valuable platforms for understanding, in a much pure and manageable system, the features, mechanisms of formation and spreading of many misfolded proteins, and the effects of different compounds against their process of aggregation. More research is also necessary to evaluate the reproducibility, sensitivity, and specificity of these technologies when applied in the field of NDs diagnosis. Nonetheless, in the near future, they could potentially support the classical diagnostic workup of many NDs, finally leading to early diagnosis and patient stratification by a simple OM, blood or urine analysis. This is the key to the development of effective and targeted therapies for many neurodegenerative diseases.

References

1. Ponjoan A, Garre-Olmo J, Blanch J, et al. Epidemiology of dementia: prevalence and incidence estimates using validated electronic health records from primary care. *Clin Epidemiol.* 2019;11:217–228.
2. Dickson DW. Neuropathology of Parkinson disease. *Parkinsonism Relat Disord.* 2018;46(Suppl 1):S30–S33.
3. Levin J, Kurz A, Arzberger T, Giese A, Hoglinger GU. The differential diagnosis and treatment of atypical Parkinsonism. *Dtsch Arztebl Int.* 2016;113(5):61–69.
4. Coon EA, Singer W, Low PA. Pure autonomic failure. *Mayo Clin Proc.* 2019;94(10):2087–2098.
5. Gao YL, Wang N, Sun FR, Cao XP, Zhang W, Yu JT. Tau in neurodegenerative disease. *Ann Transl Med.* 2018;6(10):175.
6. Bodea LG, Eckert A, Ittner LM, Piguet O, Gotz J. Tau physiology and pathomechanisms in frontotemporal lobar degeneration. *J Neurochem.* 2016;138 (Suppl 1):71–94.
7. Ikeda C, Yokota O, Miki T, et al. Pick's disease with neuronal four-repeat tau accumulation in the basal ganglia, brain stem nuclei and cerebellum. *Neuropathology.* 2017;37(6):544–559.
8. Irwin DJ. Tauopathies as clinicopathological entities. *Parkinsonism Relat Disord.* 2016;22(Suppl 1):S29–S33.
9. Yamada M. Cerebral amyloid angiopathy: emerging concepts. *J Stroke.* 2015; 17(1):17–30.
10. Shimada H, Shinotoh H, Hirano S, et al. β-Amyloid in Lewy body disease is related to Alzheimer's disease-like atrophy. *Mov Disord.* 2013;28(2):169–175.
11. Lünemann JD, Schmidt J, Schmid D, et al. Beta-amyloid is a substrate of autophagy in sporadic inclusion body myositis. *Ann Neurol.* 2007;61(5):476–483.
12. Feneberg E, Gray E, Ansorge O, Talbot K, Turner MR. Towards a TDP-43-based biomarker for ALS and FTLD. *Mol Neurobiol.* 2018;55(10):7789–7801.
13. Prusiner SB. Novel proteinaceous infectious particles cause scrapie. *Science.* 1982;216 (4542):136–144.
14. Prusiner SB. Molecular biology of prion diseases. *Science.* 1991;252(5012): 1515–1522.
15. Höglinger GU, Respondek G, Stamelou M, et al. Clinical diagnosis of progressive supranuclear palsy: the movement disorder society criteria. *Mov Disord.* 2017; 32(6):853–864.

16. Sabbagh MN, Lue LF, Fayard D, Shi J. Increasing precision of clinical diagnosis of Alzheimer's disease using a combined algorithm incorporating clinical and novel biomarker data. *Neurol Ther.* 2017;6(Suppl 1):83–95.
17. Postuma RB, Berg D, Stern M, et al. MDS clinical diagnostic criteria for Parkinson's disease. *Mov Disord.* 2015;30(12):1591–1601.
18. Rizzo G, Arcuti S, Copetti M, et al. Accuracy of clinical diagnosis of dementia with Lewy bodies: a systematic review and meta-analysis. *J Neurol Neurosurg Psychiatry.* 2018;89(4):358–366.
19. Balasa M, Gelpi E, Martín I, et al. Diagnostic accuracy of behavioral variant frontotemporal dementia consortium criteria (FTDC) in a clinicopathological cohort. *Neuropathol Appl Neurobiol.* 2015;41(7):882–892.
20. Gilman S, Wenning GK, Low PA, et al. Second consensus statement on the diagnosis of multiple system atrophy. *Neurology.* 2008;71(9):670–676.
21. Armstrong MJ, Litvan I, Lang AE, et al. Criteria for the diagnosis of corticobasal degeneration. *Neurology.* 2013;80(5):496–503.
22. Frost B, Diamond MI. Prion-like mechanisms in neurodegenerative diseases. *Nat Rev Neurosci.* 2010;11(3):155–159.
23. Guo JL, Lee VM. Cell-to-cell transmission of pathogenic proteins in neurodegenerative diseases. *Nat Med.* 2014;20(2):130–138.
24. Aulić S, Le TT, Moda F, et al. Defined α-synuclein prion-like molecular assemblies spreading in cell culture. *BMC Neurosci.* 2014;15:69.
25. Ayers JI, Giasson BI, Borchelt DR. Prion-like spreading in tauopathies. *Biol Psychiatry.* 2018;83(4):337–346.
26. Desplats P, Lee HJ, Bae EJ, et al. Inclusion formation and neuronal cell death through neuron-to-neuron transmission of alpha-synuclein. *Proc Natl Acad Sci U S A.* 2009;106 (31):13010–13015.
27. Holmes BB, Diamond MI. Prion-like properties of Tau protein: the importance of extracellular Tau as a therapeutic target. *J Biol Chem.* 2014;289(29):19855–19861.
28. Gousset K, Schiff E, Langevin C, et al. Prions hijack tunnelling nanotubes for intercellular spread. *Nat Cell Biol.* 2009;11(3):328–336.
29. Gousset K, Zurzolo C. Tunnelling nanotubes: a highway for prion spreading? *Prion.* 2009;3(2):94–98.
30. Frost B, Jacks RL, Diamond MI. Propagation of tau misfolding from the outside to the inside of a cell. *J Biol Chem.* 2009;284(19):12845–12852.
31. Danzer KM, Krebs SK, Wolff M, Birk G, Hengerer B. Seeding induced by alpha-synuclein oligomers provides evidence for spreading of alpha-synuclein pathology. *J Neurochem.* 2009;111(1):192–203.
32. Soria FN, Pampliega O, Bourdenx M, Meissner WG, Bezard E, Dehay B. Exosomes, an unmasked culprit in neurodegenerative diseases. *Front Neurosci.* 2017;11:26.
33. Pan KM, Baldwin M, Nguyen J, et al. Conversion of alpha-helices into beta-sheets features in the formation of the scrapie prion proteins. *Proc Natl Acad Sci U S A.* 1993;90(23):10962–10966.
34. Caughey BW, Dong A, Bhat KS, Ernst D, Hayes SF, Caughey WS. Secondary structure analysis of the scrapie-associated protein PrP 27-30 in water by infrared spectroscopy. *Biochemistry.* 1991;30(31):7672–7680.
35. Pastrana MA, Sajnani G, Onisko B, et al. Isolation and characterization of a proteinase K-sensitive PrPSc fraction. *Biochemistry.* 2006;45(51):15710–15717.
36. Kim C, Haldiman T, Cohen Y, et al. Protease-sensitive conformers in broad spectrum of distinct PrPSc structures in sporadic Creutzfeldt-Jakob disease are indicator of progression rate. *PLoS Pathog.* 2011;7(9):e1002242.
37. Parchi P, Chen SG, Brown P, et al. Different patterns of truncated prion protein fragments correlate with distinct phenotypes in P102L Gerstmann-Sträussler-Scheinker disease. *Proc Natl Acad Sci U S A.* 1998;95(14):8322–8327.

38. Lau AL, Yam AY, Michelitsch MM, et al. Characterization of prion protein (PrP)-derived peptides that discriminate full-length PrPSc from PrPC. *Proc Natl Acad Sci U S A*. 2007;104(28):11551–11556.
39. Soto C. Unfolding the role of protein misfolding in neurodegenerative diseases. *Nat Rev Neurosci*. 2003;4(1):49–60.
40. Bieschke J, Weber P, Sarafoff N, Beekes M, Giese A, Kretzschmar H. Autocatalytic self-propagation of misfolded prion protein. *Proc Natl Acad Sci U S A*. 2004;101 (33):12207–12211.
41. Noble GP, Wang DW, Walsh DJ, et al. A structural and functional comparison between infectious and non-infectious autocatalytic recombinant PrP conformers. *PLoS Pathog*. 2015;11(6):e1005017.
42. Vázquez-Fernández E, Young HS, Requena JR, Wille H. The structure of mammalian prions and their aggregates. *Int Rev Cell Mol Biol*. 2017;329:277–301.
43. Richardson JS, Richardson DC. Natural beta-sheet proteins use negative design to avoid edge-to-edge aggregation. *Proc Natl Acad Sci U S A*. 2002;99(5):2754–2759.
44. Wille H, Requena JR. The structure of PrP. *Pathogens*. 2018;7(1):20.
45. Caughey B. Interactions between prion protein isoforms: the kiss of death? *Trends Biochem Sci*. 2001;26(4):235–242.
46. Watts JC, Prusiner SB. Mouse models for studying the formation and propagation of prions. *J Biol Chem*. 2014;289(29):19841–19849.
47. Bueler H, Fischer M, Lang Y, et al. Normal development and behaviour of mice lacking the neuronal cell-surface PrP protein. *Nature*. 1992;356(6370):577–582.
48. Cassard H, Torres JM, Lacroux C, et al. Evidence for zoonotic potential of ovine scrapie prions. *Nat Commun*. 2014;5:5821.
49. Bruce ME, Will RG, Ironside JW, et al. Transmissions to mice indicate that 'new variant' CJD is caused by the BSE agent. *Nature*. 1997;389(6650):498–501.
50. Collinge J. Prion diseases of humans and animals: their causes and molecular basis. *Annu Rev Neurosci*. 2001;24:519–550.
51. Hill AF, Collinge J. Prion strains and species barriers. *Contrib Microbiol*. 2004; 11:33–49.
52. Hill AF, Desbruslais M, Joiner S, et al. The same prion strain causes vCJD and BSE. *Nature*. 1997;389(6650):448–450. 526.
53. Collinge J, Sidle KC, Meads J, Ironside J, Hill AF. Molecular analysis of prion strain variation and the aetiology of 'new variant' CJD. *Nature*. 1996;383(6602):685–690.
54. Hagiwara K, Hara H, Hanada K. Species-barrier phenomenon in prion transmissibility from a viewpoint of protein science. *J Biochem*. 2013;153(2):139–145.
55. Parchi P, Giese A, Capellari S, et al. Classification of sporadic Creutzfeldt-Jakob disease based on molecular and phenotypic analysis of 300 subjects. *Ann Neurol*. 1999;46(2): 224–233.
56. Telling GC, Parchi P, DeArmond SJ, et al. Evidence for the conformation of the pathologic isoform of the prion protein enciphering and propagating prion diversity. *Science*. 1996;274(5295):2079–2082.
57. Bessen RA, Kocisko DA, Raymond GJ, Nandan S, Lansbury PT, Caughey B. Non-genetic propagation of strain-specific properties of scrapie prion protein. *Nature*. 1995; 375(6533):698–700.
58. Tanaka M, Chien P, Naber N, Cooke R, Weissman JS. Conformational variations in an infectious protein determine prion strain differences. *Nature*. 2004;428(6980): 323–328.
59. Brown P, Preece M, Brandel JP, et al. Iatrogenic Creutzfeldt-Jakob disease at the millennium. *Neurology*. 2000;55(8):1075–1081.
60. Mead S, Mahal SP, Beck J, et al. Sporadic—but not variant—Creutzfeldt-Jakob disease is associated with polymorphisms upstream of PRNP exon 1. *Am J Hum Genet*. 2001; 69(6):1225–1235.

61. Mackay GA, Knight RS, Ironside JW. The molecular epidemiology of variant CJD. *Int J Mol Epidemiol Genet.* 2011;2(3):217–227.
62. Marciniuk K, Taschuk R, Napper S. Evidence for prion-like mechanisms in several neurodegenerative diseases: potential implications for immunotherapy. *Clin Dev Immunol.* 2013;2013:473706.
63. Angot E, Steiner JA, Lema Tomé CM, et al. Alpha-synuclein cell-to-cell transfer and seeding in grafted dopaminergic neurons in vivo. *PLoS One.* 2012;7(6):e39465.
64. Luk KC, Song C, O'Brien P, et al. Exogenous alpha-synuclein fibrils seed the formation of Lewy body-like intracellular inclusions in cultured cells. *Proc Natl Acad Sci U S A.* 2009;106(47):20051–20056.
65. Nonaka T, Watanabe ST, Iwatsubo T, Hasegawa M. Seeded aggregation and toxicity of {alpha}-synuclein and tau: cellular models of neurodegenerative diseases. *J Biol Chem.* 2010;285(45):34885–34898.
66. Hansen C, Angot E, Bergström AL, et al. α-Synuclein propagates from mouse brain to grafted dopaminergic neurons and seeds aggregation in cultured human cells. *J Clin Invest.* 2011;121(2):715–725.
67. Freundt EC, Maynard N, Clancy EK, et al. Neuron-to-neuron transmission of α-synuclein fibrils through axonal transport. *Ann Neurol.* 2012;72(4):517–524.
68. Krejciova Z, Carlson GA, Giles K, Prusiner SB. Replication of multiple system atrophy prions in primary astrocyte cultures from transgenic mice expressing human α-synuclein. *Acta Neuropathol Commun.* 2019;7(1):81.
69. Yamasaki TR, Holmes BB, Furman JL, et al. Parkinson's disease and multiple system atrophy have distinct α-synuclein seed characteristics. *J Biol Chem.* 2019;294(3): 1045–1058.
70. Luk KC, Kehm VM, Zhang B, O'Brien P, Trojanowski JQ, Lee VM. Intracerebral inoculation of pathological α-synuclein initiates a rapidly progressive neurodegenerative α-synucleinopathy in mice. *J Exp Med.* 2012;209(5):975–986.
71. Luk KC, Kehm V, Carroll J, et al. Pathological α-synuclein transmission initiates Parkinson-like neurodegeneration in nontransgenic mice. *Science.* 2012;338(6109): 949–953.
72. Zhang B, Kehm V, Gathagan R, et al. Stereotaxic targeting of alpha-synuclein pathology in mouse brain using preformed fibrils. *Methods Mol Biol.* 2019;1948:45–57.
73. Watts JC, Giles K, Oehler A, et al. Transmission of multiple system atrophy prions to transgenic mice. *Proc Natl Acad Sci U S A.* 2013;110(48):19555–19560.
74. Kfoury N, Holmes BB, Jiang H, Holtzman DM, Diamond MI. Trans-cellular propagation of Tau aggregation by fibrillar species. *J Biol Chem.* 2012;287(23):19440–19451.
75. Guo JL, Lee VM. Seeding of normal Tau by pathological Tau conformers drives pathogenesis of Alzheimer-like tangles. *J Biol Chem.* 2011;286(17):15317–15331.
76. Iba M, Guo JL, McBride JD, Zhang B, Trojanowski JQ, Lee VM. Synthetic tau fibrils mediate transmission of neurofibrillary tangles in a transgenic mouse model of Alzheimer's-like tauopathy. *J Neurosci.* 2013;33(3):1024–1037.
77. Liu L, Drouet V, Wu JW, et al. Trans-synaptic spread of tau pathology in vivo. *PLoS One.* 2012;7(2):e31302.
78. Gibbons GS, Lee VMY, Trojanowski JQ. Mechanisms of cell-to-cell transmission of pathological Tau: a review. *JAMA Neurol.* 2019;76(1):101–108.
79. Narasimhan S, Guo JL, Changolkar L, et al. Pathological Tau strains from human brains recapitulate the diversity of tauopathies in nontransgenic mouse brain. *J Neurosci.* 2017;37(47):11406–11423.
80. Baker HF, Ridley RM, Duchen LW, Crow TJ, Bruton CJ. Evidence for the experimental transmission of cerebral beta-amyloidosis to primates. *Int J Exp Pathol.* 1993; 74(5):441–454.

81. Walker LC, Schelle J, Jucker M. The prion-like properties of amyloid-β assemblies: implications for Alzheimer's disease. *Cold Spring Harb Perspect Med.* 2016;6(7):a024398.
82. Walker LC, Jucker M. Neurodegenerative diseases: expanding the prion concept. *Annu Rev Neurosci.* 2015;38:87–103.
83. Kane MD, Lipinski WJ, Callahan MJ, et al. Evidence for seeding of beta-amyloid by intracerebral infusion of Alzheimer brain extracts in beta-amyloid precursor protein-transgenic mice. *J Neurosci.* 2000;20(10):3606–3611.
84. Walker LC, Callahan MJ, Bian F, Durham RA, Roher AE, Lipinski WJ. Exogenous induction of cerebral beta-amyloidosis in betaAPP-transgenic mice. *Peptides.* 2002; 23(7):1241–1247.
85. Meyer-Luehmann M, Coomaraswamy J, Bolmont T, et al. Exogenous induction of cerebral beta-amyloidogenesis is governed by agent and host. *Science.* 2006;313 (5794):1781–1784.
86. Nath S, Agholme L, Kurudenkandy FR, Granseth B, Marcusson J, Hallbeck M. Spreading of neurodegenerative pathology via neuron-to-neuron transmission of β-amyloid. *J Neurosci.* 2012;32(26):8767–8777.
87. Jaunmuktane Z, Mead S, Ellis M, et al. Evidence for human transmission of amyloid-β pathology and cerebral amyloid angiopathy. *Nature.* 2015;525(7568):247–250.
88. Boone C, Goodwin CR, Anderson WS. Iatrogenic Alzheimer disease? Amyloid-β protein transmission between humans. *Neurosurgery.* 2016;78(6):N17–N18.
89. Nonaka T, Masuda-Suzukake M, Arai T, et al. Prion-like properties of pathological TDP-43 aggregates from diseased brains. *Cell Rep.* 2013;4(1):124–134.
90. Shimonaka S, Nonaka T, Suzuki G, Hisanaga S, Hasegawa M. Templated aggregation of TAR DNA-binding protein of 43 kDa (TDP-43) by seeding with TDP-43 peptide fibrils. *J Biol Chem.* 2016;291(17):8896–8907.
91. Porta S, Xu Y, Restrepo CR, et al. Patient-derived frontotemporal lobar degeneration brain extracts induce formation and spreading of TDP-43 pathology in vivo. *Nat Commun.* 2018;9(1):4220.
92. Angeli A, Mencacci NE, Duran R, et al. Genotype and phenotype in Parkinson's disease: lessons in heterogeneity from deep brain stimulation. *Mov Disord.* 2013;28(10): 1370–1375.
93. Olivares GH, Olguín P, Klein AD. Modeling Parkinson's disease heterogeneity to accelerate precision medicine. *Trends Mol Med.* 2019;25(12):1052–1055.
94. Perry RH, Irving D, Blessed G, Fairbairn A, Perry EK. Senile dementia of Lewy body type. A clinically and neuropathologically distinct form of Lewy body dementia in the elderly. *J Neurol Sci.* 1990;95(2):119–139.
95. Carol LF. *Dementia With Lewy Bodies: What Is Its Relationship to Alzheimer's Disease?* Tokyo: Springer; 2001:190–194.
96. Roncevic D, Palma JA, Martinez J, Goulding N, Norcliffe-Kaufmann L, Kaufmann H. Cerebellar and parkinsonian phenotypes in multiple system atrophy: similarities, differences and survival. *J Neural Transm (Vienna).* 2014;121(5):507–512.
97. Sanders DW, Kaufman SK, DeVos SL, et al. Distinct tau prion strains propagate in cells and mice and define different tauopathies. *Neuron.* 2014;82(6):1271–1288.
98. Mailliot C, Sergeant N, Bussière T, Caillet-Boudin ML, Delacourte A, Buée L. Phosphorylation of specific sets of tau isoforms reflects different neurofibrillary degeneration processes. *FEBS Lett.* 1998;433(3):201–204.
99. Dickson DW, Kouri N, Murray ME, Josephs KA. Neuropathology of frontotemporal lobar degeneration-tau (FTLD-tau). *J Mol Neurosci.* 2011;45(3):384–389.
100. Jebarupa B, Muralidharan M, Arun A, Mandal AK, Mitra G. Conformational heterogeneity of tau: implication on intrinsic disorder, acid stability and fibrillation in Alzheimer's disease. *Biophys Chem.* 2018;241:27–37.

101. Fitzpatrick AWP, Falcon B, He S, et al. Cryo-EM structures of tau filaments from Alzheimer's disease. *Nature*. 2017;547(7662):185–190.
102. Cohen M, Appleby B, Safar JG. Distinct prion-like strains of amyloid beta implicated in phenotypic diversity of Alzheimer's disease. *Prion*. 2016;10(1):9–17.
103. Laferrière F, Maniecka Z, Pérez-Berlanga M, et al. TDP-43 extracted from frontotemporal lobar degeneration subject brains displays distinct aggregate assemblies and neurotoxic effects reflecting disease progression rates. *Nat Neurosci*. 2019;22(1):65–77.
104. Eichner T, Radford SE. A diversity of assembly mechanisms of a generic amyloid fold. *Mol Cell*. 2011;43(1):8–18.
105. Tycko R. Amyloid polymorphism: structural basis and neurobiological relevance. *Neuron*. 2015;86(3):632–645.
106. Iadanza MG, Jackson MP, Hewitt EW, Ranson NA, Radford SE. A new era for understanding amyloid structures and disease. *Nat Rev Mol Cell Biol*. 2018; 19(12):755–773.
107. Guo JL, Covell DJ, Daniels JP, et al. Distinct α-synuclein strains differentially promote tau inclusions in neurons. *Cell*. 2013;154(1):103–117.
108. Goedert M, Eisenberg DS, Crowther RA. Propagation of Tau aggregates and neurodegeneration. *Annu Rev Neurosci*. 2017;40:189–210.
109. Peng C, Gathagan RJ, Covell DJ, et al. Cellular milieu imparts distinct pathological α-synuclein strains in α-synucleinopathies. *Nature*. 2018;557(7706):558–563.
110. Falcon B, Zhang W, Murzin AG, et al. Structures of filaments from Pick's disease reveal a novel tau protein fold. *Nature*. 2018;561(7721):137–140.
111. Condello C, Lemmin T, Stöhr J, et al. Structural heterogeneity and intersubject variability of Aβ in familial and sporadic Alzheimer's disease. *Proc Natl Acad Sci U S A*. 2018;115(4):E782–E791.
112. Di Fede G, Catania M, Maderna E, et al. Molecular subtypes of Alzheimer's disease. *Sci Rep*. 2018;8(1):3269.
113. Lim KH. Diverse misfolded conformational strains and cross-seeding of misfolded proteins implicated in neurodegenerative diseases. *Front Mol Neurosci*. 2019;12:158.
114. Clinton LK, Blurton-Jones M, Myczek K, Trojanowski JQ, LaFerla FM. Synergistic Interactions between Abeta, tau, and alpha-synuclein: acceleration of neuropathology and cognitive decline. *J Neurosci*. 2010;30(21):7281–7289.
115. Irwin DJ, Lee VM, Trojanowski JQ. Parkinson's disease dementia: convergence of α-synuclein, tau and amyloid-β pathologies. *Nat Rev Neurosci*. 2013;14(9):626–636.
116. Nonaka T, Masuda-Suzukake M, Hasegawa M. Molecular mechanisms of the co-deposition of multiple pathological proteins in neurodegenerative diseases. *Neuropathology*. 2018;38(1):64–71.
117. Lippa CF, Schmidt ML, Lee VM, Trojanowski JQ. Antibodies to alpha-synuclein detect Lewy bodies in many Down's syndrome brains with Alzheimer's disease. *Ann Neurol*. 1999;45(3):353–357.
118. Lippa CF, Fujiwara H, Mann DM, et al. Lewy bodies contain altered alpha-synuclein in brains of many familial Alzheimer's disease patients with mutations in presenilin and amyloid precursor protein genes. *Am J Pathol*. 1998;153(5):1365–1370.
119. Forman MS, Schmidt ML, Kasturi S, Perl DP, Lee VM, Trojanowski JQ. Tau and alpha-synuclein pathology in amygdala of Parkinsonism-dementia complex patients of Guam. *Am J Pathol*. 2002;160(5):1725–1731.
120. Saborio GP, Permanne B, Soto C. Sensitive detection of pathological prion protein by cyclic amplification of protein misfolding. *Nature*. 2001;411(6839):810–813.
121. Morales R, Duran-Aniotz C, Diaz-Espinoza R, Camacho MV, Soto C. Protein misfolding cyclic amplification of infectious prions. *Nat Protoc*. 2012;7(7):1397–1409.
122. Saá P, Castilla J, Soto C. Ultra-efficient replication of infectious prions by automated protein misfolding cyclic amplification. *J Biol Chem*. 2006;281(46):35245–35252.

123. Eckland TE, Shikiya RA, Bartz JC. Independent amplification of co-infected long incubation period low conversion efficiency prion strains. *PLoS Pathog*. 2018;14(10): e1007323.
124. Shikiya RA, Ayers JI, Schutt CR, Kincaid AE, Bartz JC. Coinfecting prion strains compete for a limiting cellular resource. *J Virol*. 2010;84(11):5706–5714.
125. Franz M, Eiden M, Balkema-Buschmann A, et al. Detection of PrP(Sc) in peripheral tissues of clinically affected cattle after oral challenge with bovine spongiform encephalopathy. *J Gen Virol*. 2012;93(Pt 12):2740–2748.
126. Kramm C, Pritzkow S, Lyon A, Nichols T, Morales R, Soto C. Detection of prions in blood of cervids at the asymptomatic stage of chronic wasting disease. *Sci Rep*. 2017; 7(1):17241.
127. Bucalossi C, Cosseddu G, D'Agostino C, et al. Assessment of the genetic susceptibility of sheep to scrapie by protein misfolding cyclic amplification and comparison with experimental scrapie transmission studies. *J Virol*. 2011;85(16):8386–8392.
128. Johnson CJ, Aiken JM, McKenzie D, Samuel MD, Pedersen JA. Highly efficient amplification of chronic wasting disease agent by protein misfolding cyclic amplification with beads (PMCAb). *PLoS One*. 2012;7(4):e35383.
129. Saá P, Sferrazza GF, Ottenberg G, Oelschlegel AM, Dorsey K, Lasmézas CI. Strain-specific role of RNAs in prion replication. *J Virol*. 2012;86(19):10494–10504.
130. Gonzalez-Montalban N, Makarava N, Ostapchenko VG, et al. Highly efficient protein misfolding cyclic amplification. *PLoS Pathog*. 2011;7(2):e1001277.
131. Levavasseur E, Privat N, Martin JC, et al. Molecular modeling of prion transmission to humans. *Viruses*. 2014;6(10):3766–3777.
132. Redaelli V, Bistaffa E, Zanusso G, et al. Detection of prion seeding activity in the olfactory mucosa of patients with Fatal Familial Insomnia. *Sci Rep*. 2017;7:46269.
133. Castilla J, Morales R, Saá P, Barria M, Gambetti P, Soto C. Cell-free propagation of prion strains. *EMBO J*. 2008;27(19):2557–2566.
134. Fernández-Borges N, de Castro J, Castilla J. In vitro studies of the transmission barrier. *Prion*. 2009;3(4):220–223.
135. Igel-Egalon A, Béringue V, Rezaei H, Sibille P. Prion strains and transmission barrier phenomena. *Pathogens*. 2018;7(1):5.
136. Green KM, Castilla J, Seward TS, et al. Accelerated high fidelity prion amplification within and across prion species barriers. *PLoS Pathog*. 2008;4(8):e1000139.
137. Meyerett C, Michel B, Pulford B, et al. In vitro strain adaptation of CWD prions by serial protein misfolding cyclic amplification. *Virology*. 2008;382(2):267–276.
138. Raymond GJ, Hope J, Kocisko DA, et al. Molecular assessment of the potential transmissibilities of BSE and scrapie to humans. *Nature*. 1997;388(6639):285–288.
139. Bossers A, Belt PBGM, Raymond GJ, Caughey B, de Vries R, Smits MA. Scrapie susceptibility-linked polymorphisms modulate the in vitro conversion of sheep prion protein to protease-resistant forms. *Proc Natl Acad Sci U S A*. 1997;94(10):4931–4936.
140. Li J, Browning S, Mahal SP, Oelschlegel AM, Weissmann C. Darwinian evolution of prions in cell culture. *Science*. 2010;327(5967):869–872.
141. Katorcha E, Gonzalez-Montalban N, Makarava N, Kovacs GG, Baskakov IV. Prion replication environment defines the fate of prion strain adaptation. *PLoS Pathog*. 2018;14(6):e1007093.
142. Baskakov IV. The many shades of prion strain adaptation. *Prion*. 2014;8(2):169–172.
143. Bistaffa E, Moda F, Virgilio T, et al. Synthetic prion selection and adaptation. *Mol Neurobiol*. 2019;56(4):2978–2989.
144. Moda F, Le TN, Aulić S, et al. Synthetic prions with novel strain-specified properties. *PLoS Pathog*. 2015;11(12):e1005354.
145. Mahal SP, Browning S, Li J, Suponitsky-Kroyter I, Weissmann C. Transfer of a prion strain to different hosts leads to emergence of strain variants. *Proc Natl Acad Sci U S A*. 2010;107(52):22653–22658.

146. Ghaemmaghami S, Ahn M, Lessard P, et al. Continuous quinacrine treatment results in the formation of drug-resistant prions. *PLoS Pathog*. 2009;5(11):e1000673.
147. Bian J, Kang HE, Telling GC. Quinacrine promotes replication and conformational mutation of chronic wasting disease prions. *Proc Natl Acad Sci U S A*. 2014;111(16): 6028–6033.
148. Oelschlegel AM, Weissmann C. Acquisition of drug resistance and dependence by prions. *PLoS Pathog*. 2013;9(2):e1003158.
149. Shorter J. Emergence and natural selection of drug-resistant prions. *Mol BioSyst*. 2010;6 (7):1115–1130.
150. Lyon A, Mays CE, Borriello F, Telling GC, Soto C, Pritzkow S. Application of PMCA to screen for prion infection in a human cell line used to produce biological therapeutics. *Sci Rep*. 2019;9(1):4847.
151. Lucassen R, Nishina K, Supattapone S. In vitro amplification of protease-resistant prion protein requires free sulfhydryl groups. *Biochemistry*. 2003;42(14):4127–4135.
152. Deleault NR, Lucassen RW, Supattapone S. RNA molecules stimulate prion protein conversion. *Nature*. 2003;425(6959):717–720.
153. Nishina K, Jenks S, Supattapone S. Ionic strength and transition metals control PrPSc protease resistance and conversion-inducing activity. *J Biol Chem*. 2004;279(39): 40788–40794.
154. Orem NR, Geoghegan JC, Deleault NR, Kascsak R, Supattapone S. Copper (II) ions potently inhibit purified PrPres amplification. *J Neurochem*. 2006;96(5):1409–1415.
155. Murayama Y, Yoshioka M, Horii H, et al. Protein misfolding cyclic amplification as a rapid test for assessment of prion inactivation. *Biochem Biophys Res Commun*. 2006;348(2):758–762.
156. Barria MA, Lee A, Green AJ, Knight R, Head MW. Rapid amplification of prions from variant Creutzfeldt-Jakob disease cerebrospinal fluid. *J Pathol Clin Res*. 2018; 4(2):86–92.
157. Moda F, Gambetti P, Notari S, et al. Prions in the urine of patients with variant Creutzfeldt-Jakob disease. *N Engl J Med*. 2014;371(6):530–539.
158. Concha-Marambio L, Pritzkow S, Moda F, et al. Detection of prions in blood from patients with variant Creutzfeldt-Jakob disease. *Sci Transl Med*. 2016;8(370):370ra183.
159. Bougard D, Brandel JP, Bélondrade M, et al. Detection of prions in the plasma of presymptomatic and symptomatic patients with variant Creutzfeldt-Jakob disease. *Sci Transl Med*. 2016;8(370):370ra182.
160. Llewelyn CA, Hewitt PE, Knight RS, et al. Possible transmission of variant Creutzfeldt-Jakob disease by blood transfusion. *Lancet*. 2004;363(9407):417–421.
161. Peden AH, Head MW, Ritchie DL, Bell JE, Ironside JW. Preclinical vCJD after blood transfusion in a PRNP codon 129 heterozygous patient. *Lancet*. 2004;364 (9433):527–529.
162. Bishop MT, Diack AB, Ritchie DL, Ironside JW, Will RG, Manson JC. Prion infectivity in the spleen of a PRNP heterozygous individual with subclinical variant Creutzfeldt-Jakob disease. *Brain*. 2013;136(Pt 4):1139–1145.
163. Diack AB, Head MW, McCutcheon S, et al. Variant CJD. 18 years of research and surveillance. *Prion*. 2014;8(4):286–295.
164. Cali I, Lavrich J, Moda F, et al. PMCA-replicated PrP. *Sci Rep*. 2019;9(1):5191.
165. Cosseddu GM, Nonno R, Vaccari G, et al. Ultra-efficient PrP(Sc) amplification highlights potentialities and pitfalls of PMCA technology. *PLoS Pathog*. 2011;7(11): e1002370.
166. Moudjou M, Sibille P, Fichet G, et al. Highly infectious prions generated by a single round of microplate-based protein misfolding cyclic amplification. *mBio*. 2013;5(1): e00829–00813.

167. Gao C, Han J, Zhang J, et al. Protein misfolding cyclic amplification cross-species products of mouse-adapted scrapie strain 139A and hamster-adapted scrapie strain 263K with brain and muscle tissues of opposite animals generate infectious prions. *Mol Neurobiol.* 2017;54(5):3771–3782.
168. Shikiya RA, Bartz JC. In vitro generation of high-titer prions. *J Virol.* 2011;85 (24):13439–13442.
169. Fernández-Borges N, Elezgarai SR, Eraña H, Castilla J. Animal models for testing anti-prion drugs. *Curr Top Med Chem.* 2013;13(19):2504–2521.
170. Murakami-Kubo I, Doh-Ura K, Ishikawa K, et al. Quinoline derivatives are therapeutic candidates for transmissible spongiform encephalopathies. *J Virol.* 2004;78(3): 1281–1288.
171. Korth C, May BC, Cohen FE, Prusiner SB. Acridine and phenothiazine derivatives as pharmacotherapeutics for prion disease. *Proc Natl Acad Sci U S A.* 2001;98(17): 9836–9841.
172. Panegyres PK, Armari E. Therapies for human prion diseases. *Am J Neurodegener Dis.* 2013;2(3):176–186.
173. Herrmann US, Schütz AK, Shirani H, et al. Structure-based drug design identifies polythiophenes as antiprion compounds. *Sci Transl Med.* 2015;7(299), 299ra123.
174. Berry DB, Lu D, Geva M, et al. Drug resistance confounding prion therapeutics. *Proc Natl Acad Sci U S A.* 2013;110(44):E4160–E4169.
175. Herva ME, Zibaee S, Fraser G, Barker RA, Goedert M, Spillantini MG. Anti-amyloid compounds inhibit α-synuclein aggregation induced by protein misfolding cyclic amplification (PMCA). *J Biol Chem.* 2014;289(17):11897–11905.
176. Meyer V, Dinkel PD, Rickman Hager E, Margittai M. Amplification of Tau fibrils from minute quantities of seeds. *Biochemistry.* 2014;53(36):5804–5809.
177. Atarashi R, Wilham JM, Christensen L, et al. Simplified ultrasensitive prion detection by recombinant PrP conversion with shaking. *Nat Methods.* 2008;5(3):211–212.
178. Atarashi R, Satoh K, Sano K, et al. Ultrasensitive human prion detection in cerebrospinal fluid by real-time quaking-induced conversion. *Nat Med.* 2011;17(2):175–178.
179. Caughey B, Orru CD, Groveman BR, et al. Amplified detection of prions and other amyloids by RT-QuIC in diagnostics and the evaluation of therapeutics and disinfectants. *Prog Mol Biol Transl Sci.* 2017;150:375–388.
180. Colby DW, Zhang Q, Wang S, et al. Prion detection by an amyloid seeding assay. *Proc Natl Acad Sci U S A.* 2007;104(52):20914–20919.
181. Orrú CD, Groveman BR, Raymond LD, et al. Bank vole prion protein as an apparently universal substrate for RT-QuIC-based detection and discrimination of prion strains. *PLoS Pathog.* 2015;11(6):e1004983.
182. Sano K, Atarashi R, Ishibashi D, Nakagaki T, Satoh K, Nishida N. Conformational properties of prion strains can be transmitted to recombinant prion protein fibrils in real-time quaking-induced conversion. *J Virol.* 2014;88(20):11791–11801.
183. McGuire LI, Peden AH, Orrú CD, et al. Real time quaking-induced conversion analysis of cerebrospinal fluid in sporadic Creutzfeldt-Jakob disease. *Ann Neurol.* 2012; 72(2):278–285.
184. ISS. *Criteri diagnostici per leencefalopatie spongiformi trasmissibili umane (EST).* Istituto Superiore di Sanità; 2017.
185. Luers L, Bannach O, Stöhr J, et al. Seeded fibrillation as molecular basis of the species barrier in human prion diseases. *PLoS One.* 2013;8(8):e72623.
186. Davenport KA, Henderson DM, Mathiason CK, Hoover EA. Assessment of the PrPc amino-terminal domain in prion species barriers. *J Virol.* 2016;90(23):10752–10761.
187. Bistaffa E, Rossi M, De Luca CMG, Moda F. Biosafety of prions. *Prog Mol Biol Transl Sci.* 2017;150:455–485.

188. Wilham JM, Orrú CD, Bessen RA, et al. Rapid end-point quantitation of prion seeding activity with sensitivity comparable to bioassays. *PLoS Pathog.* 2010;6(12): e1001217.
189. Cramm M, Schmitz M, Karch A, et al. Characteristic CSF prion seeding efficiency in humans with prion diseases. *Mol Neurobiol.* 2015;51(1):396–405.
190. Hughson AG, Race B, Kraus A, et al. Inactivation of prions and amyloid seeds with hypochlorous acid. *PLoS Pathog.* 2016;12(9):e1005914.
191. Mori T, Atarashi R, Furukawa K, et al. A direct assessment of human prion adhered to steel wire using real-time quaking-induced conversion. *Sci Rep.* 2016;6:24993.
192. Schmitz M, Cramm M, Llorens F, et al. Application of an in vitro-amplification assay as a novel pre-screening test for compounds inhibiting the aggregation of prion protein scrapie. *Sci Rep.* 2016;6:28711.
193. Hyeon JW, Kim SY, Lee SM, et al. Anti-prion screening for acridine, dextran, and tannic acid using real time-quaking induced conversion: a comparison with PrPSc-infected cell screening. *PLoS One.* 2017;12(1):e0170266.
194. Kondru N, Manne S, Greenlee J, et al. Integrated organotypic slice cultures and RT-QuIC (OSCAR) assay: implications for translational discovery in protein misfolding diseases. *Sci Rep.* 2017;7:43155.
195. Broadwell RD, Salcman M, Kaplan RS. Morphologic effect of dimethyl sulfoxide on the blood-brain barrier. *Science.* 1982;217(4555):164–166.
196. Ferreira NC, Marques IA, Conceição WA, et al. Anti-prion activity of a panel of aromatic chemical compounds: in vitro and in silico approaches. *PLoS One.* 2014;9(1): e84531.
197. Vieira TC, Cordeiro Y, Caughey B, Silva JL. Heparin binding confers prion stability and impairs its aggregation. *FASEB J.* 2014;28(6):2667–2676.
198. Park JH, Choi YG, Lee YJ, et al. Real-time quaking-induced conversion analysis for the diagnosis of sporadic Creutzfeldt-Jakob disease in Korea. *J Clin Neurol.* 2016; 12(1):101–106.
199. Orrú CD, Groveman BR, Hughson AG, Zanusso G, Coulthart MB, Caughey B. Rapid and sensitive RT-QuIC detection of human Creutzfeldt-Jakob disease using cerebrospinal fluid. *mBio.* 2015;6(1):e02451-14.
200. McGuire LI, Poleggi A, Poggiolini I, et al. Cerebrospinal fluid real-time quaking-induced conversion is a robust and reliable test for sporadic Creutzfeldt-Jakob disease: an international study. *Ann Neurol.* 2016;80(1):160–165.
201. Cramm M, Schmitz M, Karch A, et al. Stability and reproducibility underscore utility of RT-QuIC for diagnosis of Creutzfeldt-Jakob disease. *Mol Neurobiol.* 2016;53(3): 1896–1904.
202. Sano K, Satoh K, Atarashi R, et al. Early detection of abnormal prion protein in genetic human prion diseases now possible using real-time QUIC assay. *PLoS One.* 2013;8(1): e54915.
203. Franceschini A, Baiardi S, Hughson AG, et al. High diagnostic value of second generation CSF RT-QuIC across the wide spectrum of CJD prions. *Sci Rep.* 2017; 7(1):10655.
204. Green AJE. RT-QuIC: a new test for sporadic CJD. *Pract Neurol.* 2019;19(1):49–55.
205. Orrú CD, Bongianni M, Tonoli G, et al. A test for Creutzfeldt-Jakob disease using nasal brushings. *N Engl J Med.* 2014;371(6):519–529.
206. Orrù CD, Soldau K, Cordano C, et al. Prion seeds distribute throughout the eyes of sporadic Creutzfeldt-Jakob disease patients. *mBio.* 2018;9(6):e02095-18.
207. Orrú CD, Yuan J, Appleby BS, et al. Prion seeding activity and infectivity in skin samples from patients with sporadic Creutzfeldt-Jakob disease. *Sci Transl Med.* 2017;9(417):eaam7785.

208. Salvadores N, Shahnawaz M, Scarpini E, Tagliavini F, Soto C. Detection of misfolded Aβ oligomers for sensitive biochemical diagnosis of Alzheimer's disease. *Cell Rep.* 2014;7(1):261–268.
209. Fairfoul G, McGuire LI, Pal S, et al. Alpha-synuclein RT-QuIC in the CSF of patients with alpha-synucleinopathies. *Ann Clin Transl Neurol.* 2016;3(10):812–818.
210. Shahnawaz M, Tokuda T, Waragai M, et al. Development of a biochemical diagnosis of Parkinson disease by detection of α-synuclein misfolded aggregates in cerebrospinal fluid. *JAMA Neurol.* 2017;74(2):163–172.
211. Groveman BR, Orrù CD, Hughson AG, et al. Rapid and ultra-sensitive quantitation of disease-associated α-synuclein seeds in brain and cerebrospinal fluid by αSyn RT-QuIC. *Acta Neuropathol Commun.* 2018;6(1):7.
212. Manne S, Kondru N, Hepker M, et al. Ultrasensitive detection of aggregated α-synuclein in glial cells, human cerebrospinal fluid, and brain tissue using the RT-QuIC assay: new high-throughput neuroimmune biomarker assay for Parkinsonian disorders. *J NeuroImmune Pharmacol.* 2019;14(3):423–435.
213. van Rumund A, Green AJE, Fairfoul G, Esselink RAJ, Bloem BR, Verbeek MM. α-Synuclein real-time quaking-induced conversion in the cerebrospinal fluid of uncertain cases of parkinsonism. *Ann Neurol.* 2019;85(5):777–781.
214. Garrido A, Fairfoul G, Tolosa ES, Martí MJ, Green A, Group BLS. Synuclein RT-QuIC in cerebrospinal fluid of LRRK2-linked Parkinson's disease. *Ann Clin Transl Neurol.* 2019;6(6):1024–1032.
215. Bongianni M, Ladogana A, Capaldi S, et al. α-Synuclein RT-QuIC assay in cerebrospinal fluid of patients with dementia with Lewy bodies. *Ann Clin Transl Neurol.* 2019; 6(10):2120–2126.
216. Shahnawaz M, Mukherjee A, Pritzkow S, et al. Discriminating α-synuclein strains in Parkinson's disease and multiple system atrophy. *Nature.* 2020;578(7794):273–277.
217. De Luca CMG, Elia AE, Portaleone SM, et al. Efficient RT-QuIC seeding activity for α-synuclein in olfactory mucosa samples of patients with Parkinson's disease and multiple system atrophy. *Transl Neurodegener.* 2019;8:24.
218. Manne S, Kondru N, Jin H, et al. α-Synuclein real-time quaking-induced conversion in the submandibular glands of Parkinson's disease patients. *Mov Disord.* 2019;35(2): 268–278.
219. Saijo E, Ghetti B, Zanusso G, et al. Ultrasensitive and selective detection of 3-repeat tau seeding activity in Pick disease brain and cerebrospinal fluid. *Acta Neuropathol.* 2017; 133(5):751–765.
220. Saijo E, Metrick MA, Koga S, et al. 4-Repeat tau seeds and templating subtypes as brain and CSF biomarkers of frontotemporal lobar degeneration. *Acta Neuropathol.* 2019; 139(1):63–77.
221. Falcon B, Zivanov J, Zhang W, et al. Novel tau filament fold in chronic traumatic encephalopathy encloses hydrophobic molecules. *Nature.* 2019;568(7752):420–423.
222. Swuec P, Lavatelli F, Tasaki M, et al. Cryo-EM structure of cardiac amyloid fibrils from an immunoglobulin light chain AL amyloidosis patient. *Nat Commun.* 2019;10(1):1269.
223. Cao Q, Boyer DR, Sawaya MR, Ge P, Eisenberg DS. Cryo-EM structures of four polymorphic TDP-43 amyloid cores. *Nat Struct Mol Biol.* 2019;26(7):619–627.
224. O'Nuallain B, Williams AD, Westermark P, Wetzel R. Seeding specificity in amyloid growth induced by heterologous fibrils. *J Biol Chem.* 2004;279(17):17490–17499.
225. Yan J, Fu X, Ge F, et al. Cross-seeding and cross-competition in mouse apolipoprotein A-II amyloid fibrils and protein A amyloid fibrils. *Am J Pathol.* 2007;171(1):172–180.
226. Krebs MR, Morozova-Roche LA, Daniel K, Robinson CV, Dobson CM. Observation of sequence specificity in the seeding of protein amyloid fibrils. *Protein Sci.* 2004;13(7):1933–1938.

227. Katorcha E, Makarava N, Lee YJ, et al. Cross-seeding of prions by aggregated α-synuclein leads to transmissible spongiform encephalopathy. *PLoS Pathog*. 2017; 13(8):e1006563.
228. Zanusso G, Monaco S, Pocchiari M, Caughey B. Advanced tests for early and accurate diagnosis of Creutzfeldt-Jakob disease. *Nat Rev Neurol*. 2016;12(7):427.
229. Saá P, Castilla J, Soto C. Presymptomatic detection of prions in blood. *Science*. 2006;313 (5783):92–94.
230. Hwang S, Tatum T, Lebepe-Mazur S, Nicholson EM. Preparation of lyophilized recombinant prion protein for TSE diagnosis by RT-QuIC. *BMC Res Notes*. 2018; 11(1):895.
231. Langkilde AE, Morris KL, Serpell LC, Svergun DI, Vestergaard B. The architecture of amyloid-like peptide fibrils revealed by X-ray scattering, diffraction and electron microscopy. *Acta Crystallogr D Biol Crystallogr*. 2015;71(Pt 4):882–895.

CHAPTER FOURTEEN

The Principles of protein surgery and its application to the logical drug design for the treatment of neurodegenerative diseases

Kazuo Kuwata*
United Graduate School of Drug Discovery and Medical Information Sciences, Gifu University, Gifu, Japan
*Corresponding author: e-mail address: kuwata@gifu-u.ac.jp

Contents

Progress in Molecular Biology and Translational Science, Volume 175
ISSN 1877-1173
https://doi.org/10.1016/bs.pmbts.2020.08.004

Abstract

Here we describe the principles of protein surgery and its application to the molecular design for regulating the protein conformation. We initially describe the Poincare duality that defines the basis of complementarity in the time-dependent geometrical space. Next we introduce the theory of protein surgery consisting of "dissection" and "suture," which correspond to differentiation and integration in a phase space, respectively. Then we introduce the surgical (pan-manifold) differential equation and solve several simple cases. In this way, we constructed "medical quantization" strategy which makes a bridge between quantum mechanics and molecular biology. As the application of this theory, here we describe a logical drug design methodology in detail, and its application to anti-prion drug design. Finally, we propose a plan for logical drug (and medical device) design center, in which all the necessary procedures could be done in one place.

1. Anfinsen's hypothesis

Anfinsen's hypothesis[1] dictates that the native protein structure is determined solely by the primary amino acid sequence of a protein. This implies that the native conformation is kinetically accessible (no prominent barrier) even if the free energy minimum could not be attained. Proteins that cause prion, Alzheimer's, Parkinson's and other neurodegenerative diseases are all exceptions to the Anfinsen's hypothesis. In fact, almost all proteins form amyloids[2] or aggregates depending on the solvent environments suggesting that Anfinsen's hypothesis is valid only at the marginal region of the native state in phase space and the new world is spreading outside of the native ensembles.[3] Detailed information on the interactions between atoms may not necessarily determine the global shape of a protein in a phase space. Unique native conformation may be selected by the other global macroscopic mechanisms.[4]

2. Protein surgery and medical quantization

Protein surgery aims to stabilize the protein conformations in native ensemble. In fact, the regulation of protein conformation by external designed molecule is analogous to the surgical operation in low dimensional space. In medical treatment, surgical operation is conducted essentially based on the human experiences even if assisted by a robot, such as Da Vinci,[5] which is still a medical device. A future robot may move their hands according to the trajectory defined in the internally embedded codes, which seems to be less empirical, but it does not necessarily mean that the robot understands what he does. In fact, during the laparotomy, if the robot would eventually drop the dissected tissue in an operative field, which is a totally exceptional event in the surgical procedure, he would be required to understand what is going on, i.e., the meaning of the dissection and the reason why tissue dropped, before appropriately coping with this exceptional situation.

Since robot can understand only well-posed words which can be coded and translated into movements without conflicts, first we need to define the meaning of dissection and the suture. Sometimes it is harder to define operators in the visible macroscopic system than in the invisible microscopic system because of the detailed geometric constraints in former case. For the latter, we frequently use the homogeneous differential operators, such as Laplacian,[6] but for the former, we need to introduce geometrical operator which acts on the specific geometrical surface. Here we propose the correspondence of the dissection and the suture with the differentiation and the integration, respectively, on the geometrical objects.

2.1 Dissection

Static protein structure is algebraically described by 3n-3 coordinate (n: number of atoms) system, but is essentially a geometrical object of m-dimensionality, S^m (m = 6n). Of note, here we incorporate the dynamics (time-dependence) of each atom. Suppose we will conduct some action on the protein geometry, termed protein surgery.

$$\begin{aligned} S^m &= \partial\left(D^{m+1}\right) = \partial\left(D^{n+1} \times D^{m-n}\right) = \partial D^{n+1} \times D^{m-n} \cup D^{n+1} \times \partial D^{m-n} \\ &= S^n \times D^{m-n} \cup D^{n+1} S^{m-n-1} \end{aligned}$$

S and D represent the surface and disc, respectively. Thereby we can cut protein into at least two parts with different geometries, $S^n \times D^{m-n}$ and $D^{n+1} \times S^{m-n-1}$. If we remove $S^n \times D^{m-n} \in S^m$ from the original manifold, we obtain the following new manifold:

$$D^{n+1} \times S^{m-n-1} \cup D^{n+1} \times S^{m-n-1} = S^{n+1} \times S^{m-n-1}$$

when m = 1 and $n = 0$, this corresponds to a hole in the sheet (Fig. 1).

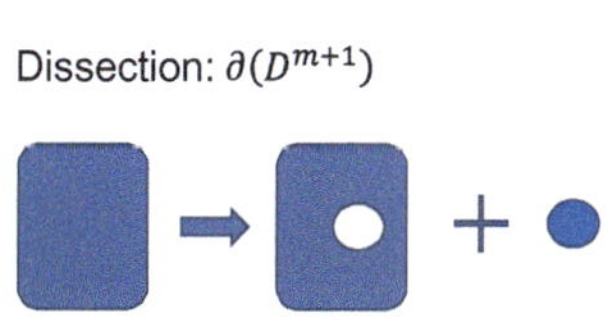

Fig. 1 The representation of "dissection" in surgery which robot can easily understand, or "dissociation" in molecular interaction.

$$\partial(D^2) = \partial(D^1 \times D^1) = \partial D^1 \times D^1 \cup D^1 \times \partial D^1 = S^0 \times D^1 \cup D^1 \times S^0$$

Then here we succeed to describe the meaning of the dissection in a well-defined manner for surgical robot in a low dimensional space. Now robot can understand dissection (differentiation) of one object generated two independent geometrical objects. Then robot succeeds to obtain the cognitive foundation, i.e., dissection generates two spatially independent objects (differentiation) and to construct the strategy for the next movements such as catching the freely dropped tissue and suture them, as shown below.

2.2 Suture

On the other hand, integration around the boundary of the objects constitutes the suture (Fig. 2), connecting at least two parts or spatially independent geometrical objects, such as body and dropped tissue using a thread. This can be done, as shown in Fig. 2, by rotating the robot arms around the boundary as defined in the "dissection" section above. Thread around the boundary can be stabilized by forming a knot which limits the random Reidemeister moves in phase space, whose trajectory traces the path of integration.

Similarly we can regulate the target protein conformation by designing new ligands, which corresponds to creation of new boundary of the complex; $\partial Y = R_+ \cup R_-$, where R_+ and R_- indicate the target protein and ligand molecule to be designed, respectively. In this case, integration boundary could be a network of such as hydrogen bonds or hydrophobic interactions.

2.3 Surgical (pan-manifold) differential equation

Since we succeeded to define the differentiation and integration in the geometrical space, now we can construct the differential equation, as follows:

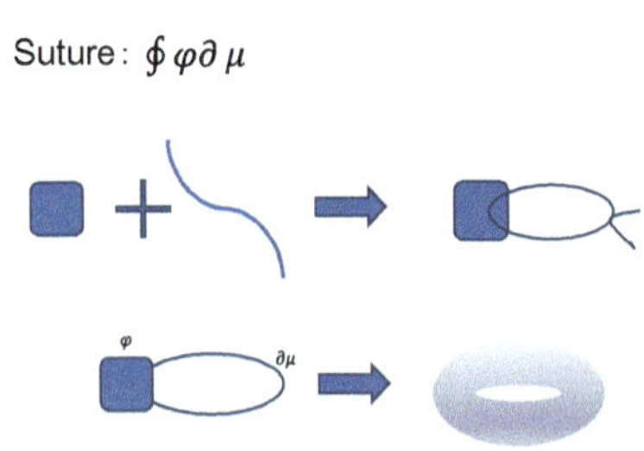

Fig. 2 The representation of "suture" in surgery which robot can easily understand, or "binding" in molecular interaction.

Example 1

$$\partial \varphi = .$$
$$\varphi = \_\_\_\_\_\_ \cdots \ \cdots$$

Example 2

$$\partial \varphi = ..$$
$$\varphi = \_\_\_\_\_\_$$

Example 3

$$\partial \varphi = \varphi$$
$$\frac{\partial \varphi}{\varphi} = 1$$
$$\oint \frac{\partial \varphi}{\varphi} = \blacksquare$$
$$\varphi = e^{\blacksquare} = \blacksquare + \blacksquare + \blacksquare + \cdots.$$

Here the solution of φ represents infinite number of geometrical objects.

2.4 Medical quantization

Logical drug design is a unification of quantum mechanics created in the early half of 20th century and molecular biology created in its late half. Thus we here describe the origin of macroscopic uncertainty essentially driven by the information entropy originated from the huge number of combinations usually occurring in biology. Embedding of quantum mechanics into the geometric macroscopic system may give us a great hint to solve many problems in theoretical biology. As shown in Table 1, medical quantization where surgical treatment (cancer surgery) and drug design are unified by a theory of protein surgery.

Is there any uncertainty in the macroscopic system? This is a definition of uncertainty in view of entropy (Kullback divergence). Poincare duality dictates that the m-dimensional homology group isomorphic to the (n-m)-dimensional co-homology group, i.e., $H^m(M) \cong H_{n-m}(M)$. The uncertainty (h) in macroscopic geometry can be obtained from the deformation quantization of the relative entropies (Kullback divergence) defined in homology and co-homology spaces, i.e.:

Table 1 Medical quantization.

Therapeutic method	**Drug design**	**Cancer surgery**	**Protein surgery**
Spread of lesion	Localizability	Intramucosal	m<<n in Poincare duality
Technical potential	Regulability and synthesizability	Resectability and sustainability	Differentiability and integrability at boundary
Purity	Specificity	Non-metastasis	Uniqueness of solution
Applicability	Safety	Complication	Solvability

$$\mathrm{h}^2 = \sum_i \mathrm{f}\left(\frac{\overrightarrow{\partial}}{\partial p_i}\frac{\overleftarrow{\partial}}{\partial q_i} - \frac{\overleftarrow{\partial}}{\partial p_i}\frac{\overrightarrow{\partial}}{\partial q_i}\right)\mathrm{g}$$

$$\mathrm{f} = \sum_i p_i(\ln \mathrm{p}_i - \ln \mathrm{q}_i)$$

$$\mathrm{g} = \sum_i \mathrm{q}_i(\ln \mathrm{q}_i - \ln \mathrm{p}_i)$$

Thus, uncertainty h is obtained as the logarithm of the likelihood ratio:

$$\mathrm{h} = \sum_i \left(\ln \frac{\mathrm{p}_i}{q_i}\right)$$

When geometrical parameters, such as quantum probabilities, are the same in both spaces, then there is no macroscopic uncertainty, but quantum uncertainty, of course.

3. Principle of logical design of a molecular chaperone

3.1 Localizability

In the logical drug design, we intend to design a molecule, which is complementary to the target protein in terms of such as geometry and electrostatic characteristics. The size of the designed molecule varies from large c.f. antibody, to small c.f. synthetic compound, both constructing the complementary space to the protein. However, here we focus on small compounds, since ab-initio antibody design is not feasible because of too many degrees of freedom at this stage. However, small compound (M.W. 500–1000) can cover only limited space in the protein, so appropriate target binding region should be carefully selected.

Followings are the required conditions for the static ligand binding, i.e., integration boundary formation:

(1) Well known strong binding site, such as substrate binding region to the active site of enzyme[7]

(2) Region with smooth boundary (static loop) formation, such as appropriate binding pocket[8]

(3) Region with strong binding interaction site, such as characteristic charge distribution[9]

(4) Other accessible cavity[10]

On the other hand, since molecular chaperone alters the conformational conversion pathway,[11] another statistical dynamical features may be considered:

(i) Hot spot in activated state (scarcely populated high-energy state)[12], in which a global boundary for integration (loop of interactions) would be formed in a phase space but at low probability. But once boundary was formed, the integration scheme (ligand binding) would stabilize the activated state and inhibit further conformational change.

(ii) Initial convertible region (ICR),[13] which is the region suffering the initial conformational alteration within the native ensemble along the misfolding pathway to the pathogenic form in case of prion. This region forms the transient boundary (loop of interactions) for integration. Once boundary was formed, the integration scheme (ligand binding) would stabilize the ICR and inhibit further conformational alteration.

Geometrical characteristics of the boundary for integration, even if it includes any knot formation, originate from the statistical dynamical features of the interaction networks (such as hydrogen bond or hydrophobic) in the target protein system.[14] If pathogenic conversion process of a prion protein involves any intermediate state or activated state, their structural characteristics originate from a statistical dynamical mode or periodicity, such as a Fourier component, or their interference. The relevant residues could be mapped as the appropriate target region on the native ensemble. When this boundary was confined within the narrow space, we may regulate it by a carefully designed small compound.

To observe the prion's conversion reaction, we constructed the in vitro conversion NMR (IVC-NMR) (Fig. 3)[13] with 800 MHz-NMR (Avance 800) equipped with a flow cell (CryoFit) connected with the ultrasonicator (QSONICA Q700). This is actually the real time PMCA reaction observed

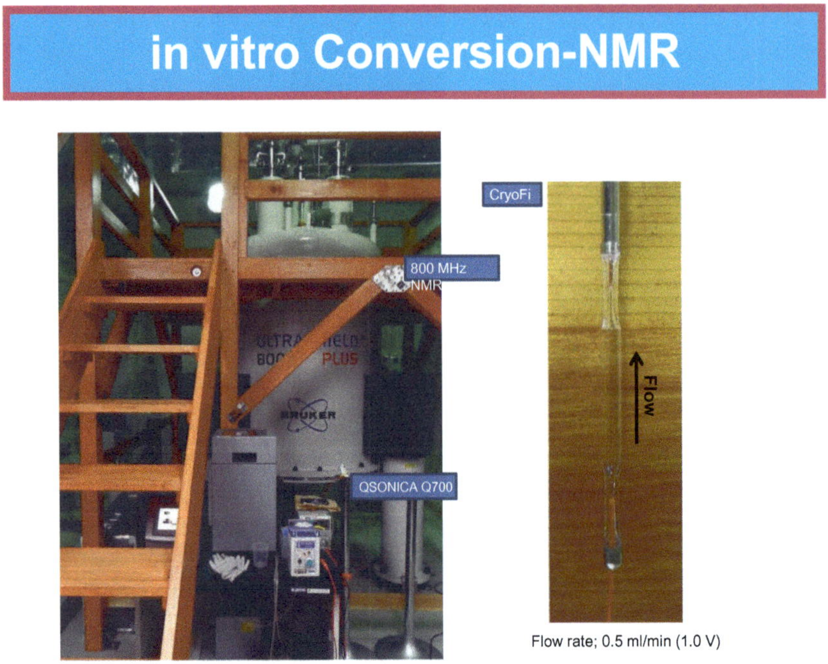

Fig. 3 The instruments used for in vitro conversion NMR (800 MHz) at Gifu University.

in NMR.[15] Normal prion protein gradually converts to the pathogenic form during cyclic flow between ultrasonicator and NMR flow cell. Flow rate was set at 0.5 mL/min. Chemical shifts of the HSQC spectra of normal prion protein alter gradually because of the conformational change of prion protein. We found the initial conformational change occurred mainly at residues 190, 192, 193, 194, 195, 196, 197 (Fig. 4), which correspond to the loop region between Helix 2 and 3 (Fig. 5). We termed this region, ICR (initial convertible region).[13]

3.2 Synthesizability and regulability

When binding region or integration boundary was already elucidated, candidate compound can be designed based on the quantum chemical calculations.[16] Before synthesis, it is analyzed how the designed compound or predicted interactions could regulate the statistical and dynamical features of a target protein. This could be done mainly by classical molecular dynamics or dynamical quantum chemistry calculation as described below. We can compare the free energy of native conformation between bound and unbound state. Although coarse grained model[17] could yield the

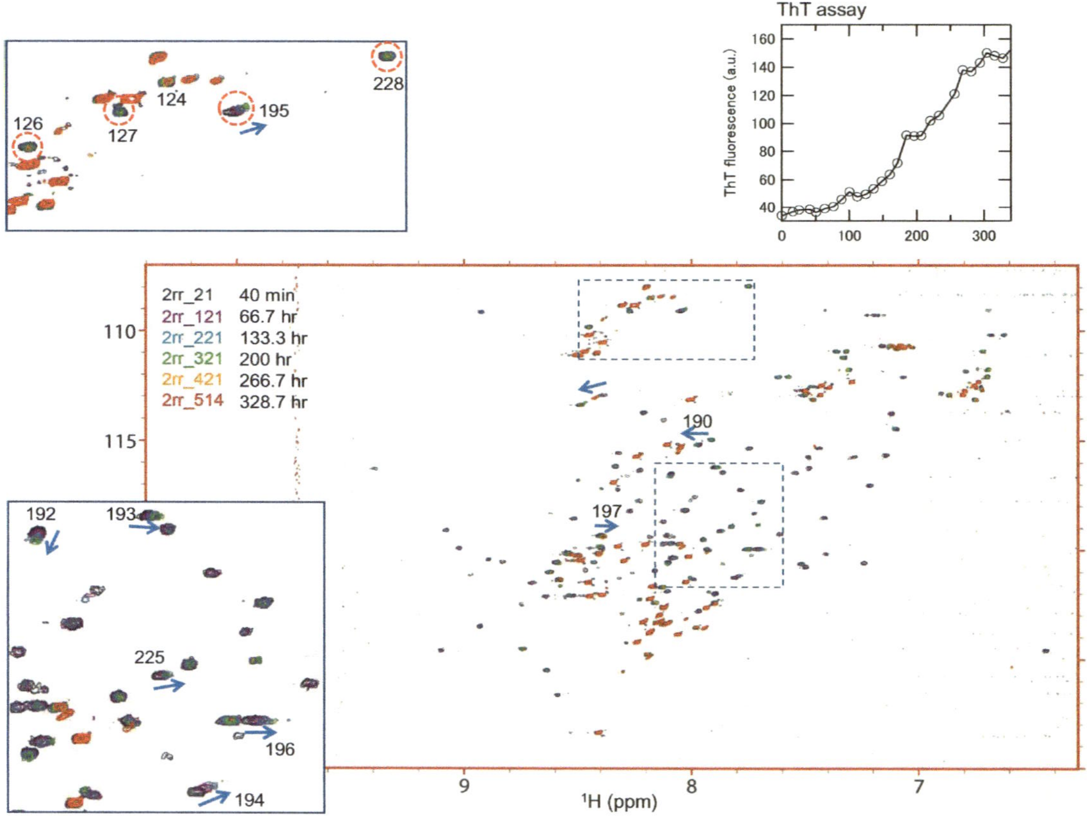

Fig. 4 Alteration of chemical shifts of normal prion protein during the pathogenic conversion in NMR.

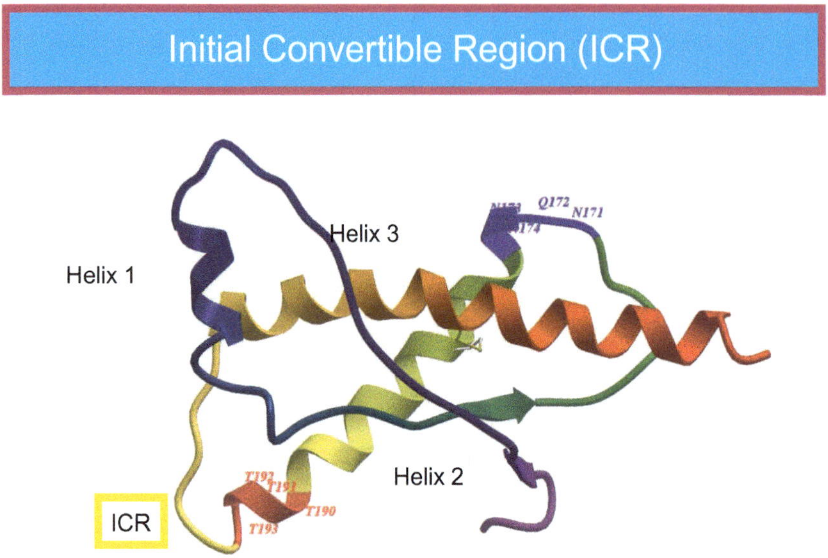

Fig. 5 ICR includes the 190–193(TTTT) region at the C-terminal of Helix 2 and 2–3 loop region.

statistical mechanical parameters easily, we found that the obtained free energy using coarse grained model was not accurate enough. Absolute statistical parameters might be critically dependent on the detailed potential functions and also on the statistical ensemble.[18] For folding simulation, MD simulation may be the most appropriate method, but we need a plenty of time to gather the statistical ensemble.

It must be emphasized that even if all the interactions between atoms were defined by the quantum chemical methodology, overall global shape or structure could be still indeterministic and unstable. So far we do not have any global geometric potential function that define the global membraneous structure (such as bubble formation). Any geometrical structure constructed only by local interactions has inherent instability. For example, we can flip over the sphere smoothly in four-dimensional space.

In general, it was shown that any ligand that binds to the native state essentially stabilizes the native state.[19] Furthermore, selective binding to the integration boundary would hamper further conformational conversion by increasing the barrier between the native and the abnormal form.[13]

However, synthesizability of the designed compound is another independent problem. Now we have accumulated a huge number of organic synthetic roots in a database. We need to find out the most optimized root for synthesis of target compound using AI. Although this process includes no essential difficulty in small compounds except for several asymmetry generations, for medium to large molecule design, the similar multiple conformation problems for drug arises as for the target protein conformation as described above.

3.3 Specificity

Specificity of the interaction between the designed molecule and the target protein is the most important to avoid the side effects due to the interaction with the off-target proteins. To attain the specificity, the interaction needs to be based on the conformational characteristics between two molecules and not on the local electronic one. If the latter is dominant, since other proteins composed of common amino acids might have a chance to interact with the designed molecule. The number of conformations that protein can take is huge (more than $\sim 10^{50}$), so that antibody can attain such a specificity. However, even for the synthetic compound (M.W. ~500), it is feasible to get it because of the induced fit interaction which significantly perturbs the conformation of the target protein.[13]

To regulate the protein conformation, it is critical to focus on the entropy (information entropy) contribution,[13] since biological complexity essentially arises from the huge number of combinations in protein conformation as well as those in the ligand binding manner.

4. General strategy to solve problems

4.1 Classical molecular dynamics simulation (CMD)

CMD assume the potential functions of molecules, and obtained free energy must be dependent on these potential functions. But we found that the classical molecular dynamics (CMD) could yield free energy corresponding to the experimentally obtained one, when exact statistical ensemble was taken.[18]

4.2 Quantum chemistry calculation

We coded a quantum chemistry program, PAICS in C-language.[8,20–23] Thus far all the quantum chemistry programs except for PAICS were written in Fortran. PAICS are coded by ~200,000 lines of C-language and sources were open in public (http://www.paics.net/index_e.html), and continuously upgraded including the time-dependent solution of Schrödinger's equation.

4.3 Quantum computing

Recently developed quantum computing is essentially fit to the drug discovery problem. Drug discovery is to find out the molecule whose dynamical space is co-homologous to that of the target molecule. That can be written in a more defined manner, i.e., suture of the topological space, i.e., $\partial Y = R_+ \bigcup R_-$ as described above.

4.4 Organic synthesis

It is desirable to use the full automatic facility for the organic synthesis of the novel compounds ever existent in the universe. Actually, number of compounds with molecular weight less than 500 is about $4^{50} \sim 2^{100} \sim 10^{30}$. if we assume four kind of atoms (C, H, N, O) with average atomic weight of ~ 10 and linear arrangement. Here the number of covalent bond topology is not taken into account so that the actual number of molecules might be much larger than 10^{30}. Thus even if a company has a stock of compound library of 10^7, which is as small as about 1/(Avogadro's number) of the real chemical space. This means that we have not encounter almost all molecules in the chemical space, and we are unaware of the functions and toxicities of the newly synthesized molecules. Therefore, it is desirable to establish the unmanned synthetic robot system. Working principle of this robotic system is essentially equivalent with that of surgical treatment described above.

(a) GMP synthesis

Once large-scale synthetic pathway of the new functional molecules was established, the pharmaceutical production must be performed according to the GMP guideline.

(b) Pharmaceutical production

For the production of the injection, we need to construct the isolator system.

4.5 Non-clinical POC

Non-clinical proof of concept includes the proofs of efficacy in cell, organ and animal level.

4.6 Non-clinical safety

Non-clinical safety tests must be done according to the GLP guideline. Also pharmacodynamics tests must be done before clinical trial. We performed

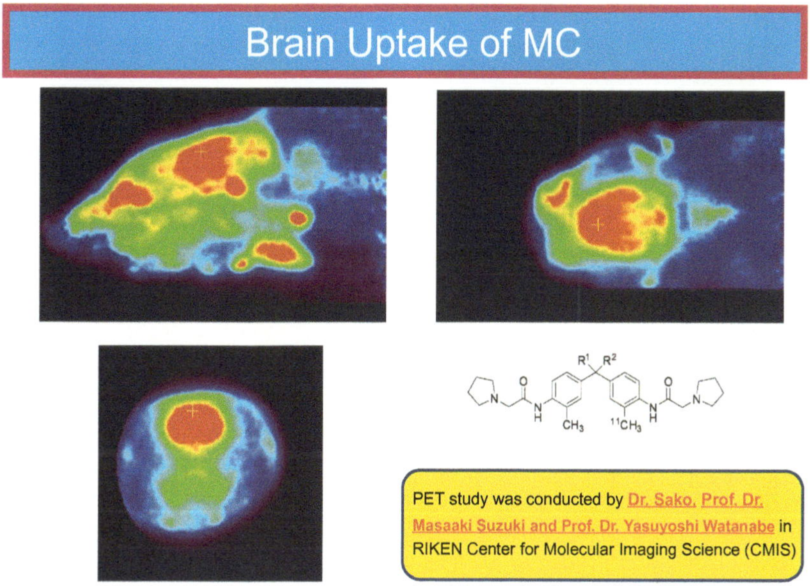

Fig. 6 GN8, an anti-prion agent modified by $C^{11}H3$ for PET observation, distributes mainly in brain in rat.

the brain uptake of a medical chaperone (MC) for prion (GN8)[12] using PET study[24] (Fig. 6), and we confirmed that MC mainly distributes in the brain.

4.7 Clinical trial

For clinical trial, clinical safety is examined in phase I/II trial and clinical POC is tested in phase II trial.

5. Logical drug design center

In order to conduct the logical drug design described above, we are planning to construct the logical drug design center (Fig. 7) where all the necessary procedure can be done at one place. This is advantageous since, in formation, materials, animals as well as human beings can move according to the logic.

a Quantum computing center

To facilitate the drug design, using a quantum computing technologies, here we propose the "Logical drug design center." Core software[16] is developed

Fig. 7 A perspective view of the logical drug design center, which includes a quantum computing center, robotic organic synthesis facilities, medical device production facilities, non-clinical facilities, and a hospital for clinical trial.

in the independently established software company and software license would be freely shared with any drug companies in the world.

b Organic synthesis facility

Quantum computer facilities would be connected with robotic organic synthesis facilities via newly established quantum cryptographic highway system. Organic synthesis is automatically performed by a quantum chemistry calculation and synthesis root database, and medical device is automatically designed by a solving surgical differential equations. These facilities are completely based on GMP (Good Manufacturing Practice) guidelines.

c Non-clinical facility

Non-clinical facilities are completely based on GLP (Good Laboratory Practice) guideline. Establishment of International collaborations on the production and usage of disease model animals. Efficient non-clinical tests are performed by using a united regulatory methodologies, and establish a unique structure-function relationships of any designed compound. Using disease model cells, functions and side effects of the compounds are automatically examined. The disease model animals are produced by CRISPR and non-clinical tests are performed also automatically. These are regulated by a quantum highway system.

d Hospital for clinical trial

Hospital for clinical trial includes 10,000 beds, 100 advanced operating rooms, 100 CT and 100 MRI, and to be established based on GCP (Good Clinical Practice) guideline, and open for drug companies in the world. Any intractable diseases in areas of neurological, circulatory, respiratory, digestive, urological, orthopedical, obstetrics and gynecological, dermatological, ophthalmological and otolaryngological, and psychic diseases can be treated in this hospital with new drugs or technologies as well as well-established treatments.

References

1. Anfinsen CB. Principles that govern the folding of protein chains. *Science*. 1973; 181:223–230.
2. Fandrich M, Fletcher MA, Dobson CM. Amyloid fibrils from muscle myoglobin. *Nature*. 2001;410:165–166. https://doi.org/10.1038/35065514.
3. Kuwata K, et al. Locally disordered conformer of the hamster prion protein: a crucial intermediate to PrPSc? *Biochemistry*. 2002;41:12277–12283.
4. Haas AL, Rose IA. Hemin inhibits ATP-dependent ubiquitin-dependent proteolysis: role of hemin in regulating ubiquitin conjugate degradation. *Proc Natl Acad Sci U S A*. 1981;78:6845–6848. https://doi.org/10.1073/pnas.78.11.6845.
5. Bodner J, Wykypiel H, Wetscher G, Schmid T. First experiences with the da Vinci operating robot in thoracic surgery. *Eur J Cardiothorac Surg*. 2004;25:844–851. https://doi.org/10.1016/j.ejcts.2004.02.001.
6. Chan YM, Hirschfelder JO. Extending the domain of the Laplacian operator with the use of delta functions. *Proc Natl Acad Sci U S A*. 1968;61:1–6. https://doi.org/10.1073/pnas.61.1.1.
7. Wilton DJ, Kitahara R, Akasaka K, Pandya MJ, Williamson MP. Pressure-dependent structure changes in barnase on ligand binding reveal intermediate rate fluctuations. *Biophys J*. 2009;97:1482–1490. https://doi.org/10.1016/j.bpj.2009.06.022.
8. Ishikawa T, et al. A theoretical study of the two binding modes between lysozyme and tri-NAG with an explicit solvent model based on the fragment molecular orbital method. *Phys Chem Chem Phys*. 2013;15:3646–3654. https://doi.org/10.1039/c3cp42761g.
9. Fukuoka M, et al. Structure-based discovery of anti-influenza virus A compounds among medicines. *Biochim Biophys Acta*. 2012;1820:90–95. https://doi.org/10.1016/j.bbagen.2011.11.003.
10. Ali F, Yamaguchi K, Fukuoka M, Elhelaly AE, Kuwata K. Logical design of an anti-cancer agent targeting the plant homeodomain in Pygopus2. *Cancer Sci*. 2016; 107:1321–1328. https://doi.org/10.1111/cas.12995.
11. Fan JQ. Pharmacological chaperone therapy for lysosomal storage disorders—leveraging aspects of the folding pathway to maximize activity of misfolded mutant proteins. *FEBS J*. 2007;274:4943. https://doi.org/10.1111/j.1742-4658.2007.06043.x.

12. Kuwata K, et al. Hot spots in prion protein for pathogenic conversion. *Proc Natl Acad Sci U S A*. 2007;104:11921–11926.
13. Yamaguchi K, et al. A designer molecular chaperone against transmissible spongiform encephalopathy slows disease progression in mice and macaques. *Nat Biomed Eng*. 2019;3:206–219. https://doi.org/10.1038/s41551-019-0349-8.
14. Kuwata K, Kamatari YO, Akasaka K, James TL. Slow conformational dynamics in the hamster prion protein. *Biochemistry*. 2004;43:4439–4446.
15. Saborio GP, Permanne B, Soto C. Sensitive detection of pathological prion protein by cyclic amplification of protein misfolding. *Nature*. 2001;411:810–813. https://doi.org/10.1038/35081095.
16. Ma B, Yamaguchi K, Fukuoka M, Kuwata K. Logical design of anti-prion agents using NAGARA. *Biochem Biophys Res Commun*. 2015;469:930–935. https://doi.org/10.1016/j.bbrc.2015.12.106.
17. Li W, Takada S. Self-learning multiscale simulation for achieving high accuracy and high efficiency simultaneously. *J Chem Phys*. 2009;130:214108. https://doi.org/10.1063/1.3146922.
18. Tran DP, Takemura K, Kuwata K, Kitao A. Protein-ligand dissociation simulated by parallel cascade selection molecular dynamics. *J Chem Theory Comput*. 2018;14:404–417. https://doi.org/10.1021/acs.jctc.7b00504.
19. Kabir A, et al. Effects of ligand binding on the stability of aldo-keto reductases: implications for stabilizer or destabilizer chaperones. *Protein Sci*. 2016;25:2132–2141. https://doi.org/10.1002/pro.3036.
20. Ishikawa T, Ishikura T, Kuwata K. Theoretical study of the prion protein based on the fragment molecular orbital method. *J Comput Chem*. 2009;30:2594–2601. https://doi.org/10.1002/jcc.21265.
21. Ishikawa T, Kuwata K. Fragment molecular orbital calculation using the R1-MP2 method. *Chem Phys Lett*. 2009;474:195–198.
22. Ishikawa T, Kuwata K. Interaction analysis of the native structure of prion protein with quantum chemical calculations. *J Chem Theory Comput*. 2010;6:538–547. https://doi.org/10.1021/ct900456v.
23. Ishikawa T, Kuwata K. RI-MP2 gradient calculation of large molecules using the fragment molecular orbital method. *J Phys Chem Lett*. 2012;3:375–379. https://doi.org/10.1021/jz201697x.
24. Kimura T, et al. Synthesis of an (11) C-labeled antiprion GN8 derivative and evaluation of its brain uptake by positron emission tomography. *ChemMedChem*. 2013;8:1035–1039. https://doi.org/10.1002/cmdc.201300167.

Index

Note: Page numbers followed by "*f*" indicate figures and "*t*" indicate tables.

Kolibri 360 GmbH
Ferdinand-Jühlke-Straße 7
99095 Erfurt, Deutschland
produktsicherheit@kolibri360.de